W9-CRE-981

This book is dedicated to my students: past, present and future

"The reasonable man adapts himself to the world: the unreasonable one persists in trying to adapt the world to himself. Therefore all progress depends on the unreasonable man."

*Man and Superman*, George Bernard Shaw

"We are all in the gutter, but some of us are looking at the stars."

*Lady Windermere's Fan*, Oscar Wilde

# ERGONOMICS, WORK AND HEALTH

**Stephen Pheasant, PhD, FErgS**

*Consulting Ergonomist*

**Aspen Publishers, Inc.**
**Gaithersburg, Maryland**
**1991**

First published 1991
Reprinted 1992

Published in the USA by
Aspen Publishers, Inc.
200 Orchard Ridge Drive
Gaithersburg, MD 20878

Printed in Hong Kong

Library of Congress Cataloging-in-Publication Data
Pheasant, Stephen.
Ergonomics, work, and health / Stephen Pheasant.
p. cm.
Includes bibliographical references and index.
ISBN 0–87189–320–7 (Aspen)
1. Industrial hygiene. 2. Human engineering. I. Title.
RC987.P525 1991
613.8'2—dc20 91–12527
CIP

# Contents

# Foreword

It is a brave man who rises from the trenches of his own discipline and moves to occupy the high ground of an ally as they advance on the broad front of Knowledge.

As its title implies, this book is intended to place Ergonomics firmly in the clinical setting in which it is very much needed; it will stand with a regrettably small number of other texts as requisite reading for all in Occupational Health. It follows logically from the author's recent *Bodyspace*, itself a considerable help to occupational physicians.

As an examiner in the Associateship examination of the Faculty of Occupational Medicine, one of my most important, and rewarding, tasks is to read the journals of the candidates encompassing a four-week period of their working lives. It is encouraging to see ergonomic concepts and ideas increasingly featured; to see a recognition of ergonomic problems and a need to seek precisely the solutions this book is designed to offer.

The fact that patterns of work, not just "industry", have changed world-wide in the last few decades illustrates that we have substituted one set of ergonomic problems for another, no less intractable if not scientifically analysed; the computer instead of the blast furnace; the human errors of Chernobyl, Three Mile Island and Zeebrugge instead of mining disasters.

The whole gamut of the work environment is covered and the author does not evade the Byzantine problems of back pain, the Waterloo of so many aspiring authors.

Many who are invited to contribute a foreword must cry "Why me?" As I write, the travails of the Channel Tunnel construction appear in the press and on our TV screens weekly and the giant machines which grind their way towards each other in climactic consummation are awe-inspiring to behold. Years ago I told the author of the intense rivalry between mining machinery manufacturers in the Soviet Union and Western Europe. Nowhere was this more intense than in tunnelling, where vast prototypes were claimed to advance through strata at a rate undreamt of. The acme was reached by one company, the last word, and duly demonstrated. Unfortunately the sole operator was positioned at the front of the machine and disappeared into the tunnel in a cloud of noise and dust. No, he did NOT turn up in Australia and, YES, the design was radically modified. The upshot was the development of a team of ergonomists who cast a critical eye over all prototypes offered to the British mining industry. The resultant modifications and improvements had much to do with the fact that the British coalmining industry has the best safety record in the world, not due solely to good ergonomics, but in which our discipline has paid a vital part.

The moral is, never tell your best stories to your best friends in case you are asked to write a foreword. But bear this story in mind as you read this book. . . .

Dr Roy Archibald
October, 1990

# Acknowledgements

Many of the ideas which go into a book like this surface for the first time in conversation with friends and colleagues—in ways which do not necessarily lend themselves to the normal conventions of academic citation. I have discussed the central topics of this book, at length and over the years, with David Stubbs, Peter Buckle and their colleagues at the Robens Institute—to the extent that we have probably all long since forgotten who said what first. I am particularly indebted to Sheila Lee for introducing me to the fascinating world of osteopathy. Anne-Marie Potts, Jackie Nicholls and Fiona Turner have also given me the benefits of their clinical knowledge; and conversations with John Long have helped me tighten up the nuts and bolts of my conceptual frameworks. The story of the chemical plant, with which I commence this book, was told to me originally by Mike Gray of the HSE. To all these people and to anybody else I have unconsciously plagiarized I extend my thanks.

Thanks are also due to David Sanchez and Andrew Pinder, who helped with the illustrations, and Pamela Dale, who patiently typed my endless revisions of the manuscript; and to my "other publishers", Taylor and Francis Ltd, for generously allowing me to reproduce so many illustrations from *Bodyspace* (1986).

# Prologue

The safety staff at a large chemical works became concerned at the number of "slipping and falling" accidents which were occurring on a particular staircase. The stairs were cleaned, then resurfaced, then removed and replaced with a new set—before the real cause of the accidents became apparent. At the top of the stairs was a tank for storing a certain chemical. It was fitted with a manually operated valve which allowed its contents to be fed down a pipe into a second tank on the floor below. This lower tank was fitted with an indicator gauge to show how full it was. To run a measured quantity of the chemical into the lower tank, somebody had to climb the stairs to open the valve; then run downstairs to check the indicator; then run up the stairs again to close the valve.

A similar design problem was involved in the complex combination of adverse circumstances which led to the capsizing of the cross-channel ferry *Herald of Free Enterprise*—the worst peacetime disaster involving a British ship since the sinking of the *Titanic* in 1912. The ferry left the harbour at Zeebrugge with her bow doors open, owing to a so-called "human error". The procedures for closing the doors were highly unsatisfactory—for which the management of the ferry company has been severely criticized. The report of the official court of enquiry states that "from top to bottom, the body corporate was infected with the disease of sloppiness". Extraordinary as it may seem (with the wisdom of hindsight), there was no warning system on the bridge to tell the captain whether the doors were open or closed—and thus whether or not it was safe to go to sea. A simple warning light would probably have been enough to prevent the catastrophe. Its cost would have been negligible—perhaps about £150. For lack of such a device, 188 lives were lost—and a cross-channel ferry worth several million pounds.

Every time you write a cheque, or use your credit card, or make a major financial transaction, somebody has to feed the data into a computer. A data entry clerk's lot is not a very happy one. In the worst cases she may spend her working day sitting with her eyes fixed on the VDU screen, keying in strings of random numbers at high speed, using one hand only. She will probably be paid by results—and the rate at which she works may be monitored by the machine. In many organizations, a rate of 12,000 keystrokes per hour is considered to be a minimum level of proficiency for an experienced worker; 20,000 or more keystrokes per hour is not uncommon.

People who work with VDUs commonly complain of headaches, eyestrain, back pain, stiff necks or pain and discomfort in the shoulder, forearm or wrist. In most cases the symptoms are of a relatively minor nature; for some people they may become more severe. When upper limb symptoms are severe, they are often referred to as *repetitive strain injuries*. People who do jobs which require them to use the keyboard and screen most intensively (particularly for high-speed data entry) tend to be the worst affected.

Consider the following hypothetical case. A young woman goes to an osteopath complaining of a stiff neck and shoulder. She might equally well consult another kind of practitioner—either orthodox or heterodox—the principle is much the same. She tells him that she works in an office—but neither of them dwells on the subject. He examines her carefully and diagnoses a lesion in her cervical spine. He gives her some soft tissue treatment to free up the muscles in the area, he performs "a successful adjustment", and she goes away feeling very much better. But the next day she goes back to work, doing the same job, staring at the same illegible VDU screen with her head and neck in the same fixed and contorted posture which caused her problem in the first place. She is pain-free for a time—but before long the symptoms return and she goes back to her osteopath, who fixes her up again—then she goes back to work and the cycle repeats itself. The pain gets steadily worse and the pain-free periods get shorter. Eventually she concludes that there is no real hope of recovery and that the recurrences are "something she has to live with". Somewhere inside she may even think that it is in some way her own fault. For some people in this kind of situation, the only solution becomes a change of employment. One young woman, who had been forced to give up her job as an audiotypist because of a "repetitive strain injury" to her forearms (and who had been unable to work since) put it to me very well. She said "I feel like a broken office machine".

The problem should never have been allowed to reach that stage. Studies have shown that improvements in the working conditions of VDU users may *both* reduce the frequency of musculoskeletal complaints *and* increase working efficiency (as measured by daily output, error rates, etc.). The improvements which are required are often quite simple: better seating, changes in the layout of the workstation, making the screen more legible, improvements in lighting, changes of activity, and so on.

What do the accidents at the chemical works, the loss of the *Herald of Free Enterprise* and the aches and pains of VDU users have in common? Each involves a working system in which human beings interact with machines. *In each case there has been a mis-match between the machine and its user.* This has compromised both the working efficiency of the system and the health and safety of the people involved. The science which deals with problems of this kind is called *ergonomics*. It is the subject of this book.

# Chapter One
# Introduction

## What is Ergonomics?

The science of ergonomics is a little more than 40 years old. In its present form it dates back to World War II and the years which followed (although it has recognizable antecedents which can be traced back much further). Its founders were a group of British scientists, who had been working for the armed services on various projects concerned with the efficiency of the fighting man. (The group included anatomists, physiologists, psychologists and engineers.) They believed that a multidisciplinary scientific approach to the study of working efficiency could be equally relevant to industry under peacetime conditions. But there did not seem to be a name for this area of research, so they had to invent one. After some deliberation they decided upon *ergonomics*. The word is derived from the Greek: *ergos*, work; *nomos*, natural law.

At about the same time a similar discipline evolved in North America, which came to be called *human factors*. Both names are now used on both sides of the Atlantic to mean much the same thing.

People define ergonomics in a number of ways. This is probably the simplest:

*Ergonomics is the scientific study of human work.*

The word "work" can be used in a broad or narrow sense. In the narrow sense it refers to the things we do for economic gain—that is, to earn a living. For most people in an industrial society, work is the portion of their life which is outside their own control—as against their leisure hours, when they may do what they please. This usage leads to odd anomalies. For example, a woman who looks after the home and family whilst her husband "works" is not regarded as a "working woman"—despite the fact that the activities involved are equally onerous, whereas if she were to do the same tasks for someone else's family, in order to earn money, then she would be "going out to work". Similarly, if a man hires himself out to dig someone else's garden, then he is "working", whereas if he digs his own, it is a "leisure activity"—despite the fact that he may be saving money by growing his own vegetables. Thus work in the narrow sense is defined by the social relationship involved, rather than by the task itself.

But in a broader sense "work" may refer to almost any kind of human activity which involves purpose or effort. Thus to speak of "working in the garden" or "working on a theory" does not necessarily mean that we do so for economic gain; and we say that walking up a hill is "hard work". The broader usage is historically older. The Old English *weark* was used in a very general sense and its subsequent specialization came with the rise of capitalism. The word "labour" has a similar history (Williams, 1976). The science of ergonomics deals with work in the broader sense—although work in the narrow sense has been central to its development.

Work generally involves the use of machines or tools. (There are exceptions; but these are not numerous.) Ergonomics is concerned with the design of tools and machines—and with the design of objects and environments for human use in general, from screwdrivers to computer systems and from easy chairs to space vehicles. In some ways it makes more sense to define ergonomics in terms of its role in the design process. This will tend to reflect more accurately what the practising ergonomist actually does.

*Ergonomics is the application of scientific information concerning human beings to the design of objects, systems and environments for human use.*

If an object is intended for human use, it must necessarily be used for some purpose. This purpose can (broadly speaking) be called work. So the two definitions of ergonomics actually mean much the same. *Ergonomics is concerned with the design of working systems in which human beings interact with machines.* (By convention we call these *man–machine systems*.) The two approaches to ergonomics may be summarized in the following phrase, which is more of a slogan than a definition as such:

*Ergonomics is the science of matching the job to the worker and the product to the user.*

An effective match is one which optimizes:

- working efficiency (performance, productivity, etc.);
- health and safety;
- comfort and ease of use.

In this book I shall argue that these criteria generally go together. Or at least that the goals and objectives are compatible: for example, that chairs which are "good for you" will also be comfortable; that products which are easy to use will be used safely and efficiently; that inefficient working systems are often unsafe and conversely that unsafe working practices are often uneconomic; and that work which is stressful and psychologically unrewarding may have a profoundly deleterious effect on the individual's mental and physical well-being. But there are doubtless exceptions to this general principle. Working practices which are "efficient" and apparently innocuous in the short term may have deleterious long-term effects on health. These in turn may have knock-on effects on the efficiency of the working system: due to sickness absence, labour turnover, etc. Where a conflict arises in the working situation between the goals of safety and productivity, its resolution is more of a political (or ethical) question than a scientific one. But in some cases such conflicts may be more apparent than real. They frequently arise from an incomplete cost–benefit analysis of the issues involved; or one which measures only those costs of ill-health which are sustained by the working organization and neglects those costs which are met by the individual or passed on to society as a whole.

## Ergonomics, Work and Health

"If work was really a good thing, the rich would have found a way of keeping it for themselves."

Haitian proverb

This book is mainly about the applications of ergonomics to preventative medicine—and in particular the role of ergonomics in the prevention of work-related musculoskeletal disorders such as back pain, neck pain and the so-called repetitive strain injuries. But we shall also touch on some other closely related problems (accidents, stress, etc.); and inasmuch as the successful resolution of work-related musculoskeletal problems requires the elimination of those factors which lead to the continued overuse of the bodily structures involved, then ergonomics is relevant not only to primary prevention in the workplace but also to those branches of orthodox and heterodox medical practice which are concerned with the management of such conditions. We could call this branch of the subject *clinical ergonomics*.

Work may affect health in a number of ways. Some of these are summarized in Table 1.1. (The list is not complete nor are the categories mutually exclusive.) The first category includes most of the classical "industrial diseases": the dust diseases (pneumoconiosis, silicosis, byssinosis, etc.), the occupational cancers, occupational dermatitis, and so on. In the advanced industrial societies these diseases are on the decrease (although this is not equally true elsewhere in the world). This is partly because hazardous industrial operations are better regulated, and partly because of changes in the nature of the work that people do. The downward trend is likely to continue—although the development of new industrial processes will doubtless bring new hazards (e.g. recombinant DNA technology).

For some purposes (particularly legal and administrative ones) it may be useful to draw a distinction between *occupational diseases* as such and a wider range of conditions which can be characterized as *work-related*. In general, the commonly

*Table 1.1 Work and Health*

1. The worker may be exposed to some injurious agent used in the working process or arising because of it—e.g. toxic chemicals, dust, micro-organisms, allergens, ionizing radiation, etc.
2. The work may be performed under adverse environmental conditions—e.g. heat, cold, noise, etc.
3. The work may involve overexertion or cumulative overuse of musculoskeletal or other bodily structures.
4. The work may be psychologically stressful because it is mentally or emotionally demanding, boring, frustrating or socially alienating—leading to mental and "psychosomatic" ill-health, alcoholism, suicide, etc.
5. The work (or the circumstances under which it is performed) may entail a high risk of accidental injury (e.g. mining, construction work, offshore operations) or physical violence (e.g. police work).
6. The work may promote an unhealthy lifestyle (e.g. obesity in sedentary workers, alcohol-related diseases in publicans, musicians, etc.).

recognized occupational diseases are rare or unknown outside the specific occupational groups who are exposed to the causative agent concerned. (Although insomuch as the harmful agent may be encountered in many walks of life, then the condition may be widespread—e.g. industrial deafness.)

Work-related disorders, on the other hand, are typically conditions of multiple aetiology in which work is a significant contributory factor in some or all cases; and they may occur in a wide variety of working populations. Low back pain, for example, is common among labourers, nurses, truck drivers and office workers; repetitive strain injuries occur in hop pickers, production line workers and keyboard operators. The identification of the underlying *risk factors* may be a complex problem—both epidemiologically and ergonomically.

Work-related musculoskeletal disorders may result from single episodes of overexertion, cumulative overuse or a combination of both. Cumulative overuse may be due to working postures, strenuous physical activity, repetitive motions or any combination of these. Conditions of sudden onset, resulting from single episodes of overexertion, are generally classified for administrative purposes as accidents—as with pack pain of acute onset associated with lifting actions, for example. But this categorization is somewhat arbitrary, since the individual concerned may have been exposed to long periods of cumulative overuse prior to the incident which triggered his current symptoms.

In Sweden, the legal definition of an occupational disease is drawn up much more widely than in most other countries—to include any harmful effect of work other than the results of an accident. Conditions resulting from overexertion and overuse (other than accidents) are referred to as *ergonomic diseases*. About half of the occupational diseases reported in Sweden each year are classified as ergonomic diseases (Figure 1.1).

The role of ergonomics, as it pertains to health and safety at work, overlaps with that of a number of other professional disciplines: occupational medicine, occupational hygiene, occupational psychology, production engineering, produc-

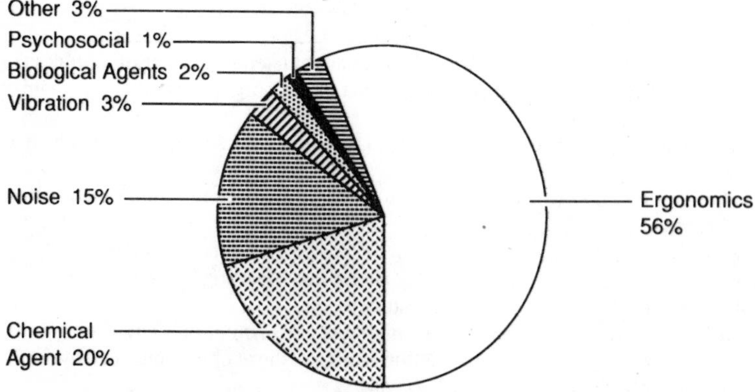


Figure 1.1 *Occupational diseases in Sweden, 1983, classified by causative agent. Data from NBOSH (1987)*

tion management, and so on (Figure 1.2). The ergonomist working in industry is principally concerned with the design of working systems and the working environment. In new technology industries the ergonomist is increasingly concerned with the design of software as well as hardware. The ergonomist's concern with the working environment may well include factors such as temperature, ventilation and noise—although we shall not deal with these in detail in this book. Toxic hazards, dust, ionizing radiation and biohazards are not generally considered to fall within the domain of ergonomics (except insomuch as they might have consequences for task design). But all of these distinctions are essentially arbitrary and the boundaries of the discipline are blurred. (This is probably a good thing.)

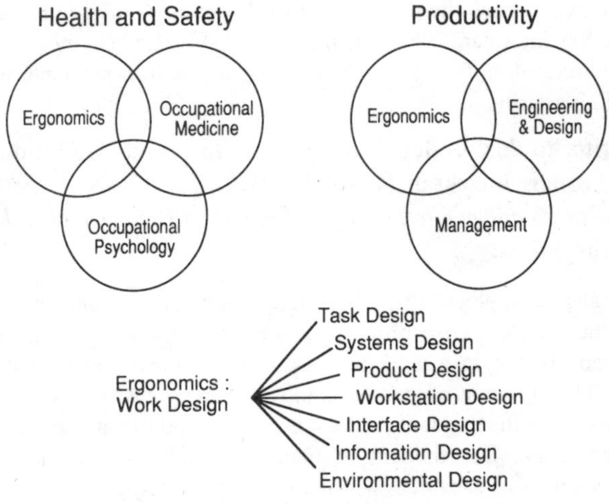


Figure 1.2 *The role of ergonomics*

## A Historical Perspective

The importance of what we would now call ergonomic factors was extensively noted by Bernardini Ramazzini—who is generally recognized as the father of occupational medicine. At the beginning of Chapter 30, about halfway through his *De Morbis Artificum* of 1713, we find the following:

> "So much for workers whose diseases are caused by the injurious qualities of the material that they handle. I now wish to turn to other workers in whom certain morbid affections gradually arise from other causes, i.e. from some particular posture of the limbs or unnatural movements of the body called for while they work. Such are the workers who all day long stand or sit, stoop or are bent double; who run or ride or exercise their bodies in all sorts of ways."

Later on, in a discussion of the diseases of scribes and notaries, he writes:

> "The maladies that afflict the clerks aforesaid arise from the three causes: first, constant sitting, secondly the incessant movement of the hand and always in the same direction, thirdly the strain on the mind from the effort not to disfigure the books by

errors or cause loss to their employers when they add, subtract, or do other sums in arithmetic."

Thus Ramazzini recognized the three principal causes of what are now referred to as *repetitive strain injuries*: fixed working posture, repetitive motions and psychological stress.

The industrial revolution, occurring in Britain from the 1760s onwards, brought fundamental changes to the nature of working life. Working conditions in early nineteenth century England were commonly brutal—as noted, for example, by the social reformer James Kay-Shuttleworth in *The Moral and Physical Condition of the Working Classes Employed in the Cotton Manufacture in Manchester* (1832):

> "Whilst the engine runs the people must work—men, women and children are yoked together with iron and steam. The animal machine—breakable in the best case, subject to a thousand sources of suffering—is chained fast to the iron machine, which knows no suffering and no weariness."

The first writer to deal in detail with the health problems of industrial workers was Charles Turner Thackrah (1795–1833) in *The Effects of the Principal Arts, Trades and Professions and of Civic States and Habits of Living on Health and Longevity* (1832), where he writes:

> "In the details of employments, I have frequently had to animadvert on the *excess of labour*. From this cause a great proportion of town operatives prematurely sink. 'Worn out' is as often applied to a work-man as a coach-horse and frequently with equal propriety. . . . The attention of masters is too exclusively engaged with the manufacture itself—the means of effecting it at the least expense—and the market for its productions. The work-people are less thought of than the machinery: the latter is frequently examined to ascertain its capabilities—the former is scarcely ever."

Friedrich Engels (1820–1895) published his account of *The Condition of the Working Class in England* in 1845. It includes the following description of the yound needlewoman who worked in the sweat shops of London's East End:

> "Enervation, exhaustion, debility, loss of appetite, pains in the shoulders, back and hips, but especially headache, begin very soon; then follow curvatures of the spine, high deformed shoulders, leanness, swollen and smarting eyes, which soon become short-sighted; coughs, narrow chests, shortness of breath, and all manners of disorders in the development of the female organism."

The plight of the seamstresses for a short time caught the imagination of Victorian England—particularly as a result of Thomas Hood's poem *The Song of the Shirt*, which first appeared in Punch in 1844 and which in the words of Engels "drew sympathetic but unavailing tears from the eyes of the daughters of the bourgeoisie".

Our Victorian forebears paid more attention than we do to matters of posture and deportment—especially in the education of young ladies. Figures 1.3 and 1.4 show good and bad positions for playing the piano and the harp—as illustrated by the physician Mathias Roth in a treatise *The Prevention of Spinal Deformities* (1861),

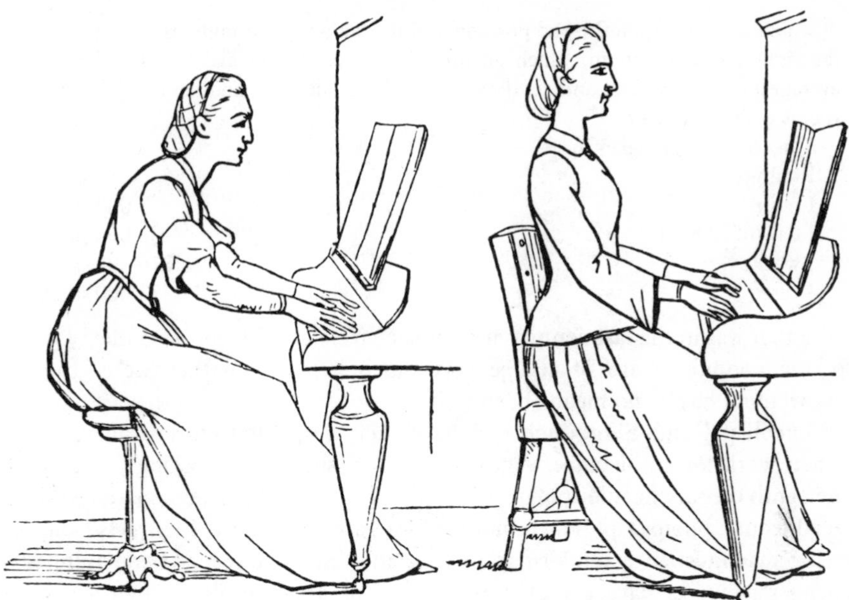

Figure 1.3 *Good and bad postures for playing the piano—from Roth (1861)*

Figure 1.4 *Good and bad postures for playing the harp—from Roth (1861)*

which also dealt *inter alia* with the evils of corsetry. Roth's treatise includes the following passage:

"The introduction of sewing machines should always be advocated by medical men, because their general use will prevent many diseases among the working class, the origin of which depends upon the bad position and the ten or eleven millions of movements of the right arm per annum, which an unhappy needlewoman makes, who works at an average of twelve hours, and six days per week, in one year, and generally confined in rooms with vitiated air.

A needlewoman makes in

| | |
|---|---|
| 1 minute | 50 stitches |
| 1 hour | 3,000 stitches |
| 1 day (of 12 hours) | 36,000 stitches |
| 1 week | 216,000 stitches |
| 1 year | 11,232,000 stitches |

But that was in the last century and things have changed—or have they? Overall working conditions are of course very much better than they were. But the improvement has by no means been uniform.

Whitechapel and Shoreditch still have their garment factories where young women work for subsistence wages. Figure 1.5 shows the working posture of a nineteenth century needlewoman as illustrated by Roth (1861), compared with one traced from a photograph taken in a modern garment factory. Figure 1.6 shows a sewing machinist in a modern factory. Roth's hopes have not been fulfilled—sewing machinists have a higher prevalence of neck problems than any other occupational group known (p. 83).

A data entry clark makes in

| | |
|---|---|
| 1 minute | 300 keystrokes |
| 1 hour | 18,000 keystrokes |
| 1 day (of 6 hours) | 108,000 keystrokes |
| 1 week | 540,000 keystrokes |
| 1 year | 25,920,000 keystrokes |

Read again the words of James Kay-Shuttleworth: the data entry operator is equally chained to the electronic machine. Read again the symptoms described by Engels: VDU users are said to suffer from most of them. In the words of Paul Branton (1988), these are "the new technology sweatshops of the latest wave of industrialization".

## The Division of Labour

The division of labour into specialized trades dates back to the new stone age or earlier. There has been a flint working industry near Brandon in Suffolk since neolithic times.

The benefits to be gained from the *subdivision* of the manufacturing process into discrete tasks were noted by Adam Smith (1723–1790), the founder of modern economic science, in *An Inquiry into the Nature and Causes of the Wealth of the Nations* (1776). He illustrates this in an account of the trade of pin making:

> "One man draws out the wire, another straights it, a third cuts it, a fourth points it, a fifth grinds the top for receiving the head; to make the head requires two or three distinct

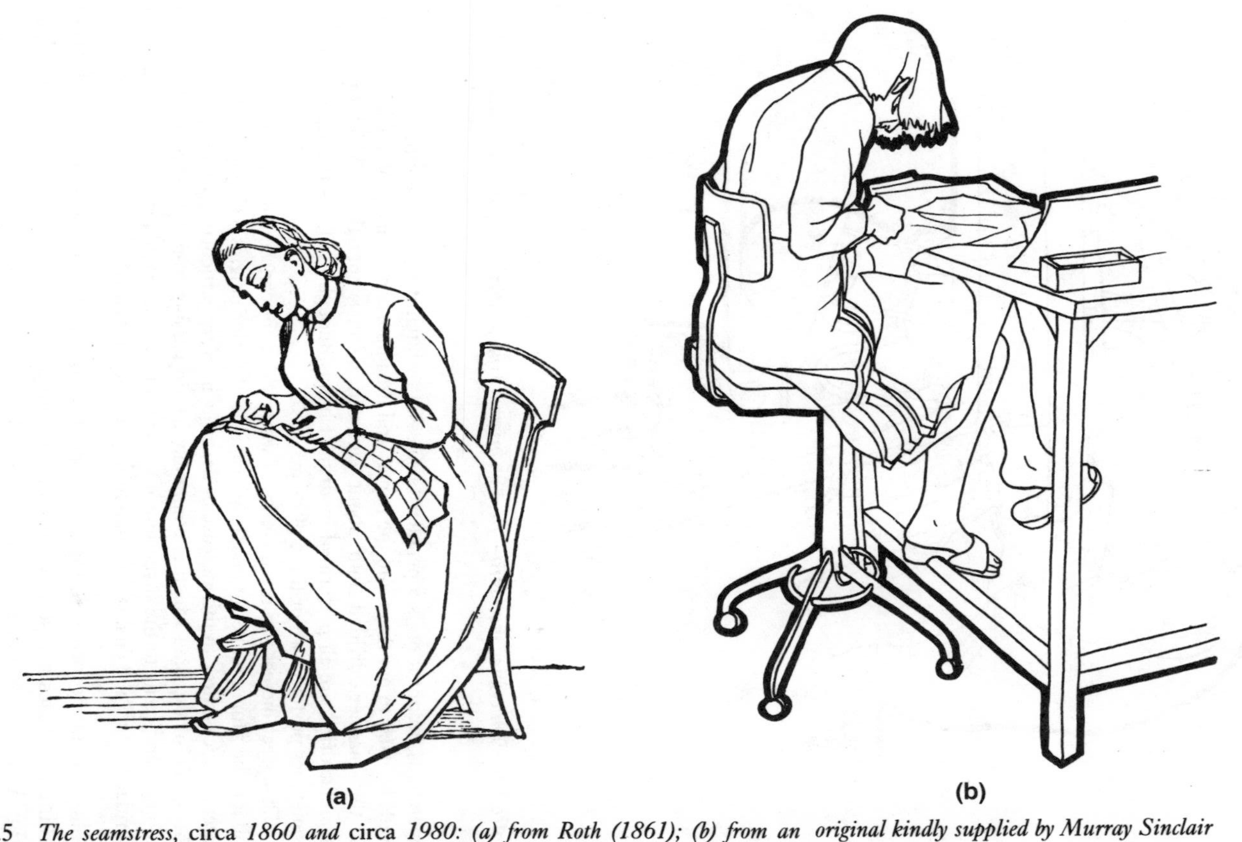


Figure 1.5 *The seamstress,* circa *1860 and* circa *1980: (a) from Roth (1861); (b) from an original kindly supplied by Murray Sinclair*

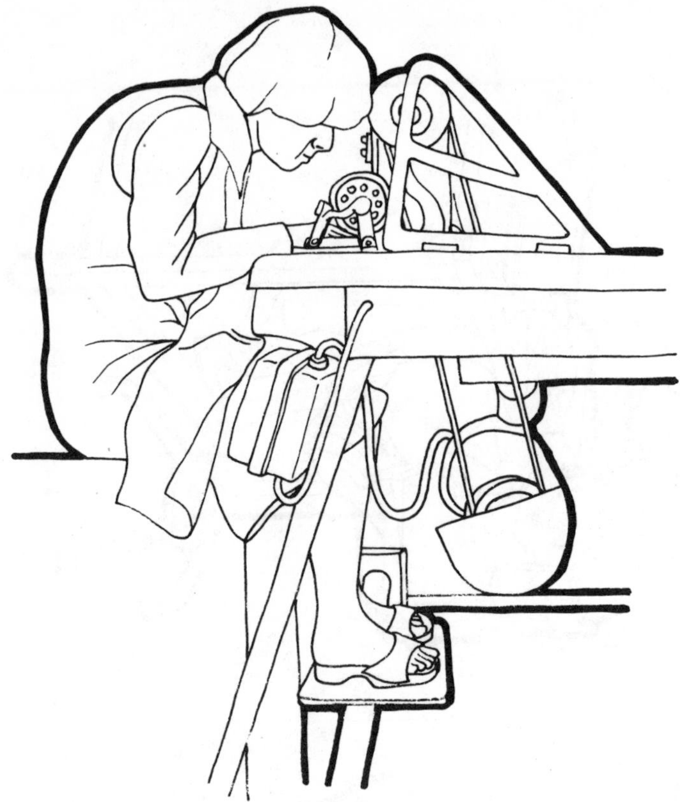

Figure 1.6 *The sewing machinist—from an original kindly supplied by Murray Sinclair*

> operations; to put it on is a peculiar business, to whiten the pins is another; it is even a trade by itself to put them into the paper; and the important business of making a pin is, in this manner, divided into about eighteen distinct operations, which, in some manufactories, are all performed by distinct hands, though in others the same man will sometimes perform two or three of them....
>
> This great increase in the quantity of work which, in consequence of the division of labour, the same number of people are capable of performing, is owing to three different circumstances; first, to the increase of dexterity in every particular workman; secondly to the saving of the time which is commonly lost in passing from one species of work to another; and lastly to the invention of a great number of machines which facilitate and abridge labour, and enable one man to do the work of many."

The systematic or "scientific" rationalization of the subdivision of industrial working processes is a somewhat later development, due in particular to the ideas of the American engineer Frederick Winslow Taylor (1856–1915), as outlined in his *Principles of Scientific Management* (1911), which includes a graphic account of his experiences in the iron and steel industry of the 1890s. Taylor's principal objective was the elimination of "the inefficient rule-of-thumb methods which are still almost universal in all trades, and in practising which our workmen waste a large part of their effort".

Taylor's studies of pig-iron handling and ore shovelling at the Bethlehem steel works in Pennsylvania have become classics. He noted, for example, the waste of effort associated with what we now call *static work*:

> "When a labourer is carrying a piece of pig-iron weighing 92 pounds in his hands, it takes him about as much effort to stand still under the load as it does to walk with it since his arm muscles are under the same severe tension whether he is moving or not."

His study of ore shovelling involved what we would now call a user trial or ergonomic experiment to determine the optimum weight of shovel load. By reorganizing the shovellers' work so as to eliminate wasted effort, Taylor was able to increase their daily average output from 16 tons per man to 59 tons; the cost to the company of handling a ton of ore was halved and the wage of an average labourer was increased by 60% (but the workforce was cut dramatically).

The system of work organization which has subsequently come to be known as *Taylorism* involves the maximum possible subdivision of the working task into discrete elements, thus minimizing the level of skill or knowledge required for the performance of any one. The manual worker becomes a pair of hands, with minimal discretionary control over his working life.

Taylorism is seen in its most extreme form in the repetitive, short-cycle tasks of the industrial assembly line—as pioneered by Henry Ford (1863–1947). In his autobiography of 1922 he describes a survey in which the jobs in his motor car assembly plants were classified according to the demands which they made upon the workers:

> "The lightest jobs were again classified to discover how many of them required the use of full faculties, and we found that 670 could be filled by legless men, 2,637 by one-legged men, two by armless men, 715 by one-armed men, and ten by blind men. Therefore, out of 7,882 kinds of job . . . 4,034 did not require full capacity."

It is widely argued that the fragmentation of the production process into discrete short-cycle tasks achieves high levels of economic performance at the cost of the *dehumanization of work*. From an ergonomic standpoint, fragmented working tasks are unsatisfactory not only because they are psychologically unrewarding and socially alienating, but also because they frequently involve fixed working postures or the repetitive loading of isolated muscle groups (neither of which is physiologically desirable).

## Ergonomics and Design

What do we mean when we say that a product (system, environment, etc.) is *ergonomically designed*? In everyday usage, the label is usually applied in an evaluative sense, to indicate various aspects of *fitness for purpose*—functional efficiency, ease of use, etc. Journalists and advertising copywriters sometimes use (or misuse) the term in slightly curious ways: as in the case of the car which is said to be "ergonomically designed yet comfortable" and the account which I once read of "ergonomic pasta" (which was designed for ease of straining and to retain its sauce). But although fitness for purpose is part of the story, it is not the whole story as the ergonomist sees it.

> "If a product (environment or system) is intended for human use, then its design should be based on the characteristics of its human users."
>
> (Pheasant, 1986, 1987)

We could call this the principle of *user-centred design*.

Ergonomics provides the scientific foundation for a user-centred approach to design—in terms *both* of methodology and research techniques *and* of a steadily growing body of descriptive data concerning the human user and concerning design solutions which have been found satisfactory (or have caused problems) in the past.

I have discussed the concept of user-centred design elsewhere (Pheasant, 1988a). I believe it has a number of characteristic features.

Firstly, user-centred design is generally *empirical*—that is, it is based upon direct observation of the way that people are and the things they do, rather than being informed by "grand theories". This is sometimes called the "bottom up" approach—as against the "top down" approach, which sets out to change things, starting from general principles. This may sound like no more than common sense. But a surprisingly large amount of bad design has been justified in terms of grand theories about what people ought to want—tower blocks and open plan offices, for example. (Although it could be argued that these supporting theories are *post facto* justifications of design approaches that were originally developed for other reasons—principally economic ones.)

The empirical user-centred approach to design is generally cyclic. It is often based upon an initial *task analysis*: that is, an investigation which sets out to provide an operational description of the actions which the user performs in order to do the job at hand with the product concerned. This may be done by means of *user trials*, in which a representative sample of users test an existing product or a mock-up of a new one. The results of such investigations suggest modifications to the design—these are then tested in further trials—and the cycle repeats itself. Traditional craft products (such as hand tools) may evolve over a period of time by a similar cyclic process of trial and error; we call this *vernacular ergonomics* (p. 269).

*The ergonomic approach to design is non-Procrustean*. Procrustes was a bandit in Greek mythology who used to invite travellers to spend the night at his house. He told them that he had a bed which would fit them exactly. To ensure that this was the case he would cut off their feet or stretch them on a rack until he achieved a satisfactory match. Procrustes' bed was made for *the average person*. Ergonomics is concerned with finding the best possible match for the greatest number of people, by adapting the product to fit the user rather than vice versa. Much of the scientific basis of ergonomics is thus concerned with defining the limits of *human adaptability and human diversity*. I have summarized some common pitfalls in this

*Table 1.2 The Five Fundamental Fallacies*

| | |
|---|---|
| No. 1 | This design is satisfactory for me—it will, therefore, be satisfactory for everybody else. |
| No. 2 | This design is satisfactory for the average person—it will, therefore, be satisfactory for everybody else. |
| No. 3 | The variability of human beings is so great that it cannot possibly be catered for in any design—but since people are wonderfully adaptable, it doesn't matter anyway. |
| No. 4 | Ergonomics is expensive—and since products are actually purchased on appearance and styling, ergonomic considerations may conveniently be ignored. |
| No. 5 | Ergonomics is extremely important. I always design things with ergonomics in mind—but I do it intuitively and rely on my common sense, so I don't need tables of data or empirical studies. |

After Pheasant (1986).

area in the *five fundamental fallacies* (Table 1.2). You might care to try applying them to some familiar area of design—clothing or motor cars, for example. A detailed refutation of the fallacies will be found in my earlier book (Pheasant, 1986).

The last fallacy is in some respects the most interesting. It concerns *empathy*—the intuitive act of putting yourself in someone else's shoes, of feeling what it is like to be someone who is different from yourself. Collectively, we do not seem to be very good at this—thus we design tower blocks that nobody wants to live in, open plan offices that nobody wants to work in, computer systems that baffle their users, and so on. Once we can overcome this problem—and that is why we need user trials and other systematic techniques for finding out about user requirements (questionnaires, interviews, etc.)—then much of ergonomics is "common sense". And much of the remainder of ergonomics is, in theory at least, fairly simple. The difficult part is putting it into practice in the real world—where other forces may be at work.

## Ergonomics and Economics

*Work is an economic process. The nature of work is therefore shaped by economic forces.*

The structure of the workforce is changing. Figure 1.7 shows the percentages of the American workforce in different occupational categories throughout this century, as given by the US Congress Office of Technology Assessment (OTA, 1985). Figures for other advanced industrialized countries in the Western bloc are likely to be similar.

The most striking feature of these figures is the decline in the number of agricultural workers—from 33% to 3% over an 80 year period. The percentage of blue collar workers has remained relatively constant by comparison, climbing slightly in the first half of the century and falling in the second. Compared with manufactured goods, the demand for food is relatively inelastic. As farming

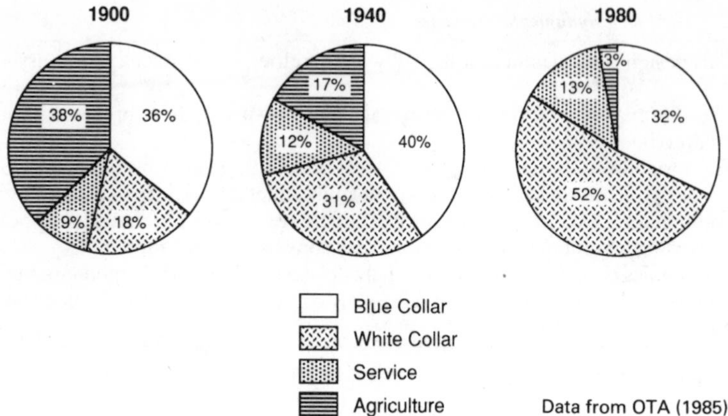


Figure 1.7 *The structure of the US workforce 1900–1980. Data from OTA (1985)*

methods become more efficient, the labour of fewer people is required to feed the population. (With increasing prosperity, people demand a greater diversity of diet; but this demand is satisfied in part by imports of foreign produce.) As manufacturing industry becomes more efficient, new demands are created. The proportion of the workforce required to meet the ever-increasing demands for manufactured goods, created by the mass markets of society, remains relatively high. As the complexity of society increases, it requires a larger proportion of its workforce for its administrative functions, whether these be in the public or private sectors. Thus, between 1900 and 1980, the percentage of the working population in white collar jobs approximately trebled. About 45% of American workers now work in offices.

Many of those classified here as white collar workers are employed in what are elsewhere referred to as the "service industries" (banking, finance, etc.), rather than in the administrative tier of the manufacturing sector of the economy. The service industries are growing in importance. There has been a steady drift away from manufacturing throughout the industrialized world for the last 30 years, due in part to pressure of competition from newly industrialized countries where labour costs are very much lower.

## Productivity

The productivity of a manufacturing operation, or some other unit of economic activity, may be defined thus:

$$\text{productivity} = \frac{\text{output}}{\text{input}}$$

The economist defines *productivity* in much the same way as the engineer defines *efficiency*: the former is a measure of economic performance, the latter is a measure of the performance of a machine.

This basic definition of productivity may be developed in a number of different ways, depending upon which particular inputs and outputs of the enterprise are taken into consideration and how they are measured. The overall productivity of an industrial enterprise could, for example, be defined thus:

$$\text{productivity} = \frac{\text{o/p}}{\text{i/p}} = \frac{V}{C+L+M+S}$$

where $V$ is the value of the goods and services produced,
$C$ is the capital costs of the production process,
$L$ is the labour costs,
$M$ is the materials costs, and
$S$ is the cost of miscellaneous supplies and services which are consumed.

(In economic theory the items in the denominator of the equation are referred to as the "factors of production".)

In practice, however, the word "productivity" is often used in a more limited sense in which only labour is taken into account. Strictly speaking, we should call this *labour productivity*. The reason for this emphasis is that in traditional manufacturing operations, labour costs are likely to consume between two-thirds and three-quarters of the added value of the goods produced; and measures of "productivity" often have implications for the process of collective bargaining between management and labour. So for many practical purposes

$$\text{productivity} = \frac{\text{o/p}}{\text{i/p}} = \frac{\text{units of output}}{\text{man-hours worked}}$$

or

$$\text{productivity} = \frac{\text{o/p}}{\text{i/p}} = \frac{\text{value of products}}{\text{labour costs}}$$

etc.

However, an increase in productivity as defined in one of these ways does not of itself indicate an increase in the overall economic performance of the enterprise. For example, investing in new machinery may increase the productivity of the labour force, but the gains may be offset by the capital costs incurred.

Ergonomic changes in the design of working systems can be shown to be economically cost-effective both *directly*, in terms of performance improvements, and *indirectly*, in terms of decreased operating costs—particularly with respect to the indirect costs of sickness absence, labour turnover, etc. In other words, ergonomic changes may both increase the numerator and reduce the denominator of the productivity equation.

## Performance

Performance of a task may be measured in terms of output of work (per unit time), error rate, or some combination of these. To take an example from traditional manufacturing technology, Corlett and Bishop (1978) described an ergonomic study of spot welding machines, in which design improvements which enabled the operators to adopt a more comfortable working posture led to an increase in output which was equivalent to a 21% improvement in productivity (as measured by the direct labour cost of making a given number of welds).

Performance improvements of a similar magnitude have been noted in VDU-based data entry tasks. Dainoff and Dainoff (1986) report an experimental study in which a VDU workstation designed according to commonly accepted ergonomic guidelines was compared with one which deliberately broke most of the rules. (The workstations differed with respect to seating, keyboard height, visual angle, lighting, glare, etc.) Subjects were recruited from an agency. They performed an experimental task involving data entry and editing under realistic conditions. A composite performance measure was used which took into account both speed and errors, and the subjects were paid by results. Performance was 25% higher at the ergonomically designed workstation; and when differences in the lighting were eliminated, there was still an 18% performance difference. The subjects preferred the ergonomically designed workstation and reported less back and shoulder pain when using it.

In practical terms, reductions in error rate may be just as important as increases in output. Accidents due to human error can be costly—although these costs are often passed on to the insurers. But human error can also result in ongoing inefficiencies in the operation of every working system—damage to capital equipment, downtime on plant and machinery, wastage of raw materials, defective output, misinformation, and so on. Data entry errors in computer systems are a case in point. Bailey (1983) cites the case of an internal investigation conducted by a large American communications company, in which it was calculated that the cost of correcting some 50 million computer errors *per annum* was of the order of $10 million.

Ergonomic improvements to the design of working systems may reduce these costs. Ong (1984) studied data entry staff at a Singapore airline terminal before and after ergonomic changes. The hardware modifications would not have involved much capital outlay. The lighting was improved and the operators were provided with document holders and footrests to improve their working postures; they were also given more rest pauses. Output (measured in keystrokes per hour) increased by 25%. But there was an even more striking reduction in the error rate—from 1.5% (i.e. 1 character in 66 incorrect) to 0.1% (1 character in 1000 incorrect). Given the magnitude of these differences, there cannot be much doubt that the capital outlay would have been rapidly repaid. Note also that although the time spent at the terminal was shorter (owing to the extra rest allowance), the daily output was higher. Before the ergonomic changes were made, a very high proportion of the workers reported musculoskeletal aches and pains. These figures were dramatically reduced—particularly the neck and shoulder problems, which were cut by more than half.

The effects of the physical environment on performance may be confounded with other factors—as illustrated by the classic series of studies conducted at the Hawthorne electrical assembly works in Chicago in the 1920s (Roethlisberger and Dickson, 1949). The original aim of these experiments was to investigate the effects of lighting on performance. It rapidly became clear, however, that other influences were at work. In one experiment, for example, output steadily increased as lighting levels were reduced—up to the point when workers complained that they could no longer see to work. It was concluded that it was the act of intervention itself—that is, the fact that the workers were taking part in an "experiment"—that was affecting performance. This has come to be known as the *Hawthorne effect*.

This finding does not mean that the physical working environment is of no importance (as some commentators have argued). In all probability the visual demands of the tasks at Hawthorne were not particularly great (Boyce, 1981). In some circumstances a well-motivated working group can maintain a high level of output despite unsatisfactory working conditions – at least on a short-term basis. But this may have long-term costs with respect to work-related ill health. Note also that in practice *negative Hawthorne effects* may be encountered: that is, there may be a resistance to change because people have adapted to the existing circumstances. And both positive and negative Hawthorne effects are very much bound up with the social climate of the organization and its labour relations.

## Costs of Musculoskeletal Disorders

The principal ways in which work-related musculoskeletal disorders may impair the economic performance of an industrial enterprise are summarized in Table 1.3. The list is based upon a brainstorming session at a workshop on RSI conducted by Peter Buckle at the Robens Institute.

*Table 1.3 Costs of Work-related Musculoskeletal Disorders*

1. *Direct Effects on Production*
   Reduced output, damage to plant and materials, defective product, etc.—leading to inability to meet production deadlines, poor customer service, overmanning, etc.
2. *Sickness Absence Costs*
   Benefit payments
   Overmanning
   Training of replacement workers
   Hiring of agency staff
3. *Labour Turnover Costs*
   Recruitment
   Training
4. *Litigation/Insurance Costs*
5. *Opportunity Costs with Respect to the Above*

Common sense tells us that musculoskeletal discomfort is likely to impair the individual's working efficiency, leading to a variety of diseconomies in the working system. In most types of organization is would be unusual for a machine to stand idle because its operator was off sick. So, in addition to sickness benefits, the costs of finding a replacement are incurred. These may include training costs, the hiring of agency staff, etc. On some types of production lines it would not be unusual to find about 10% of the workforce absent at any one time (of which half of the absences might be due to musculoskeletal problems). Thus an equivalent degree of *overmanning* will be required to keep the line running.

An annual labour turnover of 25% would not be unusual on an industrial production line. Other figures I have heard quoted include 30% for a population of hospital workers; 40% for a data processing operation; and a staggering 200% for supermarket checkout workers. A significant proportion of this labour turnover is often attributable to work-related musculoskeletal problems. Stubbs *et al.* (1986) found that 12% of nurses, intending to leave the profession for good, cited back pain as the main cause or as a contributory factor. The percentage would probably be higher in some assembly line jobs—although we do not have exact figures.

Absenteeism, labour turnover and litigation also carry *opportunity costs*. In other words, managerial staff and resources must be allocated to dealing with these problems, rather than doing something more directly productive.

As recruitment becomes more difficult, it becomes more expensive—in terms of both direct costs and opportunity costs. Industrial employers in small towns, where there are other enterprises competing for a limited labour pool, may find recruitment problems becoming particularly acute if the factory acquires a bad reputation for having "repetitive strain injuries", etc.

## The Kongsvinger Studies

The most detailed cost–benefit analysis of these issues published to date is that of Spilling *et al.* (1986), which is based upon ergonomic studies conducted at the Standard Telefon og Kabelfabrik (STK) plant at Kongsvinger in Norway. These studies have also been described by Westgaard and Arås (1984, 1985) and Aarås (1987).

Kongsvinger is a small town about 60 miles from Oslo. The STK plant employs a workforce of about 100 people (mainly women) in the assembly of parts for telephone exchanges. In 1975, workstations throughout the plant were redesigned according to ergonomic principles. Some workstations, which had previously been of a fixed height, were made adjustable, so as to enable the worker to alternate between a standing and a sitting position. Others were tilted to improve the posture of the worker's head and neck. Seats with armrests were provided and tools were counterbalanced (thus reducing the loading on shoulder muscles, etc.). Lighting and ventilation were improved.

Average musculoskeletal sick leave in the seven-year period preceding the ergonomic changes was equivalent to 5.3% of total production time. The figure climbed steadily throughout this period, reaching a peak of 10% in the final year.

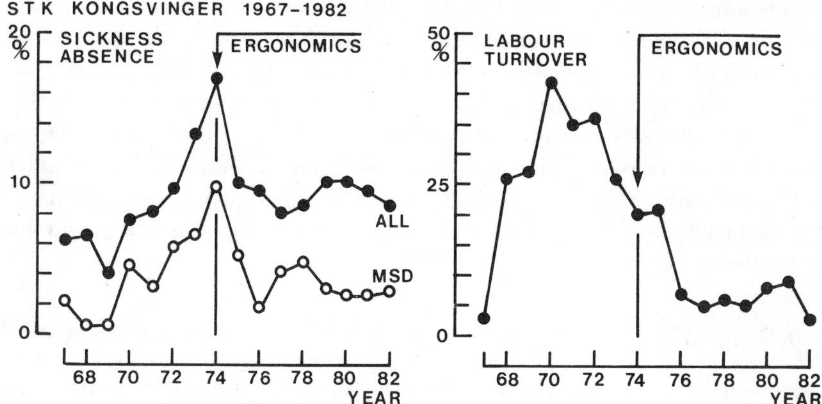


Figure 1.8 *Sickness absence and labour turnover at STK Kongsvinger before and after ergonomic improvements—data from Spilling* et al. *(1986)*

In the seven years following the ergonomic changes, the figure remained relatively steady about an average of 3.1% (Figure 1.8). Labour turnover fell from an average of 30.1% per annum in the period preceding the changes to 7.6% in the period following the changes.

The financial consequences of these differences were analysed in detail using sophisticated accounting procedures. The capital outlay required to make the changes was amortized over a twelve-year period (the projected lifetime of the installation). Reductions in recruitment costs, training costs, instructors' salaries and sickness benefits were taken into account. It was calculated that an investment of just under 340,000 Norwegian kroner (NKr) had saved the company more than 3.2 million NKr in operating costs over the twelve-year period—resulting in a net gain of close on 2.9 million NKr (11 NKr = £1).

For a further discussion of the cost-effectiveness of ergonomics, see Simpson (1988a) and Corlett (1988).

## The Distribution of Costs

Suppose you are the manufacturer of a particular product. Your production line is organized according to Taylorist principles and repetitive strain injuries are common. There is a high rate of absenteeism and labour turnover—but the work is unskilled, so there is a steady supply of replacement labour and training costs are low. You decide that the only way you could eliminate these problems would be by fundamental changes in the production process—which would raise your production costs to a point at which the selling price of your product became uncompetitive. Since the law only requires you to take such measures to guard the safety and health of your workers as are "reasonably practicable", you decide to leave things as they are.

Opting for a higher incidence of work-related disorders in the interest of keeping down your production costs is a viable strategy only if you do not have to continue to support those members of your workforce who are no longer able to work for you. But if you do not support the people you have disabled, then somebody else must. In practice this will often be the community at large. So in effect the community at large is bearing some of your production costs. This is a hidden subsidy for your firm—which keeps down the sale price of your product in other words, those members of society who do not wish to purchase your product are subsidizing those people who do. From your point of view this is a good thing, but for the rest of us it is not.

## Legislation and Common Law

If safe working practices were always cost-effective, there would be little or no need to legislate against unsafe ones. The fact that all civilized countries have felt the need to enact such legislation suggests that, in matters of health and safety, conflicts of interest between the individual worker and the enterprise for which he works are by no means uncommon. In essence, the machinery of the law is invoked, to make unsafe working practices more costly, thus shifting the point of balance of the cost–benefit equation in favour of the individual so as to achieve ethically or socially desirable ends such as the prevention of human suffering, or the reduction of costs passed on to the community at large.

A detailed discussion of health and safety legislation is beyond the scope of the present book—not least because of the radical differences which exist in the nature and provisions of such legislation in different countries.

In the UK—and in certain other countries having similar legal systems—the rights and duties of employer and employee are subject to the provisions of both statute law and common law. A worker who believes that he has been injured by his work may decide to seek redress from his employer in the civil courts. English common law, as it relates to such disputes, mainly revolves around the *tort* (French: wrong) of *negligence*, which for present purposes we could define as the failure to act with a reasonable degree of prudence, with regard to a reasonably foreseeable danger. (A tort is a breach of civil duty—in this case the duty of care, owed to one's neighbour, not to cause him injury.)

The modern common law concept of negligence dates back to the time of the industrial revolution, when, in matters related to personal injury, it replaced the medieval tort of *trespass*. This change was an essentially pragmatic response to the requirements of the times: the common law duties imposed by the older legal framework would have been incompatible with the rapid process of industrial and economic growth which was occurring at the time.

From the point of view of the plaintiff (in this case an allegedly work-injured employee) the process of litigation is a bit like trying to penetrate the defences of a fortified castle. To gain redress at law he must in general show:

(i) that he has indeed been injured;
(ii) that the injury was a direct consequence of risks to which he had been exposed in the course of his work;

(iii) that his employer had been deficient in his duty of care (i.e. that he had been negligent).

In respect of this last point, the employee must show:

(a) that the risk to which he was exposed was a *reasonably foreseeable* one;
(b) that it would have been *reasonably practical* to cirvumvent this risk.

The defendant (i.e. the employer and his legal representatives) may mount a defence at any or all of these stages. In the case of an RSI claim, he might, for example, call a medical expert, who could confidently assert that, in his experience, conditions of this sort are all in the mind (p. 80). Alternatively, the expert witness might argue that the condition could equally well have been caused by non-occupational activities—for example, peeling potatoes. In a sense, these commonly employed defences both exploit and ignore the multifactorial nature of work-related musculoskeletal problems, which in reality have *both* organic *and* psychological components and which characteristically result from a cumulative process of damage in which a number of risk factors play a part. (Although the notion that the time spent each day by a working mother in peeling potatoes could have a greater effect on the muscles of her forearm than the hours she spends at the production line would seem to tax the credulity of anyone other than a professional witness acting on behalf of her employer.) At present there seems to be something of a disparity between the legal processes by which civil disputes concerning work-related musculoskeletal injuries are resolved and the weight of the medical and scientific evidence concerning the nature of these injuries. The existence of this disparity is disturbing.

## Chapter Two

# The Physiology of Work

All observable human behaviour is mediated by muscle activity. Skeletal muscle accounts for 40–45% of body weight in an adult man of more or less average body build and 25–35% in an adult woman. A resting person consumes around 0.2 litres of oxygen per minute and has an energy expenditure of about 1 kcal/min*. During intense physical activity, the oxygen consumption of an averagely fit man can reach 3 l/min and his energy expenditure can reach 15 kcal/min. This difference is attributable almost entirely to the metabolic demands of the working muscle.

A muscle cell (or fibre) may exist in two states: passive (resting) and active (contracting). When a muscle fibre is "switched on" by its motor nerve, it briefly enters a state of excitation in which a complex sequence of electrical and chemical reactions occur, by means of which chemical energy is converted into mechanical energy and heat. During the active state, the muscle fibre is capable of exerting a tension between its ends. If the tension is unresisted, the muscle shortens and movement occurs. The tension is due to certain biochemical interactions between filamentous protein molecules (known as actin and myosin) which are arranged within the fibre in regular arrays. Shortening occurs when these filaments slide in between each other.

## Bioenergetics of Exercise

The immediate energy source for muscular contraction (as it is for most other metabolic processes) is the splitting into its component parts of a substance called adenosine triphosphate (ATP). This occurs at certain locations on the surface of the myosin filaments, known as *binding sites*, which carry an enzyme for the purpose (myosin ATPase). There is only enough ATP stored in a muscle to fuel its contraction for about 2 seconds, but as the ATP is broken down it is constantly being re-formed in a coupled chemical reaction involving another energy-rich substance called *phosphocreatine*. This is an anaerobic process—that is, it does not depend upon a supply of oxygen. There is enough phosphocreatine stored in a muscle to keep it working at full capacity for about 10 seconds at the most. When these reserves have been exhausted, energy for the resynthesis of ATP must come from other, less direct, sources.

* This is called the *basal metabolic rate*—the exact figure varies with age, sex, body build, etc. 1 kcal = 4.186 kJ.

Dietary carbohydrate is stored in the body as *glycogen*—a polymer of the simple sugar glucose ($C_6H_{12}O_6$). The glycogen molecule is built from several thousand units of glucose arranged in a branching tree-like structure. The muscles of an adult man contain about 350 g of glycogen. A reserve of about 100 g is stored in the liver—this may be called upon to maintain the blood sugar between meals and during the night, etc. (You have to keep using it if you miss breakfast.) Glucose (derived from muscle glycogen, etc.) may be broken down by two different metabolic processes: the *aerobic system* and the *anaerobic system*. The former requires oxygen, the latter does not. Both result in the synthesis of ATP but the former does it more efficiently.

The *anaerobic system* splits one molecule of glucose into two molecules of a substance called lactic acid ($C_3H_6O_3$). This process is called *anaerobic glycolysis* (or *glycogenolysis*). It is particularly important in brief episodes of intense exercise—of up to 2–3 min duration at the most. But the process of anaerobic glycolysis is self-limiting, because the accumulation of lactate in the muscles (which spills over into the bloodstream) leads to muscular fatigue—a progressive failure of the muscle's contractile system due probably to the increasing acidity. The total quantity of energy available from anaerobic glycogenolysis is of the order of 30–100 kcal, depending on the individual's capacity to tolerate the discomfort which results from lactate accumulation.

The *aerobic system* breaks down glucose completely, in the presence of oxygen, into carbon dioxide and water. In addition to glycogen and blood glucose, the aerobic system may also employ as substrates the metabolic derivatives of other foodstuffs (fats, proteins) and it may complete the breakdown of the lactate by-products of anaerobic processes. The aerobic system is the principal energy source for exercise of more than a few minutes duration. Its capacity is limited by the availability of oxygen to the working muscle—which is dependent in turn upon local muscle blood flow as well as on cardiac and respiratory capacity and on a number of other physiological factors. The maximum rate of oxygen uptake ($\dot{V}_{O_2\,max}$) which an individual is able to achieve during intense exercise is a good index of his or her overall level of physical fitness. The maximum rate of oxygen uptake is often referred to as the *maximum aerobic power* or *aerobic capacity*. (Some physiologists object to the latter term.)

The aerobic and anaerobic systems are complementary (see Table 2.1). At the beginning of exercise, the anaerobic always predominates—because the cardiorespiratory system takes time to adapt to its new level of demand. A sprint event like the 100 m dash is probably run entirely on anaerobic processes—the runner may not even take a breath. Blood lactate may reach 20 times its resting level. But the "supercharging" effect of phosphocreatine is probably the critical limiting factor. As the phosphocreatine store is depleted, the power output falls.

In exercise the lower intensity and longer duration, aerobic processes take over after the first minute or so. Once the oxygen consumption reaches a steady state level appropriate to the intensity of the exercise concerned, the anaerobic processes shut down—so that, provided the workrate stays within reasonable limits, the lactate level will remain more or less constant until the end of the work

*Table 2.1 Energy Sources in Running Events*

| | Distance (m) | Duration (s) | % Energy requirements from aerobic metabolism | Limiting factor |
|---|---|---|---|---|
| | 100 | 9.9 | 0 | Phosphocreatine depletion |
| | 200 | 19.8 | 10 | |
| | 400 | 43.9 | 65 | Phosphocreatine depletion and lactate accumulation |
| | 800 | 1:43.5 | | |
| | 1,000 | 3:36.0 | 65 | Lactate accumulation |
| | 5,000 | 13:11.7 | 87 | Glycogen depletion |
| | 10,000 | 27:21.5 | 97 | |
| Marathon | (42,000) | 130:31.6 | 100 | |

Modified from Newsholme and Leech (1988). Duration figures are winning times for the men's events in the 1988 Seoul Olympics.

period. An untrained person can exercise at around 30–50% of his maximum aerobic power without further elevating his level of blood lactate. For a top athlete, trained for an endurance event, the critical level may be as high as 85% maximum aerobic power.

A well-trained marathon runner may have a lactate level of only 2–3 times the resting level at the end of the event—so the fatigue he experiences must be due to other factors. He will probably have enough glycogen in his muscles to keep him going for about 70 min—and the liver store, if it were all used up, would be good for another 20 min. He can also get energy from the metabolism of fat. The proportion of the energy output which is derived from fat metabolism starts at zero and gradually climbs to about 50% by the end of the race—as the glycogen stores are depleted. But fat metabolism is a relatively slow process and it requires more oxygen than glycogen metabolism to generate a given quantity of ATP. So, when the glycogen stores are depleted, the pace must necessarily fall. The ultramarathon runner (as in the 53 mile London to Brighton run) will reach a state at which he is almost totally dependent on fat metabolism for his energy requirements. Low blood sugar levels now become the major problem. The brain is dependent upon blood glucose for its metabolism. The perceived effort of running gets greater (p. 158). The runner may become disorientated, his co-ordination may suffer and his consciousness may fail. Other sources of fatigue in distance running may include dehydration, hyperpyrexia (overheating) and an alteration in the balance of amino acids in the bloodstream.

In a middle distance event such as the 1500 metre or mile run, the two systems are employed more or less equally—with the anaerobic predominating at the start and in the final burst of speed, and the aerobic predominating during the middle (steady state) part of the run. The exact proportion of energy derived from the aerobic and anaerobic processes will vary from runner to runner. Even in the

10,000 metres, where anaerobic processes yield only about 3% of the energy requirements, they may still consume 30% of the glycogen store. The management of these resources is a critical part of the strategic element in middle distance running.

For a more detailed account of the bioenergetics of muscle, with representative references, see, for example, Fox and Matthews (1981) or Åstrand and Rodahl (1986); for a concise account of running in particular, Newsholme and Leech (1988).

## Muscle Fibre Types

Muscle fibres vary in their biochemical and functional properties. Two main categories are generally recognized: slow twitch or type I fibres and fast twitch or type II fibres. (Subdivisions of type II fibres are sometimes also described.) Fibre types may be recognized in microscopic preparations of autopsy or biopsy specimens by means of certain special staining techniques, which show up the enzymes which give the fibres their particular functional characteristics.

Both fast fibres and slow fibres are capable of operating both anaerobically and aerobically, but fast fibres have a greater capacity for anaerobic function and slow fibres have a greater capacity for aerobic function. *Fast fibres* are capable of contracting more rapidly at any given level of resistance (i.e. they have a greater power output) and they also probably have a greater contractile strength per unit cross-sectional area, but *slow fibres* fatigue less rapidly.

A group of muscle fibres which is innervated by a single motor nerve is called a motor unit. All of the muscle fibres in any particular motor unit are of the same functional type. Studies of glycogen depletion in biopsy specimens of human muscle, taken after different types of activity, confirm that fast fibre motor units are selectively recruited for short bursts of intense activity at a high power output, whereas slow fibre motor units are selectively recruited for endurance activities performed at lower levels of power output (Gollnick *et al.*, 1973). Slow fibre motor units tend to be smaller in size—and smaller motor units are the ones which tend to be recruited at lower levels of contractile force. So it is likely that there are many activities (such as writing, typing or walking at a gentle pace, perhaps) which will be performed almost entirely using a population of slow fibre motor units.

Overall, fast and slow fibres are probably present in the human body in approximately equal numbers, but there are considerable differences in the fibre populations of different muscles. In one study of autopsy specimens of various muscles, it was found that the majority clustered around an average value of a little under 50% slow fibres. But the soleus muscle of the calf, which has long been recognized as a "postural muscle", was predominantly (88%) made up of slow fibres, whereas the orbicularis oculi muscle (with which you blink your eyes) was predominantly (86%) made up of fast fibres (Johnson *et al.*, 1973).

There are also considerable differences in fibre populations between individuals. Fox and Matthews (1981) summarize a number of studies of the percentages of fast and slow fibres in the lower limb muscles of trained athletes.

Endurance athletes (such as marathoners, orienteers and cross-country skiers) have more slow fibres than untrained men and women, whereas power athletes (particularly sprinters, jumpers and throwers) have more fast fibres.

It is not generally believed that training can convert fast fibres into slow fibres (or vice versa)—although there is some controversy concerning this issue. It may well be that fibre populations are genetically determined, and that individuals gravitate towards those athletic events for which their fibre populations make them suited. But it does seem likely that different types of training may lead to a selective hypertrophy of fast or slow fibres, respectively, so that the percentage of the muscle's overall bulk which is made of a certain fibre type may increase. High-resistance weight training (at loads greater than 50% of the muscle's capacity) is thought to affect fast fibres to a greater extent. Endurance training increases the aerobic capacity of both fibre types—but as the levels of contractile force are low, we should expect such hypertrophy as occurs to be predominantly in the slow fibres. Walking and jogging are both in this category (Fox and Matthews, 1981).

There is a high correlation between maximum aerobic power and the percentage of slow twitch fibres in the muscles of both trained and untrained people; but for any given percentage of slow fibres the trained person will have a greater $\dot{V}_{O_2 max}$ (Bergh *et al.*, 1978). An individual's maximum rate of oxygen uptake is a function both of the capacity of his heart and lungs to deliver oxygen to the working muscles and of the capacity of the muscles to use it. Cardiac output is probably the limiting factor in most cases (at least in untrained people), but the relationship is a complex one.

## Energy Expenditure

Energy expenditure is difficult to measure directly, so, in practice, the physiological cost of work is generally measured in terms of the oxygen consumption. For most practical purposes we may assume that:

$$\text{energy expenditure (kcal/min)} = 5 \times \text{oxygen consumption (l/min)}^*$$

In most activities, energy expenditure increases linearly with body weight. For convenience, energy expenditure is often quoted for a notional "standard person" weighing 70 kg†.

The total daily energy expenditure of a typical male office worker is of the order of 2,500 kcal; for an industrial worker doing a relatively light job it would be about 3,000 kcal; and for a coal miner about 3,500 kcal. An army recruit, during his strenuous period of basic training, consumes about 4,000 kcal per day. Dockers

* Energy expenditure is also sometimes expressed in multiples of the resting level or METs, where 1 MET = 3.5 ml $O_2$/kg body weight/min.

† In the UK, at the present time, the average adult man weighs 75 kg and the average woman weighs 63 kg; in the USA the average adult man weighs 78 kg and the average woman weighs 65 kg (Pheasant, 1986).

and lumberjacks may reach 6,000 kcal on a very heavy day working overtime (Edholm, 1967).

The maximum daily energy expenditure that a fit man can be expected to maintain on a habitual basis (without deleterious long-term effects) is said to be 4,000 kcal (Edholm, 1967) or 4,800 kcal (Lehmann, 1958). Allowing for basal metabolism during the hours of sleep and energy consumed during domestic activities (about 2,000–2,300 kcal altogether), this works out at a maximum acceptable energy expenditure, averaged out over the working day, of 4.5–5 kcal/min. This is equivalent to about one-third of the maximum aerobic power of an average man.

Conversely, it is also said that, for the maintenance of physical fitness, daily energy expenditure should not be less than 3,000 kcal in men or 2,400 kcal in women (Grandjean, 1988). So the typical male office worker needs to make up a shortfall of about 500 kcal in his leisure activities. An hour's walking, at an average pace, consumes about half of this.

The physiological demands of work may be roughly classified as shown in Table 2.2. These figures (which are for an averagely fit man in his twenties) are from Åstrand and Rodahl (1986)—other sources quote slightly different figures.

Representative figures for the energy expenditure of various working and leisure activities are shown in Figure 2.1 (compiled from various sources). These are typical or mid-range values for a 70 kg person performing the task under ordinary (non-competitive) conditions. In practice, energy expenditure will vary considerably with rate of work, level of loading, and so on (and different sources may therefore quote different figures for some of these activities).

An individual's capacity to perform physically strenuous work will be determined by his or her maximum aerobic power (as measured by maximum oxygen uptake)—which is dependent on age, sex, physical conditioning, and so on. On average, the maximum aerobic power of untrained women is about 70% of that of untrained men of the same age (Åstrand, 1960). The average maximum aerobic power of a group of sedentary office workers is likely to be 10–20% lower than that of a comparable group of heavy manual workers (e.g. lumberjacks); and in normal subjects regular physical training can be expected to increase maximum aerobic power by a similar amount. The measurement of maximum oxygen uptake is a

*Table 2.2 Physiological Classification of Working Activities*

| | Oxygen consumption (l/min) | Energy expenditure (kcal/min) | Heart rate (beats/min) |
|---|---|---|---|
| Light | <0.5 | <2.5 | <90 |
| Moderate | 0.5–1.0 | 2.5–5.0 | 90–110 |
| Heavy | 1.0–1.5 | 5.0–7.5 | 110–130 |
| Very heavy | 1.5–2.0 | 7.5–10.0 | 130–150 |
| Extremely heavy | >2.0 | >10.0 | 150–170 |

After Åstrand and Rodahl (1986).

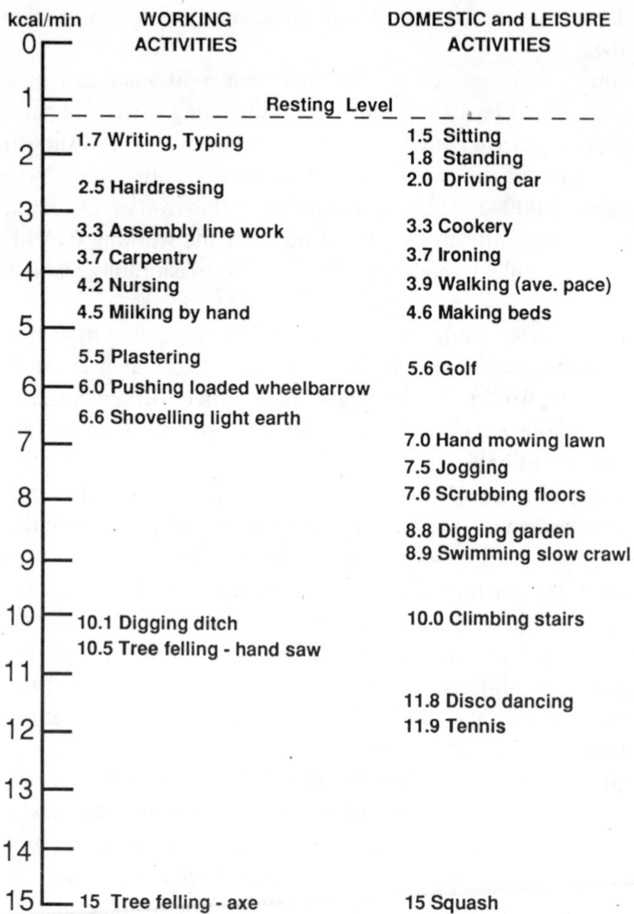


Figure 2.1 *Typical levels of energy expenditure for various activities—data from Durnin and Passmore (1967), NIOSH (1981), McArdle* et al. *(1986) and Åstrand and Rodahl (1986)*

laborious (and potentially hazardous) affair—so reliable figures for the general population are unavailable. Rough estimates for the US working population, based on data from NIOSH (1981), are given in Table 2.3. A symmetrical distribution about the average value was assumed—see p. 118 for an explanation of the statistical concepts involved. These figures are broadly comparable with those of Åstrand (1960). As a rough guide, it may be appropriate to take a "standard reference value" of 15 kcal/min in men or 10.5 kcal/min in women (NIOSH, 1981).

We noted earlier that untrained people can sustain a level of energy expenditure equivalent to 33–50% of their maximum aerobic power without the accumulation of lactic acid leading to fatigue. Åstrand (1967) found that, left to their own devices, building workers pace themselves at about 40% of maximum aerobic

*Table 2.3 Estimated Maximum Aerobic Power of US Working Population*

| Age | Maximum aerobic power | | | | | |
|---|---|---|---|---|---|---|
| | Men | | | Women | | |
| | 5 percentile | 50 percentile | 95 percentile | 5 percentile | 50 percentile | 95 percentile |
| 20 | 13 | 19 | 25 | 9 | 11 | 13 |
| 40 | 11 | 15 | 19 | 7 | 10 | 13 |
| 60 | 9 | 13 | 17 | 5 | 9 | 13 |

power. Studies of industrial workers have found somewhat lower figures (NIOSH, 1981). It has therefore been proposed that: "the average energy expenditure for continuous work (e.g. over an 8 hour working day) should not exceed 33% of the worker's maximum aerobic power" (NIOSH, 1981).

For a "standard reference man" this works out at 5 kcal/min—a figure which has been widely quoted in the ergonomics literature (Lehmann, 1958, 1962; Murrell, 1969; Grandjean, 1988; etc.). But note that insomuch as this figure is based upon the capacity of an *average* man, it will be beyond the capacity of 50% of the male population.

Consider the case of a shovelling task which requires an energy expenditure of 6 kcal/min. This would be well within the comfortable range of workloads for a fit young man who had a maximum power of 24 kcal/min and we should not expect him to find it particularly arduous. But for a 60-year-old man with a maximum power of only 9 kcal/min (i.e. in the lower part of the normal range for his age) it would be very arduous indeed—and we should not expect him to be able to maintain that level of work output for more than a short time.

Maximum oxygen uptake is usually measured in tasks like bicycle ergometry or treadmill running, which principally involve the large muscle groups of the lower limbs. Treadmill running tends to give a higher figure (5–15%) than bicycle ergometry. Tasks which mainly involve upper limb muscles give lower figures than bicycle ergometry—probably because the quantity of working muscle involved is less. These differences may be important when it comes to setting limits for lifting and handling tasks (p. 310).

In many real working tasks, periods of intense physical activity alternate with rest periods or periods of lighter work. In general, this is likely to maximize the total quantity of physical work the individual is able to perform during the working day—compared, for example, with working at a steady but lower level (Åstrand, 1960). In such cases, work should be designed so that the overall energy expenditure—averaged over the working day and taking into account both work and rest—remains within acceptable limits. A number of formulae have been proposed for calculating the rest allowances which are required in physical work (e.g. Spitzer, 1951; Murrell, 1969). Such equations have the following general form:

$$\frac{r}{t} = \frac{E-A}{E-B}$$

where $r$ = resting time (in minutes),
$t$ = total working day (in minutes),
$E$ = energy expenditure during working task,
$B$ = energy expenditure during rest (e.g. 1.5 kcal), and
$A$ = average level of energy expenditure considered acceptable (e.g. 33% maximum aerobic power, 5 kcal/min, etc.).

The weakness of this approach is that it tells you *how much* rest should be taken but not *when* it should be taken. There is no simple answer to this problem. In general, however, the recovery times which are required will be shorter if rest pauses are taken *before* the onset of the subjective symptoms of fatigue. This is particularly important where work–rest schedules are self-imposed. Recovery curves during rest are exponential in form—frequent short rest pauses are therefore likely to be more beneficial than occasional long ones.

## Static and Dynamic Work

Muscle actions may be static or dynamic. In dynamic actions, muscles change length and move loads. In static activity, muscles remain the same length and prevent movements or support loads*. Thus lifting a crate or putting it down is an example of dynamic work, whereas holding a crate in your hands is an example of static work. The task of carrying the crate across the room has both static and dynamic components.

To the physicist, static work is a contradiction in terms—since, by definition, "work is done when a force moves its point of application". The physiologist sees nothing wrong with the idea of static work—since the muscle is consuming energy despite the fact that it is not creating movement.

Static and dynamic work have a number of important physiological differences. As a muscle contracts, the pressure within it rises—which is why it becomes hard to the touch. This increases the resistance to blood flow. Vascular occlusion commences at a contractile force equivalent to about 25% of the muscle's maximum capacity, and by 70% the blood flow is cut off completely.

In dynamic work, the interruption of blood flow is transient—and the alternate phases of contraction and relaxation have a beneficial pumping effect. The blood flow to a working muscle during vigorous dynamic activity may be 20–30 times greater than it is at rest. In continuous static exertions, the blood flow is diminished. At loadings of more than about 15% of maximum, the blood supply is unable to meet the demands of the working muscle. The oxygen supply is inadequate for aerobic metabolism and there is an accumulation of waste products (especially lactate) and heat. It is generally recognized, therefore, that static work is more fatiguing than dynamic work (but see p. 47).

* The terms "isometric" and "isotonic" describe contractions in which a muscle acts at constant length and constant tension, respectively. True isotonic contractions do not occur outside the laboratory.

In dynamic exercise, the heart rate may increase to maximal levels (see below) with relatively little change in blood pressure: the systolic pressure is unlikely to exceed 175 mmHg and the diastolic 100 mmHg (as compared with a "normal" resting value of 120/80). In static work, however, the heart rate is unlikely to exceed 120 beats per minute but there is a much greater rise in blood pressure. During continuous static contractions (at more than 15% loading) the blood pressure continues to climb throughout the work period—and may reach dramatically high levels. For heavy static contractions involving large muscle groups, the systolic pressure can exceed 300 mmHg and the diastolic 150 mmHg (Åstrand and Rodahl, 1986).

In practical terms this means that work which has a high static component imposes a greater strain on the heart than purely dynamic work.

For a further discussion of the differences between static and dynamic work, see Asmussen (1981) and Hietanen (1984).

## Physical Workload: Cardiovascular Demands

For dynamic work involving large muscle groups, there is a linear relationship between heart rate and energy expenditure (or physical workload). But the relationship is task-dependent and becomes non-linear under some circumstances—see Figure 2.2. A non-linear increase in heart rate is probably more common in untrained or unfit individuals.

On average, maximum heart rate is said to be 220 beats per minute minus the person's age in years—with a standard deviation of about 10 beats/min. (The standard deviation is a measure of the degree of variability within a population—see p. 118.)

At physical workloads of moderate intensity, heart rate climbs to a plateau;

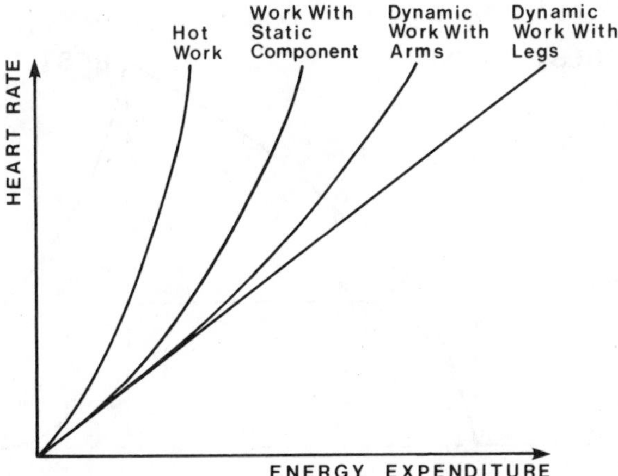


Figure 2.2 *Heart rate and energy expenditure for different types of work—after Grandjean (1988)*

remains more or less constant throughout the working period; and falls relatively rapidly to its resting level when the work ceases. At high physical workloads, the heart rate continues to climb throughout the working period and subsequent recovery times are very much longer—see Figure 2.3. The continued climb in heart rate is associated with the onset of fatigue, and occurs at levels of workload in excess of 30–50% of maximum aerobic power.

For a given level of energy expenditure, both heart rate and blood pressure are higher in work using the arms than in work using the legs. Thus heavy arm work such as shovelling snow or digging the garden may be hazardous for unfit people or cardiac patients. Heart rate is also higher where the task has a static work component—for example, in carrying tasks as compared with repetitive lifting tasks at similar levels of loading.

Working with the arms in a raised position increases the loading on the heart. Åstrand *et al.* (1968) measured oxygen consumption, heart rate and blood pressure in skilled carpenters hammering nails at bench height, into a wall at eye level and into the ceiling above the head. The number of hammer strokes per minute and the oxygen consumption were similar in all three conditions. But the work output (as measured in nails per minute) was very much less when nailing into the ceiling or the wall than when working on the bench, and the elevation of both heart rate and blood pressure was greater (although the differences were less clear for blood pressure). The hammering action has a greater static work component when performed with the arms in a raised position.

Heart rate during work is a function of ambient temperature. In a hot environment, the heart is required *both* to pump blood to working muscles *and* to perfuse the skin with blood so as to achieve the degree of heat loss which is required to prevent an excessive increase in body temperature. Thus, although temperature and humidity may have a negligible effect on energy expenditure, they may have a marked effect on heart rate (and thus a person's ability to sustain a given workload without excessive fatigue). In a study of the physiological cost of

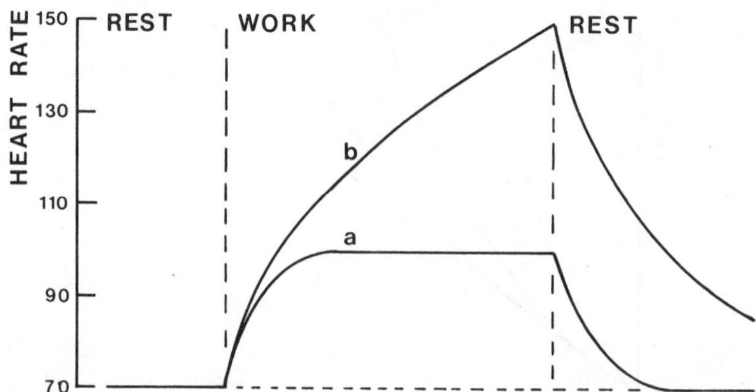


Figure 2.3 *Heart rate responses at two levels of work intensity: (a) less than 30–50% maximum; (b) more than 30–50% maximum*

carrying loads under different climatic conditions, Kamon and Belding (1971) found an increase in heart rate of about 10 beats per minute for each 10 °C increase in air temperature (above normal levels).

At low levels of physical work load, heart rate may also climb in response to psychological stress. Rodahl (1989) cites the interesting example of two bank clerks, during the period when they were balancing their accounts at the end of the working day. One clerk was experienced and balanced her accounts in 27 min, without an appreciable increase in heart rate above the resting level. The other took 36 min to balance her accounts—during which her heart rate hit peaks of 130 beats/min.

In many respects, therefore, heart rate is a better index of the overall physiological demands of work than energy expenditure. And it has the additional advantage of being very much easier to measure in the field.

It has been proposed that *for continuous work*:

(i) the heart rate should not continue to rise during the working period and should return to the resting level within about 15 min;

(ii) for men, the heart rate during work should not be more than 40 beats/min higher than the resting level, and for women it should not be more than 35 beats/min higher than the resting level (Grandjean, 1988).

For an averagely fit 30-year-old man with a resting pulse of 70 beats/min this gives a maximum acceptable working heart rate of 110 beats/min—which, if he has a maximum heart rate of 190, is equivalent to 33% of his heart rate reserve (i.e. the difference between his maximum and resting levels).

## Coronary Heart Disease

Coronary heart disease (angina pectoris, myocardial infarction, etc.) is the greatest single cause of death in the industrialized world. It has a complex multifactorial aetiology. About one-half of the total variation in risk in men (and slightly less in women) is statistically attributable to four risk factors: age, serum cholesterol, blood pressure and smoking. Physical activity has a protective effect—partly because it modifies these major risk factors and partly independently (Karvonen, 1984; Shaper, 1988). For example, Andersen *et al.* (1978) note that the prevalence of hypertension is very much less amongst forest workers than in sedentary city dwellers.

The association between coronary heart disease and the lack of physical activity at work was first demonstrated by Morris *et al.* (1953), who found a higher incidence in the drivers of London double-decker buses as compared with the conductors (who in those days used to run up and down the stairs collecting fares, etc.). Stress, obesity, etc., may well have been confounding factors in this study, but the basic connection has been confirmed elsewhere. Brunner *et al.* (1974) found that sedentary workers on kibbutzim had more than double the incidence of myocardial infarction than people in more active jobs who lived in the same environment and ate the same communally prepared food. Paffenbarger *et al.*

(1971) found that the level of physical activity on the job strongly reduced the risk of death from coronary heart disease in San Francisco longshoremen.

The protective effects of leisuretime physical activity have been shown in British executive grade civil servants by Morris *et al.* (1973) and in Harvard graduates by Paffenbarger *et al.* (1978). The former study suggested that physical activity only had a measurable protective effect at intensities beyond a threshold level of 7.5 kcal/min (or 50% maximum aerobic power for an average man). The Paffenbarger study showed that it was habitual physical activity in later life, rather than student athletics, that was critical; that relatively modest levels of exercise had a beneficial effect; and that the effect was independent of smoking, obesity, high blood pressure or the death of a parent from heart disease.

But the effects of physical activity at work on mortality from coronary heart disease may easily be masked by other factors. English and Finnish data both place miners and quarrymen close to the top of the coronary heart disease mortality table and construction workers close to the bottom (Karvonen, 1984)—despite the fact that both occupational groups have physically very demanding jobs.

## Obesity

It is estimated that 40–50% of adult city dwellers in the advanced industrial societies may be regarded as obese (Andersen *et al.*, 1978). Overweight people suffer a measurable reduction in life expectancy, and obesity is associated with various kinds of ill-health: cardiovascular disorders, musculoskeletal disorders, diabetes, psychosocial problems, etc.

Of itself, obesity is not now thought to be an independent risk factor for coronary heart disease, but it is strongly associated with other major risk factors (hypertension, serum cholesterol). So the overweight person may be around twice as likely to suffer from coronary heart disease—although the connection is mediated via these other factors (Shaper, 1988).

Weight gain occurs when caloric intake exceeds energy expenditure over a period of time. The search for a mediating physiological variable, which allegedly enables some people to eat heartily and remain thin, has so far proved inconclusive.

But the relationship between caloric intake and energy expenditure is itself complex. In a physically active person, body weight will be to some extent self-regulating—since any weight gain which occurs will be partially offset by an increase in the energy expended in any particular activity. There is evidence, however, that obese people accomplish their everyday activities with a greater economy of effort. Chirico and Stunkard (1960) compared the levels of physical activity of obese and non-obese people who had similar jobs. Subjects wore a pedometer which recorded the number of steps they took throughout the day. As measured in these terms, the non-obese subjects were around twice as physically active. (In one sense, therefore, we could say that the obese subjects were doing things more efficiently—p. 127.)

In general, we should expect people to regulate their food intake according to their energy expenditure. (Thus a brisk walk may sharpen your appetite—at least, it sharpens mine.) Mayer *et al.* (1956) studied sedentary and manual workers in India. For the various groups of manual workers, caloric intake increased in proportion to the heaviness of the job; thus there were no differences in body weight between these occupational groups. But the relationship broke down for the sedentary workers, who had an average caloric intake equivalent to that of men in a much heavier occupational category. Thus the sedentary workers had a greater average body weight—i.e. they were more obese. This suggests that there might be a maladjustment of the hunger mechanism in sedentary people.

## Fitness Training

It is generally recognized that regular periods of vigorous aerobic exercise (i.e. dynamic work involving large muscle groups) result in improvements in cardiorespiratory fitness and enhanced physical work capacity, whereas the absence of such activity leads in the long term to a steady deterioration. And although it is difficult to be precise, it is widely believed that the intensities of exercise which are required for the improvement of fitness are relatively high. (Higher, that is, than the levels associated with the onset of fatigue, etc.) This is referred to as the *overload principle* in physical training.

Åstrand and Rodahl (1986) state that in untrained individuals an energy expenditure of 50% aerobic capacity, repeated for two or three half-hour sessions per week, will result in a gradual improvement in fitness. (For an average person this would be equivalent to a heart rate of about 130 beats/min.) Similarly, studies reported by Shephard (1968) suggest that cardiovascular training effects are only significant where a heart rate of 120 beats/min or more is sustained over a period of time. (A figure of 20 min several times per week is widely quoted.) According to Goldsmith and Hale (1971), repeated short bursts of relatively intense activity lead to an increase in fitness, but more prolonged activity at a lower level has little or no demonstrable effect.

Notwithstanding this weight of expert opinion, experience suggests that a daily walk lasting 45 minutes to one hour, at a fairly brisk pace, will result in substantial long-term gains in fitness in untrained people. An hour's walking a day keeps the cardiologist away. For a middle-aged man in the lower part of the normal range of fitness, this is unlikely to represent an energy expenditure much in excess of one-third maximum. It may be, therefore, that the beneficial effects of low-intensity exercise have been underestimated. (Perhaps because people who do research in this area tend to be one-time athletes themselves.)

Given the sedentary nature of working life, there must inevitably be a decline in the number of people whose work calls for a level of physical exertion commensurate with the maintenance of fitness. Should we be designing some of the physical effort back into work? We cannot reasonably expect that the deliberate introduction of working methods which are sufficiently inefficient to be beneficial will command much support in the real world. But recreational fitness programmes based in the workplace—involving lunchtime exercise sessions, sporting activities after work, etc.—are a much better proposition.

Cox *et al.* (1981) described one such programme, in which employees in an insurance company had the option of taking part in 30 min exercise sessions three times per week. The sessions involved rhythmic callisthenics, jogging and ball games. People attending the classes showed significant improvements in fitness. Absenteeism was 22% lower in those attending the classes regularly (compared with non-participant controls), and labour turnover over a 10 month follow-up period was 1.5% in participants compared with 15% in other employees. It was estimated that if 20% of employees were to take part in regular exercise classes, changes of this magnitude would lead to a 1% reduction in company payroll costs.

## Biomechanics of Muscle

### The Length–Tension Curve

In its resting state the normal muscle has a degree of resistance to stretch—due primarily to the elastic properties of the connective tissue which surrounds the muscle and binds its fibres into bundles. The tension which is required to stretch the muscle to a given length may be plotted as the *passive tension–length curve*, as shown in Figure 2.4. Note that the slope of the curve (equivalent to the *stiffness* of the muscle) increases with length.

It is important to distinguish between this passive resistance and the property of muscle which is referred to as "tone". Until quite recently it was widely held that under normal physiological conditions there was always a low background level of motor unit activity—which would only be abolished if the muscle was deprived of its nerve supply. This concept of muscle tone has been shown to be quite erroneous. Electromyographic* studies have shown beyond all doubt that,

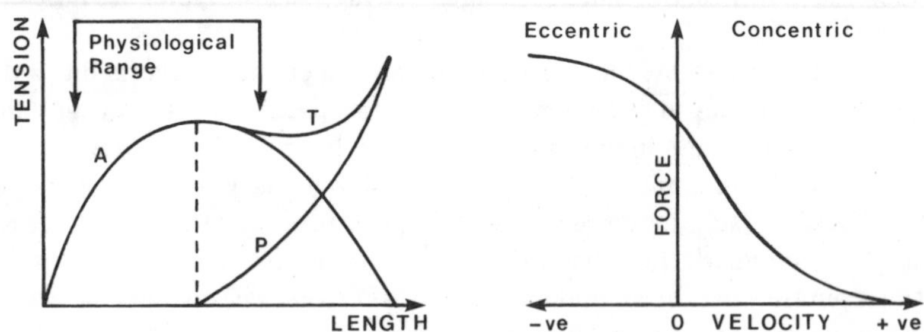


Figure 2.4 *The mechanics of muscle.* Left, *the length–tension relationship: A, active; P, passive; T, total.* Right, *the force–velocity relationship*

* Electromyography (emg) is a technique for detecting muscle activity. When a muscle fibre enters its active state, a wave of electrical depolarization, known as an action potential, passes along its surface. This electrical signal may be detected using electrodes taped to the skin, over the belly of the muscle, or by needles or wires inserted into the muscle itself. The intensity of the emg signal is a measure of the number of motor units which are active.

under normal circumstances, muscles which are not required for any active functional role relax rapidly and completely.

Some people do, however, find it difficult to relax their muscles completely and at will. In most cases, this may be regarded as a minor deviation from normal function. It is often a sign of a psychologically tense state (pages 53 and 147).

The relationship between length and tension in a fully active muscle is shown in Figure 2.4, where it is labelled "T" for total. The difference between the total tension and the passive tension gives us the *active tension–length curve*—which describes the capacity of the actual contractile mechanism of the muscle itself (see Gordon *et al.*, 1966). It so happens that the maximum active tensions are found at about the unstretched equilibrium length of the resting muscle—it is not clear why.

In real life, muscles never act singly—always in groups. They act between their points of bony insertion, across articulations (joints) of complex geometry, so as to turn these joints about their axes of rotation. (Note that any human movement, however complex, is a combination of rotations at the joints concerned.) The leverage of a muscle (that is, the perpendicular distance of its line of pull from the axis of the joints) may well change with joint angle. So in practice the nearest thing to a length–tension curve that you can measure on a living person is an *angle–torque curve.** Figure 2.5 shows the angle–torque relationship for the muscles which

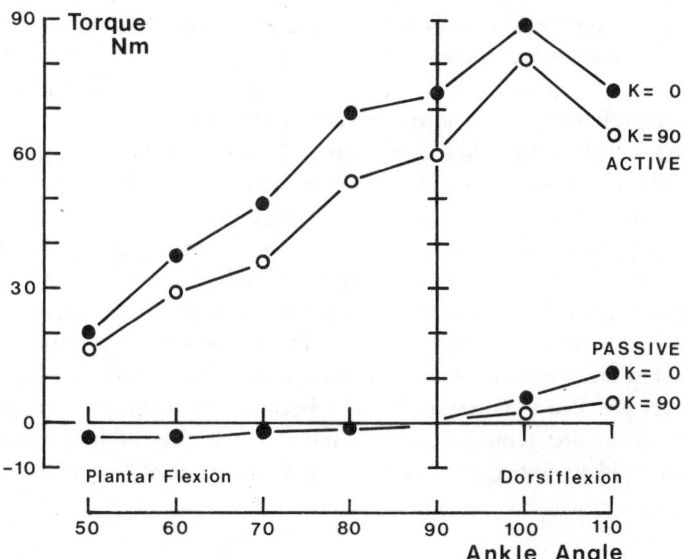


Figure 2.5 *The angle–torque relationship for the muscles which plantar flex the ankle. K = angle of knee flexion. Data from Pheasant (1977)*

* A torque or moment is the turning effect of a force acting on a lever—given by the product of the force and the perpendicular distance from the line of action of the force to the axis of rotation.

plantar flex the ankle joint (data from Pheasant, 1977). Note that active torques are dependent on knee angle—because the gastrocnemius muscle (of the calf) crosses both joints. Passive torques are only dependent on knee angle when the ankle is dorsiflexed—thus stretching the gastrocnemius. None of the anterior tibial muscles, which dorsiflex the ankle (and are tense in plantar flexion) cross the knee. (For an explanation of these anatomical terms of movement see p. 54.)

Only part of the tension–length curve seems to be used in the living person. In general, the length at which the maximum active tension is generated is somewhere near to the *muscle's maximum length in the body*. In the normal physiological range of motion, the muscle shortens to about half of its maximum length—and loses about 80% of its capacity to generate tension.

The passive resistance of resting muscle is generally small compared with the tensions which may be exerted in active muscular contractions. But it may be an important limitation on movement under certain conditions. If you are unable to touch your toes with your knees straight, it is probably the passive tension in your hamstring muscles which prevents you from doing so. (This is sometimes called "the ligamentous action of muscles".) When we talk about a person having "tight hamstrings" or "tight calf muscles", we are probably referring to a situation in which the muscles are unusually fibrous and resistant to stretch. But this is not the whole story.

When a muscle is stretched, close to the limit of its range of motion, neural signals from tissue which is under tension trigger a protective reflex—that is, the muscle is activated to prevent it from being stretched to the point of damage. Some of these neural signals reach consciousness as pain. (Note also that the resistance which is encountered at the extremes of joint motion may be due in part to ligaments and other soft tissue structures around the joint.)

If a muscle is habitually held in a shortened position, it will become more fibrous (and thus stiffer and less extensible)—and in due course it will become physically shorter. Studies on immobilized animals have shown that both the active and passive length–tension characteristics are modified (Tabary *et al.*, 1972). Conversely, if the muscle is repeatedly stretched, it will in due course adapt by increasing in length. To achieve this in practice—as in an exercise class—we have to learn to switch off the protective reflex. The process has to be managed with care: long slow stretches are very much safer than dynamic movements. "Warm-up" exercises are also important—possibly because the mechanical properties of connective tissue are temperature dependent, such that at low temperatures tearing is more likely to occur; but it may also be a matter of blood flow.

## The Force–Velocity Curve

A contraction in which a muscle shortens against resistance is described as *concentric*: the muscle is said to be doing *positive work*. A contraction in which a muscle actively resists being stretched is described as *eccentric*: the muscle is said to be doing *negative work*. Thus, when a muscle lifts a load, it is doing positive work; when it lowers the load, it is doing negative work. In the concentric action the muscle acts like a motor; in the eccentric it acts like a brake or "plays out rope".

The capacity of a muscle to exert a force in order to cause or resist the movement of a load is described by the *force–velocity curve* as shown in Figure 2.4—the tension in the muscle is plotted against its rate of shortening or lengthening. (Shortening is plotted as a positive velocity; lengthening as a negative velocity.)

The right-hand portion of the curve deals with *concentric conditions* or *positive work*. To state it simply, you can lift a heavy weight slowly or a light weight quickly—but you can't lift a heavy weight quickly. The maximum velocity of shortening is achieved when there is no resistance to the muscle contraction—in real life we never quite reach this point because our muscles always have to overcome the inertia of our limbs, air resistance, and so on. The zero velocity condition is an isometric contraction.

The left-hand portion of the curve deals with *eccentric conditions* or *negative work*. The tensions a muscle can exert under eccentric conditions are substantially greater than its isometric maximum. We are not sure exactly how much greater; the curve plotted here is a bit conservative and it might be anything up to twice as much (Grieve and Pheasant, 1982).

Under eccentric conditions, the muscle is absorbing energy. This has a number of important consequences. It is under eccentric conditions that acute muscle injuries are most likely to occur—a sudden stretch of the active muscle causes a build-up of tension which results in a tearing of the muscle tissue or its tendinous attachments. Anecdotal evidence suggests that rupture of the tendo calcaneus (Achilles tendon) characteristically occurs during an eccentric contraction of the calf muscles—such as would be caused by the foot being unexpectedly driven into dorsiflexion when running on uneven ground.

There is also some evidence that eccentric contractions are more effective than either isometric or concentric contractions when it comes to muscle training (Komi and Buskirk, 1972), although not all authorities agree on this point (Fox and Matthews, 1981). But it is certainly true that eccentric contractions are a prime cause of post-exercise soreness (p. 48).

The physiological cost of performing a given quantity of negative work is considerably less than the cost of performing an equal quantity of positive work. This was demonstrated in a classic experiment using a stationary tandem bicycle which had been rigged up so that two people could work against each other: one pedalled in the ordinary way whilst the other acted as a brake. The oxygen consumption of the former (who was doing the positive work) was very much greater than that of the latter (Abbot *et al.*, 1952; Bigland-Ritchie and Woods, 1976). This is presumably a consequence of the shape of the force–velocity curve. Each individual fibre can exert a greater force under eccentric conditions than under concentric conditions; the number of fibres required to exert a given force (and thus the energy expenditure required) will therefore be less.

## Back-swing

When we analyse the movements of people performing natural everyday activities, we commonly encounter a sequence of events in which the active shortening of a muscle is preceded by a dynamic stretch—that is, an eccentric phase (in which the muscle is absorbing energy) precedes the concentric phase (in which the muscle is performing positive work). The working muscles may be stretched by their antagonists or by an external force such as gravity.

Consider what happens when you strike a blow with a hammer, or when you punch somebody. In each case the actual blow itself is preceded by a back-swing action in which the muscles which will be called upon to do the work are dynamically stretched. Electromyographic studies of walking have shown that many of the muscle groups in the lower limb go through this same sequence, of stretch followed by shortening, at various phases in the gait cycle (Inman *et al.*, 1981).

Consider what happens when you play a backhand drive in tennis or squash. Your objective is to get the racquet moving with the maximum possible velocity at the time it hits the ball. You start by "winding yourself up", lengthening all the muscles that will be used in the shot. Then, as the back-swing is approaching completion, you commence the positive phase by "leading with the hips"—that is, you use the muscles of the lower limb to start the pelvis turning. This imposes an additional dynamic stretch on the abdominal muscles, which transfer the momentum of the pelvis to the upper trunk, which is the next part of the body to move—followed in turn by the shoulders, the upper arms and finally the forearms. In squash (but not in tennis) the shot is concluded by a final flick of the wrist. The whole action is executed as a single smooth unwinding motion. The heavier parts of the body move first and the lighter parts lag behind. Each muscle group recruited in turn is subjected to a dynamic stretch. This optimizes the efficiency of the transfer of momentum. Each successive moving part acquires a greater angular velocity.

Experimental studies have shown that the work which is done when a muscle actively shortens is greater if the shortening is immediately preceded by an isometric contraction than if it starts from rest—and greater still if the shortening is preceded by an active lengthening (Cavagna *et al.*, 1968; Asmussen and Sørensen, 1971). You can demonstrate this for yourself. The natural way to perform a standing jump is to bounce down and spring up in a single smooth action. You can jump higher this way than you can from a static squatting position.

The increased energy which becomes available for positive work may be due to the storage of energy by the deformation of elastic structures in series with the muscle (tendon, etc.)—giving a recoil effect like a bouncing ball. Or it might be due to more subtle biochemical effects, or to a mixture of both (Grieve and Pheasant, 1982).

## Strength

A person's strength may be defined as the greatest force (or torque) he (or she) is able to exert in a given test situation.

In a sample of fit young people (of both sexes) it would not be surprising to find that the tallest was half as big again as the shortest and that the heaviest was twice the weight of the lightest. But the strongest could easily have four or five times the strength of the weakest. A person's height is mainly determined by his or her genes; but weight and strength, although they have a genetic component, are much more affected by lifestyle.

It is often said that: "On average, women are about two-thirds as strong as men". This is true up to a point, but the distributions of strength in comparable samples of men and women are likely to overlap considerably and the magnitude of the sex differences in strength is task dependent.

When you look at the available data sets in which it is possible to make a valid comparison, you find that the strength of an average woman divided by that of an average man may be anywhere between about 40% and 90%—although the values do tend to cluster around an overall grand mean of 61%. Tests of upper limb muscles tend to show somewhat greater sex differences than tests of the lower limb, with tests of trunk muscle strength falling somewhere between the two. If we subdivide the upper limb category, we tend to find greater sex differences in tests of upper arm and shoulder muscles than in tests of the forearm and hand (Pheasant, 1983, 1986).

Maximum strength is reached between the ages of 20 and 25 (probably earlier in women than in men). It remains more or less constant up until about 40, after which it begins to decline, falling to somewhere between 75% and 80% of its maximum value by the age of 65 (Hettinger, 1961; Asmussen and Heebøll-Nielsen, 1962; Larsson *et al.*, 1979). The average grip strength, reported in one sample of 80-year-old men, is about 55% of what you would expect in 30-year-olds, and the range of variation is about the same (Damon and Stoudt, 1963). The widely quoted data set of Asmussen and Heebøll-Nielsen (1962) indicates that women age more rapidly than men and that "the legs give out first". The first conclusion is contradicted by Montoye and Lamphier (1977); the second by Viitisalo *et al.* (1985). The rate at which this decline occurs is a function of exercise and lifestyle.

### Ergonomic Aspects of Strength

In practice, we can ask three kinds of question concerning strength:

(i) Under what circumstances is it possible to exert the greatest force in a particular type of action?
(ii) What range of forces are the members of a particular population capable of exerting under a given set of circumstances?
(iii) What force is it reasonable to expect people to exert under a given set of circumstances?

These questions are increasingly difficult to answer. The first may sometimes be answered from first principles. The second will almost always require an empirical study to be conducted on a valid sample of subjects (unless the particular specific population concerned happens to have already been studied under these specific circumstances). The third question poses some major conceptual difficulties which we shall discuss in a later chapter (p. 306).

In the laboratory, it is relatively easy to devise tests of strength, where performance is determined by the capacity of a single muscle group: in real life, things are not this simple.

Consider a "tug-o'-war". You use leg muscles, thigh muscles, back muscles, shoulder muscles and muscles to grip the rope. Each of these groups is subject to fatigue—possibly at different rates. You also "use your body weight": we know that in a tug-o'-war the heavier team has a better chance of winning and we put the heaviest member of the team at the back as the "anchor man". Is this because of a positive correlation between weight and muscle strength, or is body weight actually employed in the pulling action? It is probably a mixture of both. We also know that it is essential to get "a good footing"—that is, you need a high coefficient of friction between your feet and the floor to prevent yourself from sliding forwards. Finally, it may well be that there comes a point at which you relax your grip because the rope begins to burn your hands. Which of these is the limiting factor? Somewhere in the chain there is a weak link and this determines the strength of the chain as a whole.

The biomechanical factors which limit the expression of strength in the performance of everyday tasks may be divided into three main categories, the last of which has two subdivisions:

- *muscular limitations*;
- *gravitational limitations*;
- *interface limitations*—which include *frictional limitations* and *discomfort limitations*.

In tests of strength which are subject to *muscular limitations*, the effects of posture are due partly to the angle–torque curve and partly to simple leverage effect. For example, the force which can be exerted in a one-handed pulling action, as shown in Figure 2.6, will decrease with increased elbow flexion: both because the torque ($T_e$) available from the elbow flexors decreases as the muscles shorten and because the reaction force ($F$) tending to straighten the limb has a greater mechanical advantage ($T_e = d \times F$). Note that as $d$ approaches zero, it becomes increasingly unlikely that the strength of the pulling action is limited by the capacity of the elbow flexors as such. When the limb is straight, the limiting factor might be the capacity of the muscles which retract the shoulder girdle (trapezius, etc.) or the back extensors (erector spinae)—or it might be a matter of body balance.

Consider now the case of a person using both hands to pull vertically downward on an immovable object located above his head. The maximum force he could exert would be equal to his body weight at the point where he raised himself off the

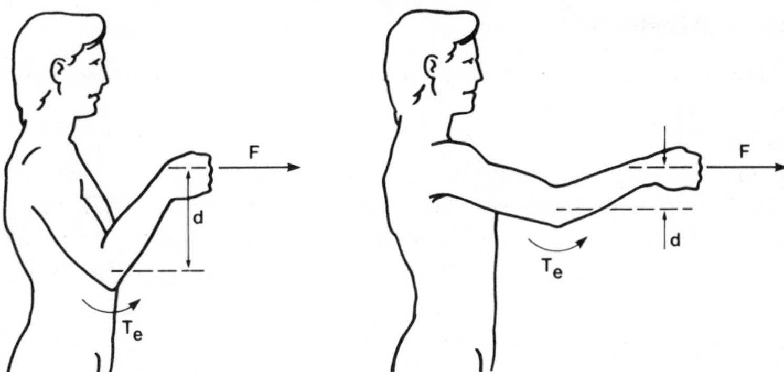


Figure 2.6 *Leverage effects in a one-handed pulling action. Reproduced from Pheasant (1986)*

floor. This is a particularly clear example of a *gravitational limitation*. Many pushing, pulling and lifting actions (in the standing position) are limited (at least in part) by the person's ability to use his body weight as a counterbalance to the force he is exerting. When you heave on a rope (as in the tug-o'-war), you generally get your feet as far forward as possible and your body weight as far back as possible so as to maximize this counterbalancing effect. Conversely, when pushing a car, you get your feet as far back as possible and your weight as far forward as possible.

*Interface limitations* are ones which result from the nature of the physical connections between the person and the objects with which he is in contact—both at the hands and at the feet. If these connections are inadequate, they may prevent the person from bringing his full muscular or gravitational strength to bear on the object concerned.

The strength of a horizontal pulling action cannot exceed the *frictional resistance* ($R$) between the feet and the floor—which is given by the equation $R = \mu C$, where $C$ is the vertical compressive force between the feet and the floor, and $\mu$ is the coefficient of limiting friction (which depends on the materials and conditions underfoot and the soles of the person's shoes). If the person is pulling exactly horizontally, then $C$ will be equal to body weight, since no other vertical forces are involved. When the horizontal pulling force equals $R$, the person begins to slip forward. Vinyl tiles, for example, have a value for $\mu$ of about 0.33, so on a vinyl floor you cannot exert a horizontal force of more than one-third of body weight without slipping.

*Discomfort limitations* are self-explanatory: we do not carry on exerting a force if it hurts us to do so. In many cases this is quite unconscious—a protective reflex is presumably involved.

These various types of limitations sometimes interact in a complex and interesting way. For example, the horizontal force you can exert on a well-designed handle varies with the height of the handle above the ground. (The limitation is probably partly gravitational and partly muscular.) But for a badly designed handle the height makes no difference—because an interface limitation has taken over.

## Fatigue and Endurance

Two varieties of fatigue are generally recognized: *local muscular fatigue* and *general fatigue*. Our concern here is with the former. For a more general discussion of fatigue see Chapter 7.

A person's capacity for continuous static work is described by the *load–endurance curve* (Rohmert, 1960; Caldwell, 1963; Monod and Scherrer, 1967), as shown in Figure 2.7. Endurance time is plotted as a function of loading, the latter being expressed as a percentage of the individual's maximum capacity as measured in an isometric strength test of brief duration. This has the effect of eliminating differences due to strength as such.

A force of anything up to 10–15% maximum may be sustained for lengthy periods of time without obvious signs of fatigue. Beyond this point, endurance time falls exponentially with loading. A force of 50% maximum can be sustained for about 1 minute and 70% maximum for about 30 seconds. Maximum forces can only be sustained for a second or so. It was originally thought that these figures were the same for all muscle groups and were independent of factors such as age and sex. Subsequent research has shown that things are not quite this simple.

The data of Poulsen and Jørgensen (1971) suggest that the trunk extensors (erector spinae, etc.) have longer endurance times than other muscle groups. This has been confirmed by Rohmert *et al.* (1986). Endurance times (at a given percentage of maximum loading) typically increase (by an average of about 20% between the ages of 20 and 60), owing to differential loss of fast motor units; and at a given level of relative loading, women have longer endurance times than men (Petrofsky, 1982).

The traditionally accepted figure of 10–15% maximum as a loading which can be sustained for an "unlimited" period of time has also subsequently been

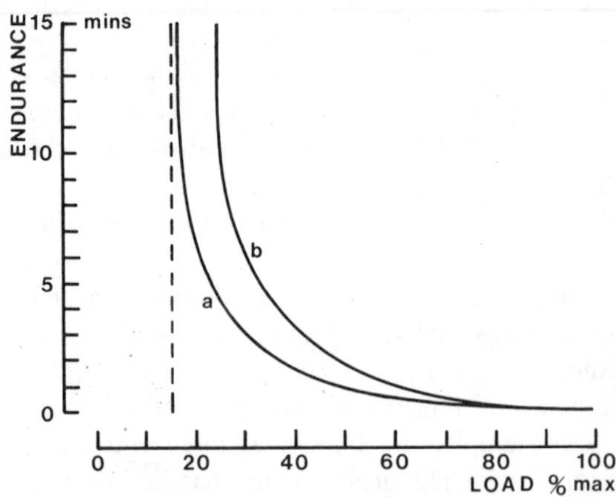


Figure 2.7 *The load–endurance curve: (a) for limb muscles, based on the data of Rohmert (1960), Caldwell (1963) and Monod and Scherrer (1967); (b) for back muscles, based on the data of Poulsen and Jørgensen (1971)*

challenged. In Rohmert's original studies, the "unlimited" period was actually set at 10–15 minutes, at which time the experiment was terminated. Both Björksten and Jonsson (1977) and Hagberg (1981) confirmed the experimental form of the load–endurance curve—but found that the endurance time for a force of 8% maximum was only about an hour. For longer periods of time (up to and including an 8 hour working day) the critical load beyond which fatigue will occur might well be as low as 3% maximum (Björksten and Jonsson, 1977).

The onset of fatigue in continuous static work is thought to be the consequence of the impairment of blood flow, the accumulation of waste products (lactate, etc.) leading to a failure of the muscle's contractile mechanism. (The biochemistry of fatigue is complex and at present it is only partly understood—for a recent review see Vøllestad and Sejersted, 1988.) The process of fatigue in dynamic work is less well understood. Monod and Scherrer (1967) have shown that for any muscle group it is possible to establish a critical rate of work which may be sustained for long periods without evidence of fatigue. They argued that it is also dependent upon blood flow, although this would be difficult to prove directly.

Hagberg (1981) showed that endurance times in a task involving alternating concentric and eccentric actions of the same muscle (without an intervening phase of relaxation) were similar to those for continuous static contractions at a similar level of loading. In most natural repetitive dynamic tasks, antagonistic muscle groups alternate in their activity—thus each muscle has alternate phases of contraction and relaxation. It may be that these relaxation phases in the work cycle (in which blood flow is re-established) are more important than the nature of the action itself (Jonsson, 1988).

To compensate for the diminishing force of muscular contraction, an increasing number of motor units have to be recruited to maintain a given level of tension—and the amplitude of the emg signal increases. There is also a downward shift in the frequency spectrum of the emg signal. This is partly due to a reduction in the rate at which the electrical changes which accompany muscle contraction are propagated along the muscle fibre. But it may also be due to a synchronization of motor unit activity—particularly at high levels of loading (Jørgensen *et al.*, 1988). This synchronization may lead to *muscle tremor*. The underlying mechanism is obscure (see Lippold, 1973).

The accumulation of metabolic waste products stimulates sensory nerve endings in the muscle. This results in a sensation of mild discomfort, which steadily mounts until it becomes intolerably painful to continue the activity concerned. It is now recognized that the substance responsible for the pain is not lactate itself. It is not clear whether the end-point of a sustained contraction is determined by the pain or by the failure of the contractile mechanism. Subjectively they seem to occur at the same time.

The performance decrements of local fatigue have a strong psychological component. There are many anecdotes of the feats of endurance that people are able to perform in times of crisis. Athletes have been found to have higher lactate levels after an important contest than after a so-called "all-out effort" in the laboratory. Given the right incentive, they are able to will their muscles to keep working at lactate levels well beyond those which would normally force them to quit.

The psychological component of fatigue has been studied experimentally. Ikai and Steinhaus (1961) required subjects to perform a long series of maximal efforts. Every so often they did things like unexpectedly firing a pistol behind the subjects' backs, or asking them to shout loudly. Both had a dramatic effect in restoring strength to the subjects' failing muscles. Asmussen and Mazin (1978a,b) conducted a very detailed series of experiments in which bouts of work were alternated with pauses in which the subjects were either completely at rest or engaged in "diverting activities". Performance was better after the latter (see also p. 164).

## Muscle Pain and Dysfunction

Muscle fatigue as such is a perfectly normal physiological state which is rapidly reversed by rest. But it has points of similarity with a number of other conditions which progress by imperceptible stages of increasing severity—via transient disturbances of normal function which lie on the borderline between physiology and pathology—to pathological states of muscular pain and dysfunction which can be exceedingly difficult to reverse.

### Post-exercise Soreness

The muscle soreness which follows unaccustomed physical activity may come on anything between 2 hours and 1 day after the activity which caused it. The muscles used in the activity become stiff and tender, and their strength is impaired. The symptoms typically reach a peak on the second day, subside thereafter, and have usually resolved completely within 4–6 days.

Muscle soreness is particularly associated with eccentric muscle contractions (Komi and Buskirk, 1972). The proximate cause is probably damage to the muscle fibres, to their tendinous attachments, or to the connective tissues within the muscle. Eccentric contractions would be most likely to cause such damage (p. 41). Fast twitch (type II) fibres are particularly affected (Newham, 1988). The subsequent release of histamine and other inflammatory substances causes oedema and pain. Muscle spasm may also be part of the picture (de Vries, 1961). The symptoms are often (but not always) most pronounced in the vicinity of the muscle tendon junction. This might be because damage is greater in this region or it might be because there is a higher concentration of nerve endings.

### Muscle Cramps

The underlying cause of these sudden, violent and intensely painful muscle spasms is unknown. Muscle cramps can occur as a consequence of disturbed electrolyte balance or during fatiguing exercise (particularly swimming), but they often seem to occur quite spontaneously—at rest or during sleep. Muscle cramps may be triggered, in a susceptible person, by using the muscle isometrically when it is in a shortened position; and they may be rapidly alleviated by stretching the muscle. This suggests a neurological explanation. Nocturnal cramps generally affect the calf muscles and the intrinsic muscles of the foot—probably because the weight of the bedclothes presses the foot into a position in which these muscles are shortened.

## Work-related Musculoskeletal Disorders

Muscle soreness and muscle cramps are transient disturbances of normal function. Muscular pain and dysfunction of a more persistent variety are known by many names, including (muscular) rheumatism, fibrositis, myalgia, myofibrositis, myofascitis, myofascial (pain) syndrome, fibromyalgia, and so on. They all refer to more or less the same range of clinical conditions. The choice of terminology partly reflects the user's conception of the underlying pathology of the condition, but it is largely a matter of fashion. In this discussion we shall mainly use the term *myalgia*—which just means muscle pain and does not imply that we know what causes it.

The examination of painful muscles often reveals "knotted" areas of palpable hardening which are tender to the touch. These may have quite definite boundaries or they may be relatively diffuse. They may be lumpy and nodular, ranging from the small pea to something approaching the size of an egg, or the muscles may feel stringy or ropey, with palpable tight bands of anything up to 2 inches in breadth, which follow the line of the muscle fibres. Squeezing the hardened area sometimes makes a crunching sound.

The precise nature of these areas of hardening has been the subject of some debate. It was once believed that they were aggregations of inflamed fibrous connective tissue—hence the term "fibrositis". Another closely related theory holds that they are areas of localized oedema. A more modern view is that the so-called "fibrositic nodules" are in reality circumscribed areas of abnormal motor activity, the palpable hardening being due to muscle spasm (perhaps with some oedema). There is some evidence for both of these views, but in neither case is it conclusive. Some biopsy studies have indeed shown abnormal quantities of fibrous connective tissue in areas of palpable hardening—but only, apparently, in severe and long-standing cases. Electromyography often shows muscle spasm—but not always (Simons, 1976; Roland, 1986). One possible explanation is that muscle spasm predominates in the early stages of the condition and fibrosis becomes significant later on (Cobb *et al.*, 1975). Another possibility, proposed by Travell and Simons (1983), is that there might be circumstances under which the contractile mechanisms of an internally damaged muscle fibre might remain in an activated state, in the absence of electrical action potentials at the fibre's surface.

Recent histological investigations of biopsy material taken from tender points in the trapezius muscles of patients with chronic myalgic symptoms have added to this picture. There are often signs of atrophy in type II (fast fibres) and type I (slow) fibres commonly acquire a "ragged" appearance (Yunnus *et al.*, 1986; Hendriksson, 1988). The latter finding is thought to result from ischaemia and to be indicative of a muscle fibre in "energy crisis". Hendriksson (1988) confirmed that subcutaneous tissues in the vicinity of trapezius tender points are indeed hypoxic and that the muscles are depleted of high-energy phosphates (ATP, phosphocreatine, etc.). Dennet and Fry (1988) have reported similar histological findings in biopsy specimens taken from the hand muscles of patients suffering from work-related "overuse syndromes" (see p. 77).

Many authorities believe that the underlying cause of persistent muscle pain is a self-perpetuating cycle in which the pain causes muscle spasm, which in turn causes further pain and so on (Travell *et al.*, 1942; Travell and Simons, 1983). The precise nature of the mechanism which maintains this cycle remains to be elucidated. The irritation of sensory nerve endings by the chemical products of tissue breakdown is doubtless part of the story. But something else is also involved. This "something" is probably neurological—an interaction between the central connections of the sensory nerves which supply the muscle and those which control motor function. We could think of it as a reverberating neural circuit which gets stuck in the "on" position (a bit like the "howl around" you get in public address systems).

If the model is correct (at least in broad terms, if not in detail), then we must regard persistent muscle pain and transient post-exercise soreness as two variants of the same basic phenomenon. (But note that palpable areas of hardening may often be detected in pain-free muscles.)

It seems likely that muscle spasm would result in an alteration in local blood flow—which could lead to hypoxia, fluid congestion and the retention of pain-producing metabolites, which in turn help to perpetuate the cycle. Disturbances of lymphatic drainage may also be part of the overall picture. The restriction of movement attendant upon the muscle spasm might in course of time lead to the laying down of fibrous tissue in the oedematous areas between the muscle fibres. So our model of muscular pain is compatible both with the older "fibrositis" concept and with the recent histological findings.

The pain of an acute muscle injury has a sharp stabbing quality. Persistent pain (in the same muscle) is more likely to be described as a dull diffuse aching or burning sensation. It may often radiate into adjacent areas or even to quite remote sites. Travell and Rinzler (1952) introduced the important concept of *trigger points* and *reference zones*. A trigger point is a circumscribed area of tenderness in the affected muscle (or in adjacent connective tissue), stimulation of which (by palpation, etc.) causes pain in the reference zone.

Trigger points have fairly consistent locations in any particular muscle. In some cases they are found near the centre of the muscle belly; in others they cluster round the points of tendinous insertion. Experience suggests that the trigger point often coincides with an area of palpable hardening, but most authorities seem to agree that this need not necessarily be the case. The patterns of referred pain associated with trigger points in any particular muscle are also fairly consistent. Some important examples, arising in the working situation, are shown in Chapter 4.

Successful treatment of the trigger point leads to an alleviation of symptoms, both locally and in the reference zone. A number of methods are available, all of which are basically aimed at breaking the cycle of pain and spasm. These include the injection of local anaesthetics into the trigger point, the use of vapo-coolant sprays (such as ethyl chloride) and the application of heat and cold. Spasm may be arrested by prolonged thumb pressure to the trigger point (as in certain acupressure techniques for pain relief) or by stretching the muscle tissue (either manually or by exercises). Some osteopathic manipulations (particularly the so-called "muscle energy techniques") also probably work in this way—as, indeed, may some methods used by acupuncturists. This rather disparate range of techniques appears to be divisible into two categories: those which operate by blocking the neural input to the muscle (anaesthetics, cold and pressure) and those which operate by feeding a new signal into the reverberating circuit, which has the effect of switching it off (stretch, acupuncture and possibly osteopathy).

What sets up the self-perpetuating cycle in the first place? In some cases there is a history of acute muscle injury. In others the process may start with a virus infection. Trigger points in muscle may be set up as a result of neural bombardment from diseased viscera (this has been reported in coronary heart disease). Some patients associate their symptoms with cold damp weather or with sitting in draughts (but these may be not so much causes as factors which exacerbate a pre-existing problem).

In many cases, however (perhaps the majority?), the condition is basically caused by overuse of the muscles concerned. A single episode of unaccustomed activity may sometimes be responsible, but more commonly it is a cumulative process operating over a period of time. The condition is very often work-related—it may be due to static loading, repetitive motions or a combination of both.

The basic course of events which is involved is summarized in Figure 2.8. We shall read this diagram from top to bottom, noting that, as we move down the page, we are dealing with conditions which are increasingly chronic and increasingly difficult to reverse.

Static loading and repetitive motions both cause local muscular fatigue. If opportunities for recovery are adequate, the discomfort is short-lived and equilibrium is restored by rest. If rest pauses are inadequate (and there is a progressive build-up of metabolic waste products) or if the workload is very intense (causing tissue damage), the reverberating neural circuit of the pain–spasm cycle may be triggered. In its benign form (muscle soreness) this condition resolves within a few days, but this resolution does not always occur.

What are the factors which determine whether the pain–spasm cycle is transient or prolonged? The prolonged static loading of muscles is undoubtedly a bad thing. But vigorous dynamic muscle activity may sometimes lead (after a transient phase of soreness) to an enhancement of strength and work capacity—whereas at other times it leads to prolonged pain and dysfunction. This seems paradoxical. The dividing line between "exercise" and "trauma" is a very fine one.

Part of the distinction is simply a matter of the intensity of the work load and of the balance between work and rest. Continued employment of an overused muscle in the same repetitive activity, before it has fully recovered from a previous work

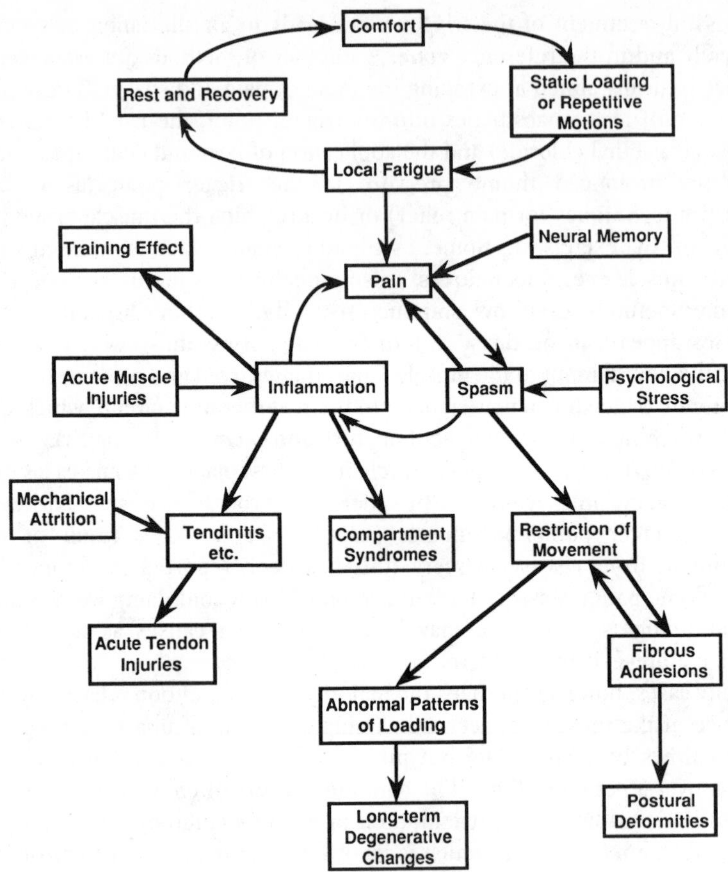


Figure 2.8 *The aetiology of work-related musculoskeletal disorders—a speculative model*

period, leads to an intensification of the pain–spasm cycle. The overall picture is predominantly one of progressive overuse and of a condition which progresses by a series of imperceptible stages, starting from the mild discomfort of ordinary muscle fatigue.

Once the pain–spasm cycle has been established, it seems to become increasingly easy to reactivate it. This underlies the recurrent nature of many musculoskeletal disorders. Attacks may be triggered by quite trivial stimuli. What causes this increased sensitivity? It could be the presence within the muscle of sensitive areas of scar tissue, but it is more likely that the explanation is to be found in the central nervous system. Current models of pain suggest that certain centres in the spinal cord may behave like "gates" which either let pain through or shut it out (Melzack and Wall, 1988). It seems that once a particular gate has been fully opened, it becomes easier to open it again. Some kind of *neural memory* is involved.

Musculoskeletal disorders often have a psychological component in their aetiology (perhaps always). The manifestation of their symptoms is often stress-related. We commonly respond to psychologically stressful situations by tensing our muscles. This was noted, for example, by Jacobson (1944), who pioneered the use of muscular relaxation techniques in stress management. Westgaard and Bjørklund (1987) have electromyographically demonstrated a diffuse muscle tension, unrelated to postural demands, in subjects performing tasks involving high mental workload and sustained visual attention at a VDU screen. Tensions were of the order of 2% maximum loading. Psychogenic muscle tension of this kind may be sufficient to trigger the pain–spasm cycle in a sensitized individual, or it may have an add-on effect with tension due to other causes (posture, etc.). But to consider any musculoskeletal condition as solely psychogenic is not to do justice to the complexity of its aetiology.

The pain–spasm cycle is likely to be associated with fluid congestion in the muscle. This may cause particularly serious problems if the muscle is located within a tight fascial compartment which cannot easily expand. The best-known of these *compartment syndromes* is the one which affects the anterior tibial muscles (or, less frequently, the peroneal muscles) in runners and walkers. A similar condition, affecting the small muscles of the hand, has been described.

Myalgic pain and dysfunction of the ordinary variety may progress to a number of other, more chronic, musculoskeletal conditions. Muscle spasm causes a restriction of movement leading to the formation of fibrous adhesions (see above). These in turn may cause a further restriction of movement—hence, perpetuating a further progressive cycle of dysfunction, and leading to a postural deformity (p. 114). Muscle spasm and the restriction of normal movement may lead to abnormal patterns of mechanical loading. Other muscles in the area will have to work harder to overcome the spasm. Joints will be subjected to greater compressive loading, and so on. The whole region becomes progressively "stiff" and there is a further impairment of blood flow. This may be an important feature of various types of long-term degenerative changes.

Some degree of inflammation of the connective tissue within the muscle is part of the basic myalgia condition. If this inflammation particularly affects the tendinous insertions, the tendon itself or the soft tissues surrounding it, we will probably refer to the condition as a *tendinitis*, a *peritendinitis* or *tenosynovitis*. It is well accepted that these are generally caused by overuse. To what extent are we to regard them as a progression of the basic myofascial syndrome or a variant in which inflammation of the tendon (etc.) happens to predominate? Or are they a separate group of conditions, resulting directly from mechanical wear and tear of the tendons and associated structures (see p. 91)?

Perhaps both are involved. But taking one consideration with another, there seems every reason to regard the Achilles tendinitis of the runner, the back pain of the manual labourer or the nurse and the "repetitive strain injury" of the keyboard user as members of the same generic class of conditions. And all of these conditions are principally caused by cumulative overuse.

In recent years, some authorities have come to regard certain long-standing forms of muscle pain as part of a more complex syndrome, which has been given the name *fibromyalgia* (Bennet, 1986, 1987; Goldenberg, 1986; etc.). The condition is defined as existing when there is chronic generalized pain and stiffness, with multiple tender points. The muscular symptoms are characteristically modulated by the weather and by stress, and the stiffness is worse in the morning. Other symptoms include chronic headaches, profound feelings of "fatigue", lethargy and exhaustion, and the symptoms of the "irritable bowel syndrome". Many patients show signs of clinical depression which apparently pre-date the muscular symptoms (Goldenberg, 1986). Sleep disturbances are common: these are considered to be a cause rather than a consequence of the condition, since muscle pain may be induced in normal subjects by interrupting their sleep (Moldofsky, 1986). The fibromyalgia concept is interesting, from the standpoint of ergonomics, in that it places some forms of persistent muscle pain within the context of a more generalized state of what could be referred to as chronic fatigue (see p. 159).

## Appendix: Anatomical Terminology

The terms which anatomists use to describe posture and movement can be a little confusing to the novice—but they are really quite simple.

The *median plane* divides the standing body, vertically down the midline, into two equal halves; the *sagittal plane* is any plane which is parallel to the median plane; and the *coronal plane* is any vertical plane which is at right angles to the median plane.

The body has two surfaces: ventral and dorsal. Broadly speaking, the ventral surface is on the front and the dorsal surface is on the back. But it is not quite this simple—and this is where it is easy to get confused. During the course of its embryological development the lower limb rotates about its own axis. So the *ventral* surface of the body is: the face; the front of the chest, abdomen, groin and thigh; the back of the knee and the calf; the sole of the foot; the palm of the hand and the surfaces of the forearm and upper arm which are continuous with the palm.

A movement occurring in the sagittal plane which brings two ventral surfaces together is called one of *flexion* and the opposite movement is called one of *extension*. A movement in the coronal plane which takes the arm or leg away from the mid-line is called one of *abduction*; and the opposite movement is called one of *adduction* (Figure 2.9). A movement in which the arm, the leg or the trunk turns about its own long axis is called one of *rotation*.

There are also some special terms used to describe the movements of particular parts of the body.

The movement which occurs at the ankle joint, by which we "point" the foot or stand on tip-toe, is called *plantar flexion* and the opposite movement is called *dorsiflexion*. (By rights they should be called flexion and extension, respectively, but physiologists seem to find this too confusing.)

In the standing position, the normal spine has a complex sinuous curve which is concave to the rear in the lumbar and cervical regions (i.e. the neck and lower back, respectively) and convex to the rear in the thoracic region (the upper back),

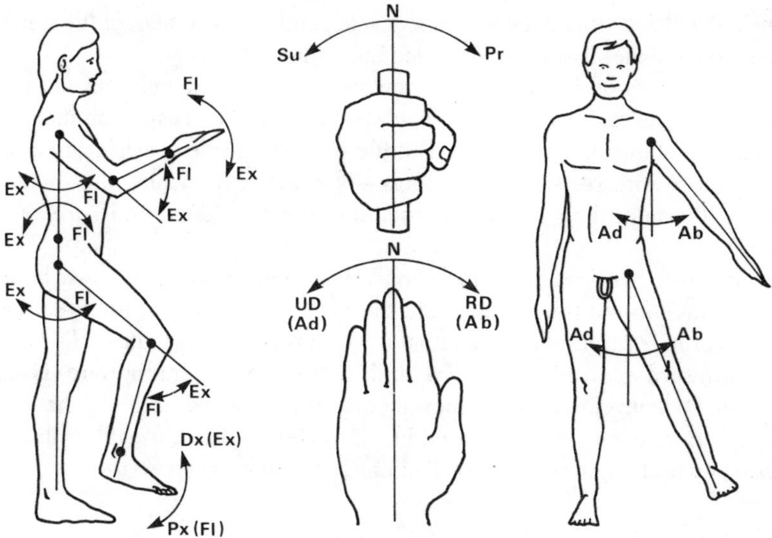


Figure 2.9 *Anatomical terms of movement. Fl = flexion; Ex = extension; Ab = abduction; Ad = adduction; Px = plantar flexion; Dx = dorsiflexion; Pr = pronation; Su = supination; UD = ulnar deviation; RD = radial deviation; N = neutral—after Pheasant (1986)*

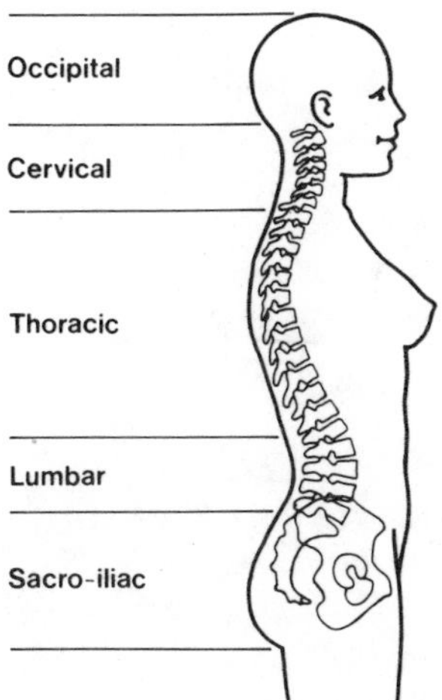


Figure 2.10 *The curves of the spine—reproduced from Pheasant (1986)*

as shown in Figure 2.10. The concavity of the cervical region is continuous with the convexity of the *occiput* (back of the head) and the lumbar concavity is continuous with the convexity of the *sacro-iliac* region and the buttocks.

A spinal curve which is concave to the rear is called a *lordosis* and a curve which is convex to the rear is called a *kyphosis*. The act of flexing the spine (as when you curl up into a ball) will flatten the lumbar lordosis and the opposite movement (extension) will accentuate the lordosis. A side bending movement of the spine (i.e. in the coronal plane) is sometimes called one of lateral flexion.

The movements of the shoulder region are complex. The humerus (arm bone) may move on the scapula (shoulder blade)—in flexion, extension, abduction and adduction (as shown in Figure 2.9). But the shoulder girdle (clavicle and scapula) may also move with respect to the thorax—in *elevation* (hunching the shoulders), *depression* (the opposite), *protraction* (rounding the shoulders) and *retraction* (the opposite).

The forearm may rotate about its own axis. The forearm has two bones in it—the radius (on the thumb side) and the ulna (on the side of the little finger). The radius rotates around the ulna (that is why it is called the radius). Turning the palm downward is called *pronation* and turning the palm upward is called *supination*. By convention, wrist movements are described as if the hand were supinated. But to save confusion, adduction of the wrist is usually called *ulnar deviation* and abduction is usually called *radial deviation*.

# Chapter Three
# Back Pain at Work

In this chapter (and the one which follows) we shall deal with the aetiology of the work-related musculoskeletal disorders and in particular with their patterns of occurrence in different occupational groups. The epidemiological database concerning the work-related musculoskeletal disorders is patchy. Low back pain has been the subject of a number of large-scale surveys of people drawn from many different walks of life. But in other cases it will be necessary to base our discussions on studies of isolated working groups or to draw upon case-historical evidence of a less structured kind.

## Epidemiology: Some Definitions

Epidemiology is the science which studies the distribution, determinants and deterrents of disease. (For a general introduction to the subject see, for example, Alderson, 1983.) The *prevalence* of a condition (or a symptom) is the percentage of people in a certain population who suffer from the condition concerned. The *point prevalence* is the percentage who are found to be suffering at a certain moment in time (i.e. when the survey is conducted). The *period prevalence* is the percentage who report having suffered at some time or another during a certain specified period leading up to the time of the survey (often 1 year). The *lifetime prevalence* is the percentage who have suffered at some time or another during their lives.

Point and period prevalences are commonly age-dependent—but need not necessarily be so. Lifetime prevalences are cumulative figures, so they are always age-dependent. The figure you get for the prevalence of a particular disorder will depend on how serious a person's symtoms have to be before you regard him as suffering from the condition concerned—so the figures you get from different surveys are not always compatible.

These difficulties are avoided in *case–control studies*—that is, investigations in which a sample of sufferers (cases) is compared with a matched sample of non-sufferers (controls). Studies of this kind enable us to identify *risk factors*—that is, characteristics which are found significantly more frequently in cases than in controls (and which therefore must affect the probability of a person suffering from the conditions concerned). We may thus calculate the *relative risk* associated with a particular factor. Thus, to say that a particular job carries a relative risk of 2 means that people in that job are twice as likely to develop the condition concerned. Note that the existence of a significant statistical association does not of itself constitute proof positive of a direct causative link—but it is strong circumstantial evidence.

The aetiology of the disorders with which we are concerned is complex and multifactorial. Some of the relevant risk factors will be work-related; others will not. In any particular working population some people will be more at risk than others, by virtue of their genetic make-up, their overall lifestyle, their prior exposure to other risks, and so on.

In a large population of diverse characteristics, engaged in diverse working tasks (like the employees of a large industrial concern, for example) we may well find that the overall prevalence of a particular condition is quite low. But this may mask the fact that such cases as there are tend to be concentrated in certain localized areas of high prevalences—like one particular assembly line, for example, or in people using a particular machine.

*Clusters* of cases of this kind are encountered quite frequently in many areas of occupational and environmental medicine. A certain amount of clustering is to be expected by chance alone—just as you can have a run of luck at roulette. So it is not always easy to determine whether a cluster should be regarded as circumstantial evidence for a causative link. The interpretation of small cluster data is a notoriously difficult epidemiological problem. Recent controversies surrounding birth defects and VDUs, and the clustering of leukaemia cases around nuclear plants and overhead power lines, are notable examples.

If we were to investigate a sizeable sample of people, all of whom were doing a similar kind of work, we might find that a large proportion of them suffered work-related aches and pains on a fairly regular basis (with prevalences in the 70–90% range); but the proportion of these who went on to develop severe clinical problems would be relatively small. We could think of this pattern of occurrence as something like a pyramid (see Figure 3.1). At the base of this pyramid are a very large number of people who suffer mild aches and pains as a result of their work and do not complain very much. At the apex are a small number who suffer work-related conditions of crippling severity. Between these two extremes there are an infinite number of gradations.

We do not know why this progression should occur in some people and not in others, but in many respects the question is academic, since the preventative measures we should apply in terms of the redesign of work would in any case be the same.

## Low Back Pain

Low back pain is the most common of the work-related musculoskeletal disorders, and in economic terms the most costly. Its epidemiology has therefore been studied in very much more detail than the other conditions with which we are concerned—so in many respects it can serve as a model for the work-related musculoskeletal disorders in general.

Pain in the lower back may be sharp, or dull; and diffuse, or localized. When the

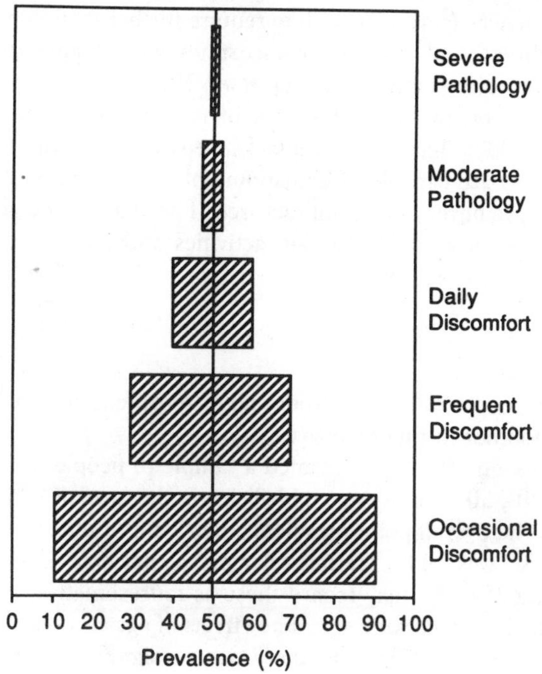


Figure 3.1 *The distribution of work-related musculoskeletal symptoms*

pain is acute in onset and/or severe, it is sometimes called *lumbago* (although diagnostically the term is quite meaningless). The pain may be limited to the mid-line or it may extend outward on either side—and in around one-third of cases it radiates into the buttock and down the back of the thigh and into the leg (*sciatica*). This may be associated with tingling in the foot and other signs suggestive of nerve root involvement. The pain may also radiate into the groin and down the front of the thigh.

Muscle spasm and limitation of movement are common findings (exactly how common depends on exactly how you define spasm). In severe cases they become pronounced. Trigger points may be palpable in the erector spinae muscle group and elsewhere (e.g. quadratus lumborum). Spasm of the psoas muscle and the hamstrings are not uncommon.

## Natural History

Most episodes of low back pain are relatively short-lived. The symptoms usually subside within a few weeks at the most, although in a minority of cases the episode may be more prolonged. *But the recurrence rate is very high.* In one follow-up study of patients who had recovered from an attack of back pain, it was found that more than 60% had another attack within 1 year (Biering-Sørensen, 1983a). In another study of patients returning to work after an attack of back pain, it was found that

49% had recurrences (severe enough to require further treatment) in the first year and 32% in the second; and recurrence rates were higher in those who had suffered several previous attacks (Troup *et al.*, 1981).

There seem to be two main patterns of recurrence. Some people (typically middle-aged or older) have regular attacks, separated by symptom-free intervals during which they are capable of demanding physical work. Others (who may be almost any age) are never quite symptom free. They have a dull ache almost all the time, which is made worse by certain activities and punctuated by episodes of intense pain (Troup *et al.*, 1981).

## Prevalence

About 70% of people will suffer from one or more episodes of low back pain at some time or another during their lives.

Biering-Sørensen (1982) investigated a sample of people which included 82% of all the 30-, 40-, 50- and 60-year-olds living in the town of Glostrup, a suburb of Copenhagen. The overall point prevalence for "pain or other trouble in the lower part of the back" was 14%; the one-year prevalence was 45%; and the lifetime prevalence was 62%. We may regard these as representative baseline figures for the adult population of working age. The effects of age and sex are shown in Table 3.1. Combining the age of groups, there was no significant sex difference in the overall prevalence of back pain but there was a sex difference in the age distribution of cases, with women having a significant excess prevalence in the oldest age group.

*Table 3.1 Prevalence of Low Back Pain in a Suburb of Copenhagen*

| Age | Point prevalence (%) | | One-year period prevalence (%) | | Cumulative lifetime prevalence (%) | |
|---|---|---|---|---|---|---|
| | Men | Women | Men | Women | Men | Women |
| 30 | 9 | 6 | 44 | 41 | 56 | 52 |
| 40 | 16 | 16 | 49 | 39 | 69 | 55 |
| 50 | 13 | 12 | 45 | 47 | 65 | 64 |
| 60 | 11 | 29 | 43 | 54 | 63 | 77 |
| All ages | 12 | 15 | 45 | 45 | 63 | 61 |

Data from Biering-Sørensen (1982).

## Sickness Absence

Sickness absence from back pain is increasing, both in absolute numbers of days and when expressed as a percentage of sickness absence from all causes. Figure 3.2 shows UK data for the years ending 1955–1982. (These figures are for certified absence, exceeding 3 days, based on DHSS statistics cited by Waddell, 1987.) Owing to changes in the reporting procedures, there are no comparable

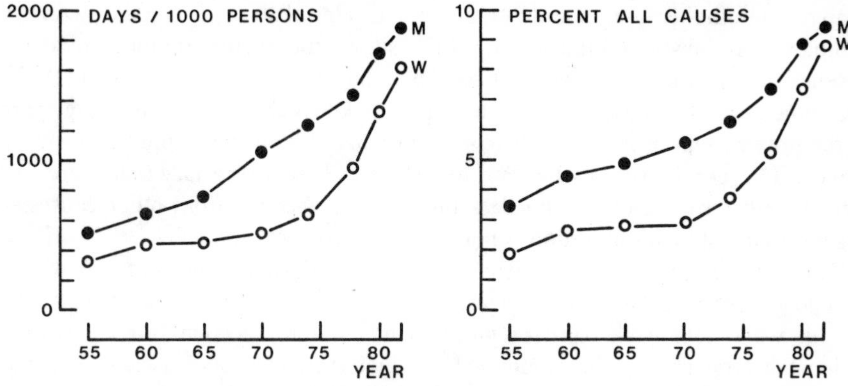


Figure 3.2 *Back pain sickness absence in the UK. M = men; W = women. Based on DHSS data quoted by Waddell (1987)*

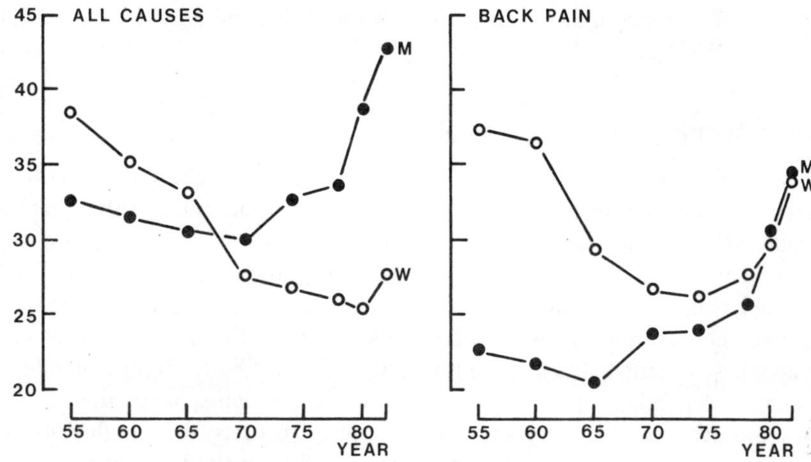


Figure 3.3 *Sickness absence in the UK—average duration of a spell. M = men; W = women. Based on DHSS data cited by Waddell (1987)*

data for subsequent years. But such figures as are available suggest that the upward trend may well be continuing.

Sickness absence is a notoriously poor indicator of morbidity as such—since a number of other factors are involved. Indeed the overall epidemiological database tells a different story: the prevalence figures recorded in various surveys over the last 30 years are, broadly speaking, comparable. Hult (1954) and Biering-Sørensen (1982), for example, recorded lifetime prevalences in men of working age of 60% and 63%, respectively. But the sickness absence data do show that the overall social and economic impact of back pain is steadily increasing. The sex difference is particularly interesting. The figures for men and women seem to be following convergent trends. The reasons are a matter for speculation.

The increases in the relative and absolute numbers of working days lost reflect changes in both the number and duration of episodes of absence. But the picture is complex—as shown in Figure 3.3, which shows the average number of days' absence per spell of sick leave—for back pain and for all causes. Waddell (1987) has argued that the upward trend in back pain sickness absence is due in large part to the prescription of increasingly lengthy periods of bed-rest as the treatment of choice. The complexity of the data in Figure 3.3—and the fact that there are similar long-term trends in the data for sickness absence from other causes—suggest that Waddell's assertion is not altogether correct.

The distribution of days per spell is heavily skewed, so the median figure would be a good deal lower that the averages shown in Figure 3.3. The 5% of cases who are off work with back pain for more than 3 months account for a disproportionately large slice of the economic loss which the condition causes—according to Frymoyer (1988), as much as 85% (although this figure is American and must therefore include compensation costs which are high by the standards of other countries). After 6 months' absence, an individual's likelihood of ever returning to work has fallen to about 50%; after 1 year about 25%; and after 2 years virtually nil (McGill, 1968).

## Applied Anatomy and Pathophysiology

The human spine is a flexible column of 24 movable vertebrae (7 cervical, 12 thoracic, 5 lumbar), plus the sacrum and coccyx. These articulate with each other and with adjacent bones at a total of 137 joints.

Each vertebra has two parts—the vertebral body in the front and the neural arch at the back (Figure 3.4). These form the vertebral canal which transmits the spinal cord. Each neural arch carries articular facets, which are the sites of synovial joints (the apophyseal joints) which permit a sliding action. The functional mobility of the spine is dependent upon the intervertebral discs which separate the vertebral bodies and act as shock absorbers. The disc has two parts: the annulus fibrosus and the nucleus pulposus, which may be translated from the Latin as the fibrous outside and the squidgy middle, respectively. The annulus is made up of concentric layers of fibrous material arranged rather like the cords of a cross-ply motor car tyre (see Figure 3.5). The chemical nature of the nucleus is such that it tends to absorb water from surrounding tissue fluids—a chemist would say that it has a high blood osmotic pressure. This causes the disc to swell until the osmotic pressure is equilibrated by the tension in the annulus. The disc is thus a pre-loaded structure—like a motor car tyre which automatically inflates itself.

Flexion of the spine causes deformation of the disc. The nucleus becomes wedge-shaped and the posterior part of the annulus comes under tension. If the flexed spine is subjected to a compressive loading, the fibres of the annulus may begin to tear, and a bulge may develop. In due course both the tear and the bulge get bigger and eventually the nuclear material may be extruded completely—pressing on nerves as they leave the vertebral canal through the intervetebral foramen and causing intense pain in the distribution of these nerves. This is called prolapsed intervertebral disc (or in lay terms "slipped disc").

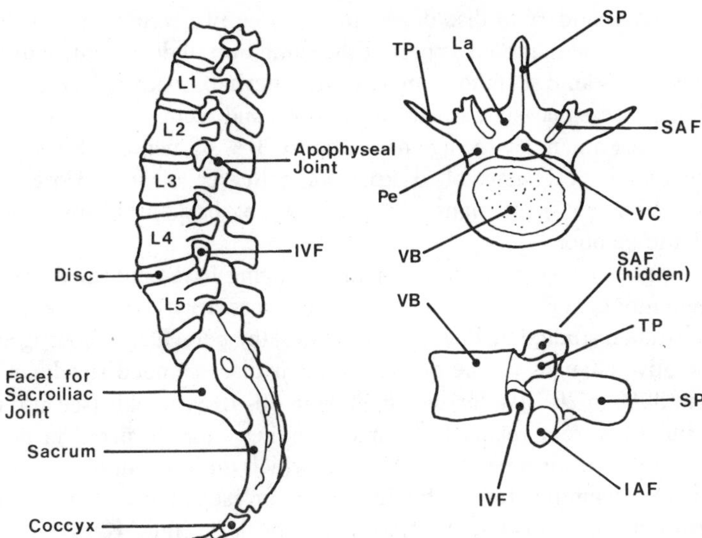


Figure 3.4 Left, *the lumbar spine and sacrum;* Right, *a single lumbar vertebra (L3). IVF = intervertebral foramen; TP = transverse process; La = lamina; SP = spinous process; SAF = superior articular facet; VC = vertebral canal; VB = vertebral body; Pe = pedicle; IAF = inferior articular facet*

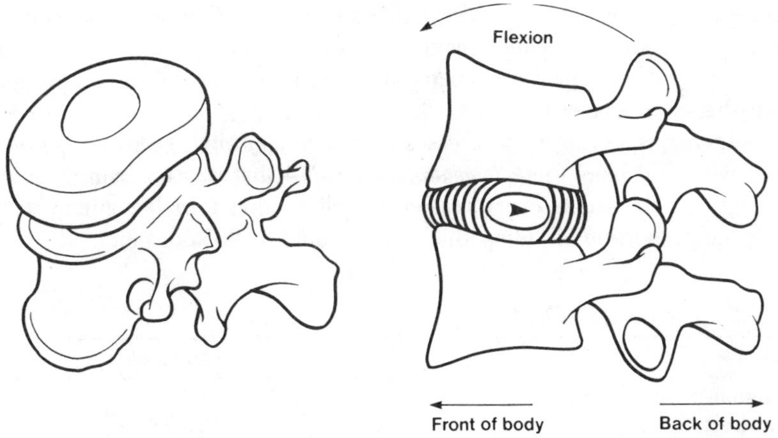


Figure 3.5 *The structure and function of the intervertebral disc. Reproduced from Pheasant (1986), based on Kapanji (1974)*

With age the disc loses its water-binding properties. The fluid content of the nucleus diminishes and the disc decreases in height. This is seen on an X-ray as a narrowing of the disc spaces. Bony growths called *osteophytes* are formed around the margin of the annulus. This is called *spondylosis.* As the vertebral bodies get closer together, the synovial facet joints may be subjected to abnormal patterns of loading, leading to further degenerative changes. *Osteoarthrosis* of the facet joints is thought to be secondary to disc degeneration. The two conditions usually occur together and may be regarded as part of the same overall degenerative process.

Changes of this kind seem to be more or less inevitable with age, at least to some extent. The process may start as early as 25 years and seems to occur more rapidly in men than in women. By the age of 65, up to 90% of men and 80% of women will show radiological signs of at least some degree of spondylosis. But the correlation between these signs and a history of back problems is generally considered to be poor.

Recent studies using magnetic resonance imaging have confirmed that neither disc degeneration nor annular bulging need necessarily cause back pain. In a sample of women studied by Powell *et al.* (1986), the percentage showing evidence of degenerative changes in one or more lumbar discs climbed steadily with age: from about 30% in 20–30-year-olds to 80% in 60–70-year-olds (see also p. 309). Annular bulges were a relatively common finding—and occurred in both discs showing evidence of degeneration and discs which did not. All of the subjects in this study were symptom-free at the time of the investigation.

In a series of back pain patients referred to a specialist clinic, Kersley (1979) was able to make a positive diagnosis in only 20% of cases (see Table 3.2). The most striking thing about these diagnoses is that, with the exception of disc lesions, they all account for a relatively small proportion of the cases. *Spondylolisthesis* is a forward displacement of a lumbar vertebra resulting from a defect in the neural arch. The defect itself, which is known as a *spondylolysis*, occurs in 3–5% of people—and progresses to spondylolisthesis in about half of these. But spondylolisthesis is commonly asymptomatic. *Osteoporotic crush fractures* are relatively common in postmenopausal women—the vertebral body collapses and appears wedge-shaped in a lateral X-ray.

This leaves us a clear 80% of cases for which *no firm diagnosis was possible.* By default we could label these *idiopathic back pain*—that is, back pain of unknown pathology. There seems every reason to believe that in a less highly selected sample the proportion of "idiopathic" cases would be 90% or more.

*Table 3.2 Diagnoses of Low Back Pain*

| Cause | % Cases |
|---|---|
| Disc lesions | 10.9 |
| Spondylosis | 2.5 |
| Gross anxiety and depression | 1.7 |
| Trauma | 1.5 |
| Spondylolisthesis | 0.7 |
| Osteoporosis with wedging | 0.7 |
| Severe postural defects | 0.7 |
| Malignancies | 0.7 |
| Total of positive diagnoses | 19.4 |
| Undiagnosed cases | 80.6 |

Source: Kersley (1979).

Are there people suffering from a form of back pain which is basically myofascial in its origin? Are their symptoms caused by the self-perpetuating cycle of pain and spasm as described at the end of the last chapter? This must be at least part of the story. Muscle spasm and limitation of movement are very common in back pain, and when these are alleviated, the pain often subsides. Back pain (as we have noted earlier) is commonly episodic and recurrent. There is a strong tendency for patients who present with definitive disc lesions to have a long history of prior episodes in which there were no obvious signs of disc involvement. This suggests that discogenic back pain is a progression of the more common idiopathic or myalgic variety. The alternative hypothesis is of course that the idiopathic cases are early stages of disc problems in which the full symptomatology has not yet become manifest. It would be difficult to prove the matter either way. The mechanism of the progression is unclear.

Osteopaths and chiropractors would not agree that 80–90% of their back pain patients are undiagnosable. In most of these cases they would find evidence of *osteopathic lesions* or *subluxations.* At the crudest level these can be regarded as misalignments of the facet joints. The articular surfaces lock together, causing a restriction of movement which may be recognized by palpation and various other techniques. The apophyseal joints contain tiny menisci (discs of fibrocartilage rather like those of the knee)—which could in principle get trapped between the joint surfaces, causing them to lock. Some clinicians (including most osteopaths and chiropractors) also believe that lesions of the sacro-iliac joint are a common cause of pain in the lower back. The fact that osteopathic and chiropractic manipulations commonly result in an alleviation of symptoms must be regarded as strong evidence for the validity of their diagnoses. But to regard the osteopathic lesions solely as a disorder of the facet joints is not to do full justice to the subtlety of the concept. Most osteopaths recognize that the restriction of movement is as much due to muscle dysfunction as to articular misalignment—and the manipulation is aimed at a restoration of normal function. So the osteopathic lesion may be regarded as a special manifestation of the complex cyclic process of myofascial pain and dysfunction which we discussed in the last chapter.

This at least is how I see it at present. From the ergonomic standpoint, these diagnostic controversies are not of much importance—since the preventative measures required would probably be much the same whether the condition is discogenic, myalgic or due to some other process as yet unidentified.

## Work-related Risk Factors

It is generally believed that people in heavy manual occupations are more likely to suffer from back pain than people whose work is less physically demanding. The classic survey of Hult (1954) confirms this. He studied over 1,000 men (aged 25–59) from various selected occupations in Vasteras and Stockholm in Sweden.

Table 3.3 shows the principal differences between the heavy and light occupational groups. The lifetime prevalence of back pain in the heavy workers

*Table 3.3 Prevalence of Low Back Pain in Heavy and Light Occupations*

a. *Lifetime prevalence of back pain and cumulative incapacity for work*

| | % with symptoms | % with disability | | |
|---|---|---|---|---|
| | | >1 day | >3 weeks | >6 months |
| Light work | 53 | 26 | 12 | 2 |
| Heavy work | 64 | 44 | 25 | 5 |
| Total | 60 | 36 | 20 | 4 |

b. *Percentage of Cases in Which Accidents or Heavy Lifting Were Considered to be the Cause of at Least One Episode*

| | Accidents | Lifting | Lifting + accidents | Neither |
|---|---|---|---|---|
| Light work | 14 | 17 | 31 | 69 |
| Heavy work | 27 | 22 | 49 | 51 |
| Total | 23 | 20 | 43 | 57 |

After Hult (1954).

was somewhat higher—64% had suffered symptoms, as against 53% for the light workers—but there were very much greater differences in the disability which these symptoms had caused, as measured by cumulative time off work. This may mean that the heavy workers' symptoms were much more severe—but it is more likely to reflect the fact that it is much more difficult to continue work when your back hurts if your job is physically demanding.

About half (49%) of the heavy workers considered at least one of their episodes to have been brought on either by an accident or by heavy lifting—as against 31% for light workers. But in a majority of cases (57% overall) neither of these was considered to have been the immediate cause of an attack on any occasion. Back pain commonly has an insidious onset, in both heavy and light workers. This suggests that the factors which cause back pain act in a cumulative fashion.

Hult's findings concerning the prevalence of back pain in light and heavy workers were confirmed almost exactly in a study of the employees of the Kodak plant at Rochester, Long Island (Rowe, 1969, 1971). He was able to analyse a working lifetime's medical records in men approaching retirement—65% of those in heavy jobs had been treated for back pain, as against 50% in moderate work and 52% in light work.

Subsequent epidemiological studies have shown, however, that to regard low back pain as predominantly a disorder of the heavy manual worker is a considerable oversimplification.

In the early 1970s, Magora published a series of very important papers concerning the effects of occupational task demands on the prevalence of low back pain (Magora 1970a, b, 1972, 1973a, b). He studied more than 3,000 Israeli men

*Table 3.4 Prevalence of Low Back Pain as a Function of Occupational Task Demands*

| | | Point prevalence (%) |
|---|---|---|
| Sitting | Often (more than 4 hours) | 12.6 |
| | Sometimes | 1.2 |
| | Rarely (less than 2 hours) | 25.9 |
| Standing | Regularly more than 4 hours | 13.8 |
| | Variable* | 2.2 |
| | Regularly less than 4 hours | 24.9 |
| Lifting (5 kg or more) | Often (more than 10 per hour) | 13.7 |
| | Variable | 2.5 |
| | Rarely or never | 23.2 |
| Sudden Maximal Physical Effort | Often (more than 15 per hour) | 18.0 |
| | Variable | 11.3 |
| | Rarely or never | 10.9 |
| Mental Concentration | Continuous | 20.6 |
| | Often | 18.2 |
| | Sometimes | 9.0 |
| Responsibility | Often | 16.8 |
| | Sometimes | 3.1 |
| | Rarely or never | 15.4 |
| Job Satisfaction | Satisfied | 11.7 |
| | Partially satisfied | 7.6 |
| | Not satisfied | 25.2 |
| | Wants change | 31.4 |
| | *Overall prevalence for all subjects:* | 12.9 |

*Free to stand or sit at will.
Data from Magora (1972, 1973a, b).

and women from eight different occupations. The principal findings of these studies are summarized in Table 3.4.

According to these data, people who frequently lift weights have a high prevalence of back pain. So do people who rarely or never lift weights at all. But people in the middle category have a very low prevalence. The same is true for bending actions. Similarly, people who either stand or sit continuously at work for long periods have a high prevalence, but people who are free to vary their posture and to stand or sit at will have a very low prevalence. The only task characteristics which have a simple one-way "dose-related" effect on the prevalence of back pain are sudden maximal physical efforts and mental concentration. It seems likely, therefore, that two quite different groups of people are at risk of back pain as a consequence of their work: those in fully sedentary occupations and those whose jobs require a great deal of lifting, bending and forceful exertion.

Is it the sitting posture itself which causes back problems or does the association arise because sedentary workers tend to be unfit? We cannot be sure—it could be a mixture of both. The loading on the spine can be high when performing sedentary working tasks (p. 106), and people with back problems often find they are worse in the sitting position (p. 72).

Does standing at work cause back pain or is the association due to other factors? Some people find that their back pain gets worse when they stand around ("cocktail party back"). This is said to be due to an accentuation of the lumbar lordosis (and the symptoms are often relieved by flexion). But hyperlordosis is not of itself a significant risk factor (p. 103). The people who do the most lifting will tend to be the ones who stand for longest. Working for long periods in a stooped position certainly causes back problems. Wickström (1978) cites studies of foundrymen and turners (lathe operators), and Buckle (1983) found that, on average, men with back problems spend longer in a stooped position than controls.

The data of Magora point to the importance of a varied pattern of working activities and the desirability of changes in posture. Vällfors (1985) confirmed that people with back problems were significantly less likely to have varied working tasks or the ability to choose their own working position. Magora also found an association between back pain and low job satisfaction. This may be partly because people who do repetitive jobs in fixed working postures are likely to have low job satisfaction. Subsequent studies have found that people with back problems are more likely to report stressful events at work (Frymoyer *et al.*, 1980) and that they have lower job satisfaction and a lower opinion of the working environment (Vällfors, 1985). We shall return to these matters in due course.

Back pain is a common condition overall, but when we divide the population into categories based on occupational task demands, we find that there are some working groups in which it is more than ten times more common than in others (and some in which it is actually quite rare). *We must therefore conclude that low back pain is predominantly a work-related condition.*

Let us try to quantify this statement. Figure 3.6 is based on Magora's data for sitting, standing and lifting. The prevalences for the categories indicative of low, middle and high physical workload have been averaged out. (Standing is taken to be indicative of high workload.) Let us assume that the 2% prevalence in the middle category figure prepresents a baseline figure for a population which is not exposed to any work-related risks. In a working population in which the three categories were represented in equal numbers, the overall prevalence would be 14%—of which 2% would be non-work-related and 12% would be work-related. Thus low back pain could be regarded as a work-related disorder in 85% of cases.

The overall picture presented by Magora's data has been confirmed in two case–control studies, of patients diagnosed as suffering from acute prolapsed discs, reported by Kelsey and co-workers (Kelsey, 1975a, b; Kelsey *et al.*, 1984a, c)—although not without certain reservations. In the first of these studies, Kelsey (1975a, b) found that sitting for half or more of the working day carried a relative risk of 1.6—the risk was greater in older workers and in drivers, and it was greatest in truck drivers (Figure 3.7). There was no appreciable increase of risk in people who had been in sedentary work for less than 5 years.

Buckle *et al.* (1980) confirmed the association with driving and also noted that drivers tended to suffer from a preponderance of right-side symptoms. In the

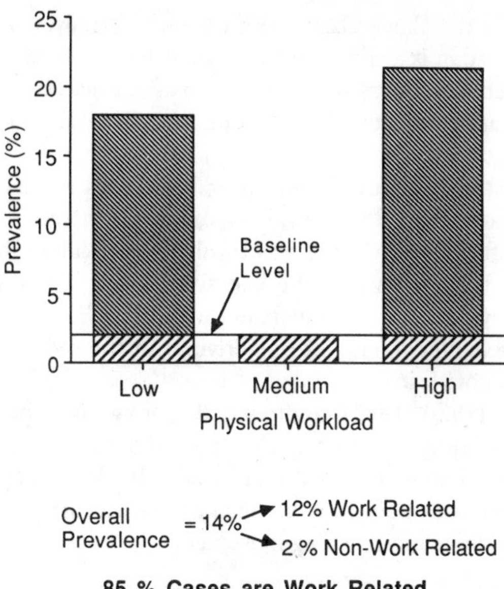


Figure 3.6 *Prevalence of low back pain as a function of physical workload. Based on data from Magora (1972)*

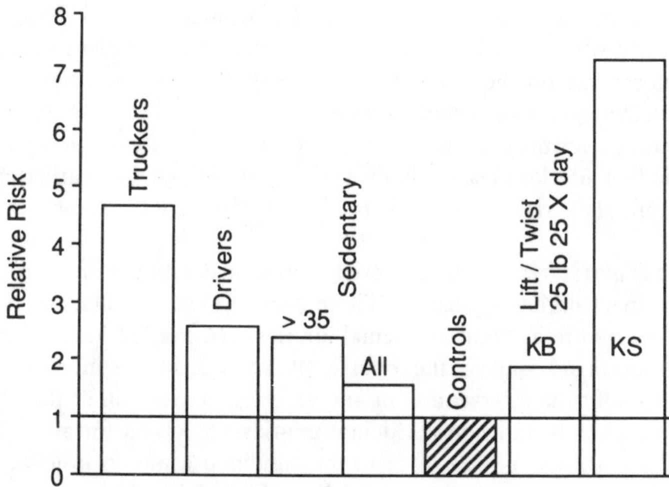


Figure 3.7 *Occupational risk factors for low back pain (diagnosed as due to prolapsed disc). Data from Kelsey and co-workers (Kelsey, 1975a, b; Kelsey* et al., *1984a, c)*

second study, Kelsey *et al.* (1984a) found that the risk of prolapsed disc increased directly with the number of hours per week spent in a motor vehicle and that people who drove older vehicles were more at risk than people who drove new ones. This is probably because of the increased vibration and jolting which occurs in older vehicles as the shock absorbers wear out. Frymoyer *et al.* (1980, 1983) confirmed that vibration is a significant risk factor for low back pain—not only for drivers of motor vehicles, but also in people who use devices such as jackhammers, chainsaws and rotary cultivators. We shall discuss the association with driving in Chapter 13.

In the second study, Kelsey (1984c) found that jobs involving lifting objects which weighed more than 25 lb (11.3 kg) carried a threefold increase of risk (which increased further if twisting was involved, particularly if the knees were straight), but the risk was lower if the knees were bent. Paradoxically, however, Kelsey found no evidence of an association with lifting in the first study and no association with sedentary work (except driving) in the second. The reason for these anomalies is not clear.

Frymoyer *et al.* (1980, 1983) confirmed that jobs involving lifting, carrying, pushing, pulling, bending and twisting all carried significantly increased risks; and in an investigation of women in the retail trade, Buckle *et al.* (1986) found an association between back pain and the need to perform frequent (more than 10 times per hour) twisting and reaching movements.

## Trigger Factors

Back pain is characteristically episodic and recurrent. Episodes of back pain sometimes have an acute onset but in other cases the onset is insidious. There is some suggestion that a higher proportion of first attacks are likely to have an insidious onset, whereas subsequent attacks are more likely to come on suddenly; but the subject has not been much investigated (Peter Buckle, personal communication). When back pain comes on suddenly, the patient is likely to regard the activity he was performing at the time as the "cause" of his condition—particularly if this is his first attack of back pain or if it is more serious than earlier episodes. Given the multifactorial aetiology of back pain, this belief may or may not be justified.

Manning *et al.* (1984) investigated events associated with the onset of episodes of back pain in a working population. These were divided into accidents *per se* and what they referred to as "non-accidental injuries". (A non-accidental injury was one which occurred during the course of an individual's normal working activities—without the intervention of any unforeseen event other than the pain itself.) The accidents and non-accidental injuries accounted for about 30% of cases each, and in about 40% of cases the condition came on insidiously. The most common accident was a loss of footing (66%), followed by an unexpected load (12%) or loss of balance (10%). Non-accidental injuries most commonly occurred during bending/straightening actions (73%), lifting and carrying (50%) and twisting (48%).

The figures of Manning *et al.* (1984) are in reasonable agreement with those of Hult (1954) given in Table 3.3. A number of other studies have addressed the same issue, including Glover (1960), Frymoyer *et al.* (1980) and Biering-Sørensen (1983a), although none of them have analysed the data in as much detail as

Manning *et al.* (1984). When we compare these sources, we find considerable variation in the percentage of cases falling into each of the statistical categories. In part, this variation is due to differences in the way the categories are defined; but sampling effects are also probably involved and it presumably makes a difference how (and why) you ask the question. The overall picture which these studies present is summarized in Figure 3.8—although we should not take the exact percentages too seriously.

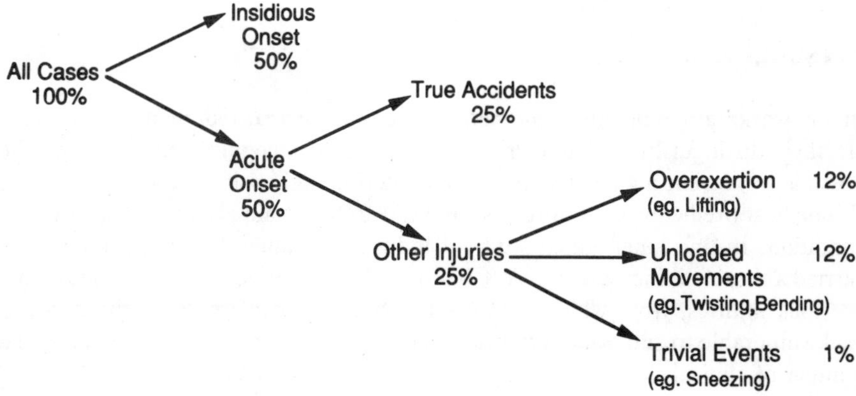


Figure 3.8 *Low back pain—mode of onset and trigger factors. Data from various sources—see text for details*

The "other injuries" category is particularly interesting: it includes overexertions such as lifting and handling injuries. However, acute attacks of back pain seem to occur equally often during the course of what seem to be perfectly normal movements, when the individual is not handling a load. (Attacks very often occur as an individual straightens up after a period in a stooped position.) And a small minority of episodes are triggered by seemingly trivial events like sneezing (Glover, 1960). We may conclude therefore that attacks of back pain may be triggered at levels of loading (on the spine and musculature) that range from very heavy to almost negligible. This implies that some people must be very much more vulnerable than others.

People who suffer from chronic or recurrent back problems can generally give a fairly good account of the sorts of things which tend to make their pain worse or better. Biering-Sørensen (1983b) studied these aggravating and relieving factors. Stooping, sitting and standing tended to aggravate symptoms, as did changes in the weather; lying and walking tended to relieve symptoms. Psychological stress commonly aggravated symptoms (much more so in women than in men), and women commonly found that their symptoms were worse during menstruation.

We are now in a position to summarize those features of the working situation which are associated with a high risk of low back pain. These are set out in Table 3.5.

*Table 3.5 Low Back Pain: Ergonomic Risk Factors*

| |
|---|
| Heavy work—lifting and handling, forceful exertions, bending, twisting, etc. |
| Prolonged sedentary work |
| Prolonged work in a stooped position |
| Vibration |
| Psychological stress |

## Personal Risk Factors

In any working population, some people must be more at risk than others. What are their distinguishing characteristics? If we could recognize these, it might be possible to prevent high-risk people from entering high-risk situations.

The best predictor of a future susceptibility to back pain is a prior history of the condition. In one longitudinal study of industrial manual workers, prior history carried a threefold increase in risk (Chaffin and Park, 1973). We may suppose that previous injuries, from which recovery has been incomplete, leave the person's back vulnerable to subsequent injuries. Lloyd and Troup (1983) have identified a number of clinical signs which are predictive of recurrence.

There is a strong clinical impression that tall people, or people with a "long back", are more prone to back problems than short people. The epidemiological evidence for this is equivocal. Some studies have shown no association (Hult, 1954; Rowe, 1969, 1971; Kelsey *et al.*, 1984a). Kelsey (1975a) found an association in women but not in men; Biering-Sørensen (1983c) found an association in men; Heliövaara (1987) found a strong association in both women and men.

Taken together, these studies point to the conclusion that there is an association between tallness and back pain, but probably a weak one. Perhaps tall people are more at risk in some jobs than in others. A tall lathe operator (for example) would have to stoop further to operate the controls of the machine. In other jobs, height could be an advantage (especially insomuch as height is significantly correlated with strength), with the two possibilities cancelling each other out in a large-scale survey.

Of the studies cited above, only Biering-Sørensen (1983c) and Heliövaara (1987) found any signs of an association between back pain and either body weight or overweight.

Postural variations and abnormalities of spinal curvatures are discussed in Chapter 5. With the exception of scoliosis—a sideways list which is commonly associated with inequalities of leg length—these are of minimal predictive value.

Standard radiological investigations of the lumbar spine are of minimal value in predicting whether a person's back is likely to stand up to the stress of industrial work. A person with spondylolisthesis would certainly be at risk in a heavy job, but the condition is rare. Structural anomalies such as spina bifida occulta and abnormal vertebral numbers are of no consequence. Spondylitic changes are so common that their predictive value is minimal. The X-ray doses involved in looking for these rare or equivocal signs are not considered to be justified in the context of pre-employment screening or personnel selection (Pope *et al.*, 1984).

The size of the vertebral canal can be measured by ultrasound. The technique is simple and safe. Patients attending a general practitioner with back pain have been found to have smaller canals than controls (Drinkall *et al.*, 1984). Miners with small canals have more sickness absence from back pain (MacDonald *et al.*, 1984). But the current consensus seems to be that the predictive value of this technique is not high enough to make it a viable option for screening purposes.

Cady *et al.* (1979) evaluated the strength, flexibility and cardiorespiratory fitness of Los Angeles firefighters. They divided them into three groups according to a composite fitness score. Over a four-year follow-up period, the incidence of back problems was 7% in the least fit group, 3% in the middle groups and only 1% in the most fit group. They also found that fitness levels were predictors of recurrence in men who had suffered back problems previously.

It is not particularly surprising to find that physical fitness is a good predictor of risk in a job which is as physically demanding as firefighting. But which components of the composite fitness score contribute most heavily to the observed differences in risk? And would they be equally good predictors for people doing less demanding jobs?

A comparison between the weight of load handled at work and the lifting strength of manual workers (measured in a comparable position) has been found to predict risk of injury (Chaffin and Park, 1973; Chaffin *et al.*, 1978—see Chapter 14 for a further discussion). But Troup *et al.* (1987) found that lifting strength, measured in standardized tests which were related to task demands, had no predictive value; and Biering-Sørensen (1983c) found no association between the strength of either back or abdominal muscles and the risk of a first-time attack, although back muscle endurance was predictive, and the strength measures were predictive of recurrences in men with pre-existing problems.

People with current problems have a significantly reduced range of spinal mobility (taking into account the effects of age and sex), but there is also a significant excess representation of people with back problems in the upper quartile of the distribution of spinal mobility (Burton *et al.*, 1989a). The findings of Biering-Sørensen (1983c) suggest that hypermobility of the spine is an independent risk factor, and tight hamstring muscles are predictive of recurrences but not first attacks.

Participation in sporting activities during childhood is associated with a reduction in risk during adult life, but adult sports participation is associated with increased risk (Burton, *et al.*, 1989b). Presumably the childhood sporting activities result in increased overall fitness but the adult activities carry an increased risk of injury. The impression I get from my students, however, is that back problems in young people (17–20 years) are very often due to sports injuries—particularly gymnastics and trampolining in girls, and rowing and rugby in boys.

Men who suffer from back problems are also more likely to suffer from a number of other medical conditions indicative of poor overall fitness—including angina pectoris, chronic bronchitis, intermittent claudication and headaches (Gyntelberg, 1974). But aerobic power does not seem to have independent predictive value (Gyntelberg, 1974; Battié *et al.*, 1989).

Smokers are more likely to suffer from back problems than non-smokers (Frymoyer *et al.*, 1980, 1983). The effects of smoking are dose-related and stronger than you would expect from the effects of coughing alone: risk increases by about 20% for every 10 cigarettes per day; but people who have given up smoking for 1 year are at no higher risk than those who have never smoked (Kelsey *et al.*, 1984a). The effect is independent of aerobic fitness (Battié *et al.*, 1989). So far, the connection has not been explained—it may be something to do with blood flow.

In women, back problems starting in pregnancy may be exacerbated by the physical trauma of childbirth and the physical workload of bringing up small children. About 20% of women with back problems attribute these to pregnancy or childbirth (Biering-Sørensen, 1983a), and both Kelsey (1975a) and Frymoyer *et al.* (1980) have found that women who have more children have more back pain.

The predictive value of the personal risk factors we have considered so far is summarized in Table 3.6.

*Table 3.6 Low Back Pain: Personal Risk Factors*

Strong Risk Factors
- Previous history of low back pain
- Low overall fitness
- Low lifting strength—compared with task demands
- Low endurance of back muscles
- Smoking
- Motherhood

Moderate Risk Factors (may be significant in extreme cases)
- Hypermobility
- Spondylosis
- Spondylolisthesis
- Scoliosis and unequal leg length
- Weak back muscles, weak abdominal muscles, tight hamstrings *(predict recurrence but not first attacks)*

Weak or Very Weak Risk Factors
- Stature
- Overweight

Factors of No Predictive Risk Value
- Lordosis or flat-back
- Abnormal vertebral numbers
- Spina bifida occulta

## Psychological Aspects

Some people believe that people with back problems (particularly chronic ones) have characteristic personality traits. Back pain patients report more episodes of anxiety and depression than controls (Frymoyer *et al.*, 1980). It is difficult to interpret these findings. The psychological states may be a precursor or a consequence of the pain, or both may be due to stress.

In its chronic forms, the underlying pathology of back pain may well be dominated by its neurological and psychological components. A combination of neural sensitization and exaggerated pain perception may lead to a limitation of activity which initiates a downward spiral of disability (Troup, 1988). Factors of this kind are characteristic of many other forms of chronic pain. Vällfors (1985) was unable to demonstrate objective clinical signs in 70% of chronic cases (defined by 3 months' sickness absence), compared with 40% in acute cases (although you might perhaps expect this from the subsidence of acute inflammation, etc.). The chronic cases were a psychologically distinguishable group: they showed more evidence of psychiatric illness, more signs of alcohol abuse and reported lower job satisfaction (although it would be difficult to distinguish cause from effects). Of these chronic cases, 31% believed that their working conditions prevented their return to work (and discussions with foremen commonly confirmed that working conditions were bad or very bad).

There are those who believe that long-term working disability due to back pain (and other comparable conditions) may involve an element of subconscious end-gaining. The medical condition may become a means of withdrawing from an intolerable situation. As a humanistic psychologist, one would ask why the experience of work is so distressing for these individuals that in their cognitive appraisal of their total life situation, pain and disability are construed as the better option?

## Summary

Back pain is common—but not uniformly so. The epidemiological data point to the conclusion that habitual patterns of working activity are more significant determinants of an individual's chances of suffering from back problems than his (or her) personal characteristics. That is, ergonomic risk factors are more important than personal risk factors.

Of the many personal risk factors which have been investigated, the ones with the greatest predictive value tend to be concerned with *lifestyle* rather than with the individual's *constitutional make-up*. In other words, back pain is mainly a consequence of what we *do*, not of what we *are*.

It is reasonable to suppose that these factors act in an additive or multiplicative fashion. Given that this is the case, it is also reasonable to suppose that any working population will include a minority of individuals whose backs are in a *vulnerable state*. The distribution of events associated with the onset of acute attacks supports this proposition. If the concept of the vulnerable back is valid, then the event which triggers the acute attack may be a more or less arbitrary end-point to a process of cumulative damage acting over weeks, months or years. We could call this the *cumulative trauma model* for the aetiology of back pain.

Throughout your life you are doing things which are potentially damaging to your back: if the rate of damage exceeds the rate of recovery, problems will arise in one context or another. But to what extent is it legitimate to regard the *context* in which the acute symptoms finally become manifest, as the *cause* of the condition?

The cumulative trauma model has important implications for work design—particularly when it comes to setting load limits for lifting and handling tasks (see Chapter 15).

## Chapter Four

# Repetitive Strain Injuries

Work-related musculoskeletal disorders of the neck, shoulder and upper limb are collectively referred to as *repetitive strain injuries* (RSI). The term originated in Australia and has been taken up in the UK. In North America the term *cumulative trauma disorders (CTD)* is applied to a similar range of conditions, and in Japan and Scandinavia they are sometimes called *occupational cervicobrachial disorders (OCD).* The term *overuse syndrome* is also used. These are all general categories, not diagnoses.

Some of the disorders which fall into the RSI category are discrete pathological entities, with relatively well defined signs and symptoms (see Table 4.1). But the same label is also applied to less well defined myalgic conditions, involving pain and dysfunction at multiple sites. Tender points are typically distributed throughout the neck, shoulder, forearm and hand (Ryan and Bampton, 1988), and histological investigations have found clear signs of muscle damage (Dennet and Fry, 1988). Prolonged pain and disability may result. This suggests that a

*Table 4.1 Work-related Musculoskeletal Disorders of the Neck, Shoulder and Upper Limb*

*Synonymous Generic Terms*
- Repetitive strain injuries (RSI)
- Cumulative trauma disorders (CTD)
- Occupational cervicobrachial disorders (OCD)
- Overuse syndrome

*Some Specific Clinical Conditions*
- Supraspinatus (or rotator cuff) tendinitis
- Thoracic outlet syndrome
- Lateral epicondylitis (tennis elbow)
- Medial epicondylitis (golfer's elbow)
- Tenosynovitis, peritendinitis (crepitans), De Quervain's disease,
- Trigger finger
- Carpal tunnel syndrome
- Ganglia

*Some Less Specific Descriptive Terms*
- Cervical syndrome
- Tension neck, stiff neck, etc.
- Frozen shoulder
- Occupational cramps (craft palsies)

*Some Occupational Variants*
- Gaoler's elbow, hopper's gout, data-processing disease, washer-woman's thumb, cowboy's thumb, writer's cramp, florist's cramp, telegraphist's cramp, musician's cramp, and very many others

neurological mechanism may well be involved and, in its chronic form at least, the condition is thought to have a strong psychogenic component. In a minority of cases, there may also be signs of vasomotor disturbance. The discrete and myalgic forms of RSI may co-exist—for example, people with tenosynovitis or carpal tunnel syndrome may also have myalgic conditions in the proximal parts of the limb.

It is widely recognized that all of these disorders principally result from cumulative overuse. (Although, in the case of carpal tunnel syndrome, systemic factors may also play a part.) They are particularly associated with the repetitive short-cycle tasks of the industrial assembly line, although they can also occur in other contexts. The causative factors most commonly implicated are:

- fixed working posture;
- repetitive motions;
- psychological stress.

These often occur in combination and were recognized almost three centuries ago by Ramazzini (see p. 7).

There are three good reasons for treating the work-related musculoskeletal disorders of the neck, shoulder and upper limb as a group:

(i) The nerves which supply the upper limb arise from cervical spinal segments (C5–T1), and form a neurovascular bundle which passes from the thoracic outlet and the root of the neck into and through the axilla. Pathology of the cervical spine (or other kinds of cervical dysfunction) or pressure on the neurovascular bundle may therefore cause symptoms referred to distal parts of the limb.

(ii) Trigger points in the muscles of the shoulder girdle may have reference zones distal in the limb: for example, the supraspinatus (which is commonly affected in people who work with their arms in a raised position) may refer pain down the outer edges of the limb as far as the thumb (see below).

(iii) Working activities which result in an excessive static (i.e. postural) loading of the neck and shoulder muscles commonly also involve repetitive contractions of the forearm muscles: for example, assembly tasks performed with the arms in a raised position, or keying actions on a keyboard which is too high. Thus neck and shoulder problems may occur in combination with forearm problems both in the individual patient and in the members of a working group. A large proportion of RSI sufferers who have clinical signs and symptoms in distal parts of the upper limb also show evidence of cervical spine dysfunction (David Tasker, personal communication).

People who suffer from neck, shoulder and upper limb problems are also more likely to have trouble with their backs: about 50% more likely, according to the data of Hult (1954). And it is worth noting that people with chronic and/or recurrent back problems often have intermittent symptoms at a number of sites:

one day in the lumbar region; another time between the shoulder blades; and so on. A number of studies of VDU operators and other office workers have shown that those who report musculoskeletal pain and discomfort are statistically more likely to suffer from headaches and eyestrain, to report glare problems, and so on (p. 233). So, statistically at least, we could regard RSI as part of a more extensive syndrome of work-related ill-health. It seems particularly ironic that low back pain is not generally classed as a cumulative trauma disorder—since it is commonly caused by cumulative trauma. The list of symptoms which Maeda (1977) included in his original account of occupational cervicobrachial disorder (OCD) was extensive, including not only musculoskeletal, visual and neurovascular problems, but also a number of stress-related symptoms such as insomnia, irritability, gastrointestinal dysfunction and menstrual disturbance. We might regard these latter symptoms as manifestations of chronic fatigue (see p. 159; compare also with the condition called fibromyalgia, p. 53).

## RSI: The Australian Epidemic

The Australian epidemic of RSI commenced in 1980 or thereabouts and peaked in 1985/1986, following which the number of cases reported began to decline. So far the phenomenon does not seem to have been repeated elsewhere, although the high prevalence of writer's cramp in nineteenth century London invites comparison (see below).

Annual incidence rates for New South Wales are shown in Figure 4.1. The figures were kindly supplied by Mike Stevenson. They show spells of disability lasting more than 3 days due to musculoskeletal problems other than back pain.

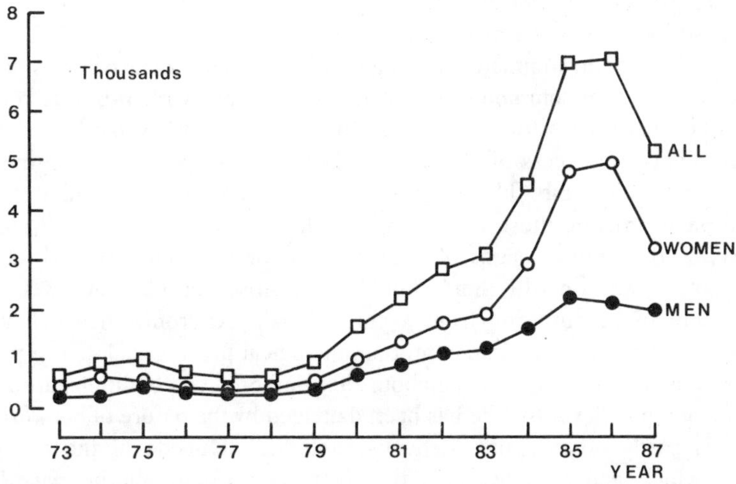


Figure 4.1 *Annual incidence of RSI in New South Wales*

(Thus they will include a small minority of lower limb problems.) The rate of increase has been greatest in office workers—specifically in users of electronic keyboards. But incidence rates in office workers remain low compared with some other occupational groups: in 1987 there were six new cases per thousand office workers, compared with 78 per thousand in production workers in the metal working and electrical industries.

In the early stages of the epidemic, there was a degree of confusion as to the nature of these conditions. Diagnoses of tenosynovitis were common—but it soon became clear that patients were suffering from a condition of a different and more complex nature. In the small minority of cases where clear signs of tenosynovitis were present (see p. 9), there were generally other problems as well, and many patients with no signs of tenosynovitis fell into the most severe and chronic category.

Media coverage of Australia's "office epidemic" was sensationalist in the extreme. Emotive terms like "keyboard cripples" were used and comparisons were drawn with AIDS. Large sums were paid out in compensation. As it became increasingly clear that the symptoms of the condition or conditions involved did not entirely match those of any specific known clinical entity, there were many who were ready to assert that RSIs were in some way "unreal"—that the symptoms were psychogenic, hysterical or frank malingering. The reporting of the issue was basically confrontational. The antithesis of "psychological" versus "real" is entirely false. But no one seemed much interested in the possibility that the condition could be *both* organic *and* psychological. (I should like to thank Leon Straker, who ran a support group for RSI victims in Perth during this period, for his graphic first-hand account of these events.)

As the epidemic began to subside, and people began to take stock of what had happened, a new consensus began to emerge: that the epidemic was "a complex psychosocial phenomenon with elements of mass hysteria" (Ferguson, 1987); that it was the product of "social iatrogenesis" (Cleland, 1987); and so on. Willis (1986) described RSI as "a metaphor for alienation".

It is worth trying to summarize the main points at issue. Tenosynovitis has long been recognized as a common disorder of assembly line workers and is known to occur in keyboard users (but not often). Some samples of keyboard users have a high prevalence (in excess of 50%) of work-related aches and pain in the neck, shoulder and upper limb. These points have not been seriously challenged. The controversy centres on those cases in which there is severe, chronic disability in the absence of definitive signs of tenosynovitis (or any other similarly discrete clinical entity), and on why there should be an epidemic of such cases. These people seem to be suffering from a chronic pain syndrome in which muscle damage, neurological sensitization and psychological processes all play a part. Of itself there is nothing very strange about this—other such conditions are known. The RSI victim believes that he has been damaged by the nature of his work. The damage is partly organic and partly psychological. Outside of the adversarial situation which generally pertains, the distinction would not be regarded as particularly important.

## Neck and Shoulder Pain

Pain in the neck is almost as common as low back pain and a similar range of pathologies are probably involved. Myalgic conditions predominate—particularly in younger age groups. Neck pain due to muscular dysfunction usually comes on gradually—often overnight. Like low back pain, it tends to be recurrent. Muscle spasm and limitation of movement are often pronounced, so the condition is often referred to as *stiff neck* or *tension neck.* The spasm is often unilateral, giving the neck a twisted appearance; thus the condition may be called *wryneck* or *torticollis.* Any neck muscle may be affected, but sternomastoid, the sub-occipital muscles and the upper fibres of trapezius are affected particularly frequently (Figures 4.2 and 4.3).

*Cervical spondylosis* is common with age: it is similar in its pathology to the equivalent condition affecting the lumbar spine. Secondary osteoarthritic changes also occur in the unco-vertebral joints. (These are small synovial articulations which are found between the bodies of cervical vertebrae but not elsewhere in the spine.)

Neck pain commonly radiates into the shoulder region and down the arm (*cervicobrachialgia*). Pain from the muscles which connect the cervical spine to the shoulder girdle is commonly referred to their insertions. Pain from the scapulo-humeral muscles (supraspinatus, infraspinatus and subscapularis) is felt around their points of origin and referred down the arm (Figure 4.3). Muscular pains in the neck, the shoulder and the upper part of the back (between the shoulder blades) thus tend to merge into one another, to the extent that there may be little point in trying to separate them in epidemiological studies. A deep diffuse aching, reaching from beneath the scapula to about the level of the elbow, is a common finding. More circumscribed distributions of pain, suggestive of nerve root

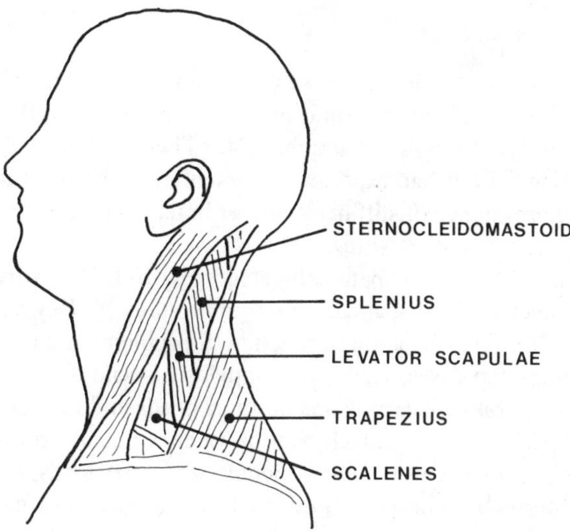


Figure 4.2 *The muscles of the neck*

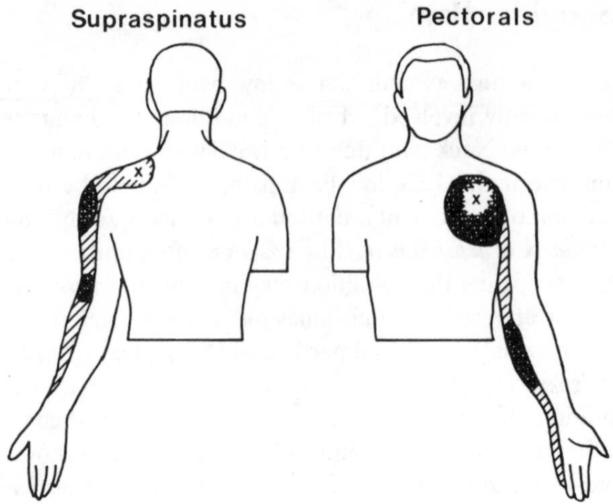


Figure 4.3 *Trigger points and reference zones for some muscles of the shoulder region—based on Travell and Rinzler (1952)*

involvement (sometimes with numbness and tingling in the fingers), are less common and generally occur at a later stage (Hult, 1954).

Some people with neck problems experience episodes of vertigo, dizziness, fainting or tinnitus when the neck is held in a position of extreme extension or rotation—for example, when changing a light bulb or painting the ceiling. This may be due to partial occlusion of the vertebral artery, or the neck muscles may perhaps supply the brain with false postural information.

## Prevalence and Risk Factors

The definitive epidemiological study of neck pain remains that of Hult (1954), in which 44% of subjects reported one or more attacks of "stiff neck" and 23% reported one or more attacks of brachialgia. (The definition of brachialgia was quite a broad one.) A few had experienced brachialgia without stiff neck, giving an overall lifetime prevalence of stiff neck and/or brachialgia of 51%, as against 60% for low back pain in the same study.

The light and heavy occupational categories in Hult's survey had similar prevalences of neck problems; and except for one study (Kelsey *et al.*, 1984b) there is little evidence for an association between neck problems and lifting and handling tasks except in special cases. Cervical spondylosis is said to be more common in populations who habitually carry loads on their heads and to occur at earlier ages: thus Bremner *et al.* (1968) found a higher prevalence in black rural Jamaicans than in whites. Neck problems also occur in people who carry loads on their shoulders, such as slaughterhouse workers who carry heavy carcasses of meat (Hult 1954); and in television cameramen, who may carry a heavy outside broadcast camera for an hour or more at a time (Sheila Lee, personal communication). The weight of

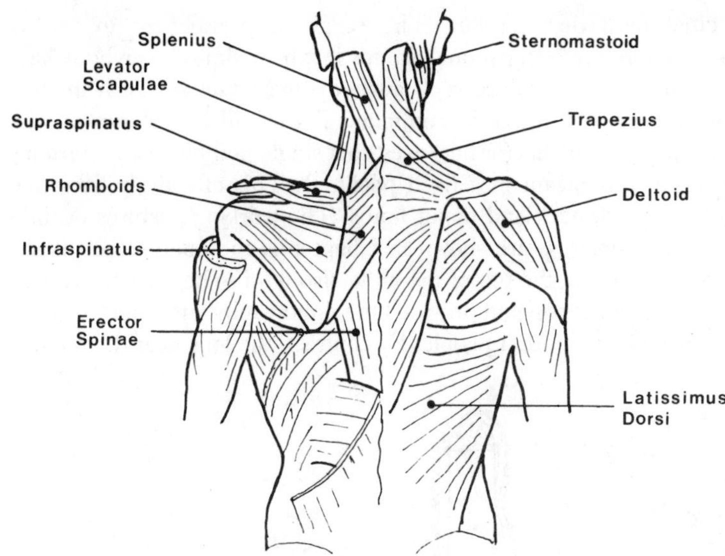


Figure 4.4 *Muscles of the neck and shoulder. Right-hand side shows first layer; left shows next layer*

the load when carried in this way is supported by tension in the muscles which join the neck and shoulder girdle: trapezius, levator scapulae and rhomboids (Figure 4.4). The tension has a high static component.

Static muscle tension may also cause neck problems in violinists and in people who habitually cradle a telephone between their shoulder and their ear so as to leave their hands free for writing. But examples of this kind are relatively rare. The most common cause of work-related neck problems is the static muscle loading which results from faulty working posture—perhaps exacerbated by psychological stress. Consider the data in Table 4.2, which were compiled from a number of

*Table 4.2 Prevalence of Neck Problems in Various Occupational Groups*

| | | % Prevalence |
|---|---|---|
| Sewing machinists | | 98 |
| Teleprinter operators | | 69 |
| Microscopists | | 66 |
| Seamstresses | | 60 |
| VDU operators: | data entry tasks | 11–24 |
| | conversational tasks | 4–12 |
| | computer-aided design | 3 |
| Key punchers | | 20 |
| Cash register operators | | 14 |
| Traditional office work (inc. typists) | | 1–8 |
| Telephone operators | | 6 |
| Nursery teachers | | 4 |

Data from various sources compiled by Grieco (1986).

sources cited by Grieco (1986). The range of prevalences in this table is remarkable. There are occupational groups in which almost everybody has a neck problem and others in which scarcely anybody does. The jobs near the top of the table have important features in common. They tend to be ones in which the worker is required to fix his (or her) eyes on a single point whilst performing some task with his hands (again in a fixed position). This effectively fixes the entire posture of the head, neck and upper limbs. The working postures of the sewing machinist and seamstress are shown in Chapter 1; the microscopist is shown in Figure 4.5. The jobs at the top of the table also tend to be ones which demand a great deal of concentration and provide little opportunity either for moving around or for socialization with other people. By contrast, the jobs near the bottom of the

Figure 4.5 *The microscopist—from an original in the author's collection*

table tend to be ones in which there is more freedom of movement and in which a more diverse range of tasks are performed, having different postural requirements; and they tend to be ones in which there is more direct contact with other people.

Neck and shoulder problems also occur in people who work for long periods with their arms in a raised position, again causing static tension in the upper fibres of the trapezius, levator scapulae, deltoid, and so on. High prevalences have been noted in assembly line workers and packers (Maeda, 1977; Waris, 1980), and in people who perform electrical wiring tasks on large frames (Westgard and Aarås, 1984, 1985).

## Headache

Headaches are often due to muscle tension. These are called tension headaches or myogenic headaches. Sometimes the neck muscles are involved. In a study of patients attending a specialist clinic because of severe headaches, muscle tension was diagnosed as the cause of 40% of cases (Lance, 1969). In patients attending their GP or coping with self-medication, the proportion would doubtless be higher—a figure of 90% is sometimes quoted.

Tension headaches are usually stress-related. Other factors which commonly precipitate attacks include noise, glare, flickering lights and the need for visual concentration. Various muscle groups may be involved. Each has a characteristic pain distribution. People who habitually clench their jaws or grind their teeth (temporalis and masseter muscles) characteristically have a pain in front of the ear and up over the temple. People who wrinkle their brows (corrugator supercilii and occipito frontalis muscles) have bilateral pain above the eyes.

Trigger points in the upper fibres of trapezius refer pain to the temple and eyebrow, and to deep behind the eyes (Figure 4.6). The sternomastoid refers pain to the ear, eyebrow and forehead and may cause red and tearful eyes, and occasionally a runny nose. The splenius muscle refers pain to the crown of the

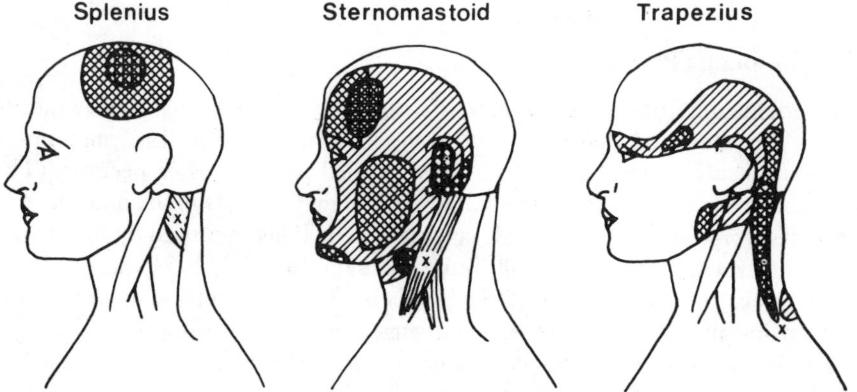


Figure 4.6 *Headache due to neck muscle dysfunction—based on Travell and Rinzler (1952)*

head (Travell and Rinzler, 1952; Travell, 1967). The similarity of these symptoms to those of both "eyestrain" and the so-called "sick building syndrome" is striking (pp. 210, 161). The mechanism of this reference pattern is of interest to the neuroanatomist. The main relay centre for pain from the head and face is the descending or spinal nucleus of the trigeminal nerve (CN V). This is co-extensive with the substantia gelatinosa in the upper part of the spinal cord, which receives sensory inputs from cervical structures.

## Humeral Tendinitis

The rotator cuff muscles (supraspinatus, infraspinatus, subscapularis and teres minor) arise from the scapula and pass just outside the capsule of the shoulder joint to insert on the upper part of the humerus. *Humeral tendinitis* or *rotator cuff tendinitis* is an inflammatory condition affecting the tendons of these muscles and surrounding soft tissues—particularly the subacromial bursa which separates the tendon of supraspinatus from the acromion and the deltoid muscle. The supraspinatus tendon is the one most commonly affected, along with the tendon of origin of the long head of biceps brachii. In older people the tendons of supraspinatus and biceps may calcify—and in advanced cases may be ruptured by muscular exertion.

Humeral tendinitis (and shoulder problems in general) are found in assembly line workers and others whose work entails sustained abduction or flexion of the arms (to more than 60°) or forceful repetition of these movements (Putz-Anderson, 1988). Frictional attrition is a favoured explanation—the tendons may rub against the head of the humerus during movement, and the supraspinatus tendon in particular may be trapped between the head of the humerus and the acromion during abduction. But other explanations have been put forward. Herberts and Kadefors (1976) found rotator cuff tendinitis in elderly welders, which they attributed to ischaemic changes resulting from the static muscle tension required by prolonged overhead work.

## Pseudo-angina Pectoris

Pain originating from chest wall muscles (especially pectoralis major) may radiate across the chest and down the (left) arm, so as to resemble the symptoms of coronary insufficiency (Figure 4.3). This is called pseudo-angina pectoris. (The resemblance is not that close: the real thing generally extends to the mid-line and becomes worse on exertion; pseudo-angina is if anything improved by activity.)

This alarming condition is really quite common. In Hult's 1954 study, 14% of the subjects who had a history of brachialgia (or 3% of the total sample) reported chest pains amongst its symptoms, and some had been off work with suspected heart disease as a result. Pseudo-angina was somewhat more common amongst subjects in heavy occupational categories. Problems with pectoralis major on the left side have been noted in workers who steady a workpiece with the left hand whilst working on it with a tool held in the right (Tichauer, 1978)—as you do when wiring a plug, for example.

## Thoracic Outlet Syndrome

This condition is caused by compression of the subclavian vessels and the brachial plexus (specifically the C8 and T1 roots) as they leave the upper aperture of the thoracic cavity (Figure 4.7). (Anatomists call this aperture the thoracic *inlet.*) The neurovascular bundle may be compressed at three points: where it emerges between the anterior and middle scalene muscles; where it passes between the clavicle and the first rib; and where it passes below and behind the insertion of the pectoralis minor muscle on to the coracoid process. (These lead to three distinct varieties of the syndrome which may be distinguished by certain diagnostic test manoeuvres, details of which are discussed in Waris, 1980.)

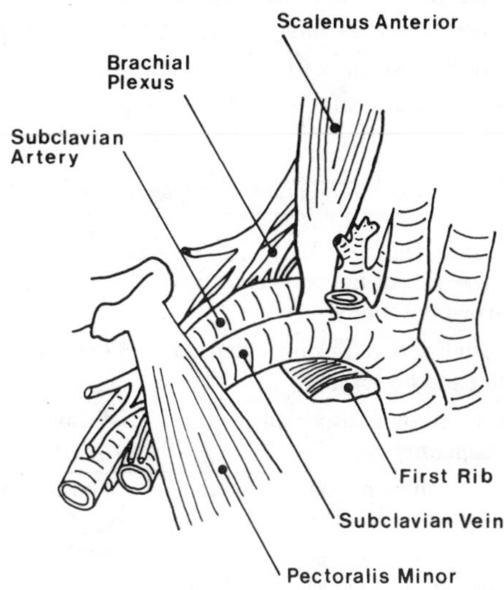


Figure 4.7 *The thoracic inlet*

The principal symptoms are pain and paraesthesiae, radiating down the underside of the arm towards the little finger (i.e. in the distribution of the C8 and T1 nerve roots). These will be made worse by carrying weights in the hands (particularly suitcases, etc.), retracting the shoulders or reaching overhead—depending upon the site of neurovascular entrapment. Arterial compression may lead to weakness, fatiguability and finally wasting of upper limb muscles; to blanching of the fingers on exertion; and very occasionally to more severe problems. Venous symptoms are rare, because the subclavian vein passes in front of scalenus anterior.

The condition is said to be up to ten times more common in women than in men, and more common on the right side than the left—owing in both cases to the angle of the shoulders. Anatomical anomalies of the thoracic inlet are also important, including cervical ribs (which occur in four men and eight women per thousand, but are asymptomatic in 90% of cases).

The condition occurs in people who carry weights on their shoulders, either directly or suspended from straps in satchels, etc. (e.g. letter carriers). Extreme postures or repetitive motions of the shoulder may also be implicated—particularly reaching overhead (as in house painters and certain industrial assembly and maintenance tasks) and repetitively reaching behind the body (as at certain supermarket checkouts).

## Disorders of the Elbow, Forearm, Wrist and Hand

The disorders in this category are all particularly associated with the repetitive, short-cycle tasks of the industrial assembly line (although this is by no means exclusively the case). Prevalences have been found to be highest in jobs where repetition rates are high and in which forceful gripping actions are required (Silverstein *et al.*, 1986). A number of specific associations and ergonomic risk

*Table 4.3 Upper Limb Musculoskeletal Disorders: Ergonomic Risk Factors*

| Task demands | Disorder |
|---|---|
| Repeated extension of wrist and/or fingers—e.g. repeated backhanded throwing actions | Tennis elbow |
| Repeated "clothes wringing" action (flexion/extension with supination/pronation and power grip) | Tenosynovitis, esp. De Quervain's |
| Repeated radial and ulnar deviation, especially with forceful grip, e.g. using a spanner | Tenosynovitis, esp. De Quervain's |
| Repeated pronation/supination with ulnar deviated wrist, e.g. twisting with pliers | Tenosynovitis, esp. De Quervain's; tennis elbow; carpal tunnel syndrome |
| Repeated gripping actions with flexed wrist | Tenosynovitis of finger flexors (trigger finger) |
| Repeated flexion/extension of wrist, especially if combined with pinch grip or power grip | Carpal tunnel syndrome |
| Prolonged pressure on elbow, especially if elbow is flexed | Ulnar nerve entrapment at elbow |
| Repeated application of force with hand, with wrist in extended position | Ulnar nerve entrapment at wrist |
| Tools causing radial deviation of wrist, especially if combined with extension and pronation | Tennis elbow |
| Tools with triggers—especially if handle is too large so that proximal interphalangeal joint is extended | Tenosynovitis of finger flexors (trigger finger) |

Sources: Tichauer (1978); Kurppa *et al.* (1979a, b); Armstrong and Chaffin (1979); Waris (1980); Armstrong *et al.* (1982, 1986); Viikari-Juntura (1984); Wallace and Buckle (1987); Putz-Anderson (1988).

factors have been described—these are summarized in Table 4.3. But in practice, almost any kind of hand-intensive work may cause these conditions. In general, awkward actions involving extremes of joint range are most likely to cause problems.

It is not possible to specify exactly how high repetition rates have to be before they pose a significantly increased risk—too many other factors are involved. (The problem is comparable with specifying a safe weight for lifting—p. 306.) As a rough guide, it has been suggested (Putz-Anderson, 1988) that there is likely to be an increased incidence of these disorders in jobs which have:

- more than 1500–2000 repetitions per hour;
- a cycle time of less than 30 seconds—particularly if more than half is taken up by a single sequence of repeated actions.

A cycle time of 30 seconds is equivalent to about 900 production units per shift.

The fact that these conditions commonly occur together (both in the individual and in members of a working population) can in many cases be predicted on anatomical grounds. For example, tennis elbow results from cumulative damage at the point of origin of the muscles which extend the wrist and hand; the tendons of these muscles are a common site for tenosynovitis; ganglia are often found at the points of insertion of these tendons; and the fleshy parts of the muscles are a common site for myalgic trigger points (Figure 4.8). The factors which cause one particular condition to occur in a given individual are likely to be idiosyncratic to that person.

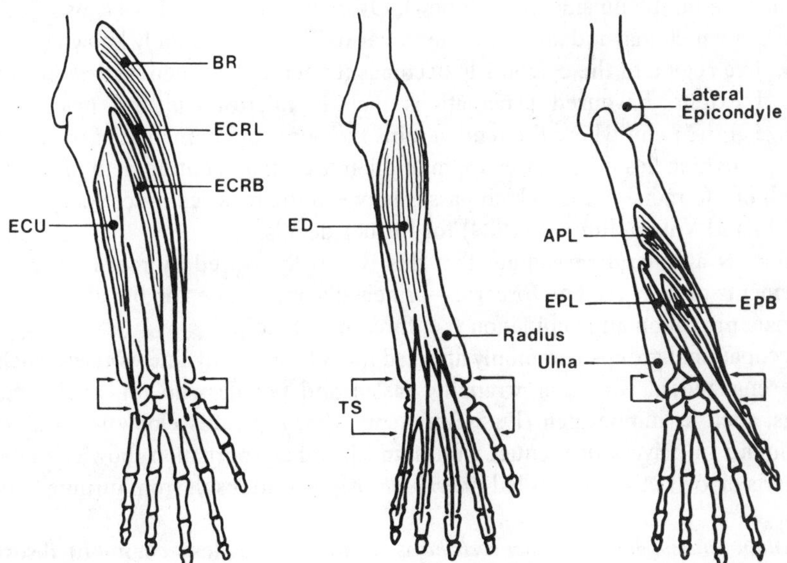


Figure 4.8 *The muscles of the extensor compartment of the forearm. Br = brachioradialis; ECRL = extensor carpi radialis longus; ECRB = extensor carpi radialis brevis; ECU = extensor carpi ulnaris; ED = extensor digitorum; APL = abductor pollicis longus; EPL = extensor pollicis longus; EPB = extensor pollicis brevis. TS = approximate extent of the tendon sheaths*

Women tend to be affected more often than men. In part this is due to the fact that women often do the more repetitive jobs. But when women and men in similar jobs are compared, the relative risk may still be as high as 3 : 1 (Silverstein *et al.*, 1986). In data based on GP samples, Buckle and Stubbs (1990) found clear signs of a sex difference for carpal tunnel syndrome, but not for tenosynovitis. These conditions may occur at any age, although they are relatively rare in people under 20. The peak incidence is widely held to be in the 40–50 year age group—but the effects of age and exposure to occupational factors may well be confounded.

Body build and hand size are not significant risk factors; neither, in general, is muscle strength, except perhaps in certain particularly srenuous jobs (Wallace and Buckle, 1987). There tends to be a high incidence amongst novices and in people returning to work after a lay-off—but seasoned workers may also be affected. In some industries the incidence of tenosynovitis is said to double after the summer holidays (Kurppa *et al.*, 1979a). A significantly higher prevalence of upper limb disorders has been found in people who show a stronger degree of hand preferences (Kucera and Robins, 1989). No other personal characteristics have been shown to have predictive value (except for the case of carpal tunnel syndrome).

## Lateral Epicondylitis: Tennis Elbow

The muscles which extend the wrist and fingers arise from a common tendinous origin which is attached to the lateral epicondyle of the humerus (the bony prominence on the outside of the elbow). Overuse of these muscles causes pain at the origin which may radiate down the forearm. This is commonly known as *tennis elbow*. The region of the epicondyle becomes tender and swollen—sometimes to a marked extent. Favoured explanations include microscopic and macroscopic damage at the point where the tendon joins the bone, inflammation of the annular ligament (which lies deep to the origin of the muscles) and entrapment of the deep branch of the radial nerve (which passes between the muscles)—see Kurppa *et al.* (1979b) and Viikari-Juntura (1984) for further details.

There is a general agreement that this is usually caused by repeated forceful movements of the wrist and forearm—principally wrist extension but also probably alternate pronation and supination (i.e. forearm rotation).

Occupational groups commonly affected include assembly line workers (such as those engaged in wiring or wrapping tasks) and people who repeatedly throw things, such as lumbermen (Putz-Anderson, 1988). Tennis elbow is, of course, also found in players of racquet games and in other sportsmen (bowler's elbow, pitcher's elbow). A variant called *gaoler's elbow* is said to result from turning keys in locks.

*Median epicondylitis* or *golfer's elbow* is a similar (but less common) disorder affecting the origin of the flexor muscles (on the medial side of the elbow). It is caused by repetitive movements of wrist flexion and forearm rotation.

## Ulnar Nerve Compression

The ulnar nerve may be compressed as it passes through a ligamentous tunnel behind the medial epicondyle, giving pain and tingling down the medial side of the forearm to the little finger. This is sometimes called cubital tunnel syndrome. It occurs in people who rest their elbows on hard wooden benches. Prolonged elbow flexion may also be a factor.

The ulnar nerve may also be compressed where it crosses the wrist (passing through the so-called Guyon's canal), or in the palm of the hand. The latter may be due to direct pressure on the handle of a badly designed tool. This may cause weakness of the interosseus muscles of the hand.

## Tenosynovitis

Strictly speaking, the terms *tenosynovitis* and *tenovaginitis* should be applied to conditions in which there is an inflammation of the synovial sheaths which surround certain tendons (including those of the forearm, which concern us here), whereas *tendinitis* and *peritendinitis* refer to inflammation of the tendon and surrounding tissue, where no such sheaths are present. But the terms are often used interchangeably.

In tenosynovitis proper, fluid exudate seeps into the tendon sheath. Fibrosis may occur and adhesions may form. The tendon may be thickened or frayed. When the movement of the tendon within the sheath is impaired, the condition is called *stenosing tenosynovitis.* In peritendinitis, the oedema and fibrosis tend to be concentrated around the muscle–tendon junction. Pathological changes are found in both peritendinous connective tissue and the muscle, but the tendon itself is relatively unaffected.

It is generally acknowledged that both conditions are caused by muscular overuse—circulatory changes are thought to be important (Kurppa *et al.*, 1979a). Peritendinitis is the more common of the two: in one series of cases in industry the ratio was 3:1 (Thompson *et al.*, 1951). For the sake of simplicity, we shall use the more familiar term "tenosynovitis" to describe both variants, whilst noting that in the majority of cases it will not be strictly accurate.

The tendons most commonly affected are those in the extensor compartment (i.e. the back) of the forearm on the radial (i.e. thumb) side: the abductor pollicis longus and extensor pollicis brevis tendons (which run round the radial styoid and form the outer margin of the "anatomical snuff box") and the tendons of extensor carpi radialis longus and brevis. Stenosing tenosynovitis of the former two tendons is called *De Quervain's disease.* The extensors of the finger may also be affected.*

* You can find most of these tendons quite easily on your hand. Form the "anatomical snuff box" by extending and slightly abducting your thumb, i.e. stretching it out backwards and away from the palm of the hand. The indentation (which may be used for taking snuff) is caused by the tendons of the muscles: on the very outer edge of the wrist the abductor pollicis longus and extensor pollicis brevis tendons are most prominent at the base of the thumb; and further over the extensor pollicis longus is prominent as it runs obliquely up the back of the thumb after running round a bony pulley on the back of the radius (the dorsal tubercle of Lister). The tendons of finger extensors (extensor digitorum and extensor indicis) are prominent on the back of the hand when you spread out your fingers.

The tendons of the finger flexors are affected less frequently—stenosing tenosynovitis of these leads to a condition known as "trigger finger".

The first and most common symptom is pain over the structures involved—a dull ache at rest which is greatly exaggerated by movement, when it may acquire a sharp "neuralgic" quality. The pain causes weakness of grip, etc. Swelling subsequently develops. (In tenosynovitis proper this is circumscribed and fusiform; in peritendinitis it is more diffuse.) The swelling is accompanied by an audible crepitus, which is described as sounding like the crunching of dry snow or the crumpling of Cellophane.

Tenosynovitis is principally caused by repetitive movements of the hand and wrist—although trauma and sprain may be involved in some cases—or almost any kind of hand-intensive work.

The connection between tenosynovitis and the repetitive, short-cycle tasks of the industrial assembly line has been recognized since the 1930s. Tenosynovitis also occurs in agricultural workers (such as cane cutters) and in traditional craftwork (such as net making). Amongst Londoners who at one time used to go hop-picking in Kent (to pay for their summer holidays), tenosynovitis was known (incorrectly) as *hopper's gout* (Hunter, 1978). Another variant is called *washerwoman's thumb.*

Hunter (1978) noted the occurrence of tenosynovitis in typists. There is a strong impression that it has become more common since the introduction of fast-action electronic keyboards; but there is little epidemiological evidence either to support this belief or to refute it. English *et al.* (1989) found a significant over-representation of VDU operators and other keyboard users in a sample of patients attending orthopaedic clinics with a range of upper limb disorders.

Tenosynovitis is one of the few musculoskeletal disorders to be classified as a prescribed industrial disease in the UK. In the early 1980s it accounted for about 30% of all reported cases of prescribed industrial diseases, and was the second most common (after dermatitis).

Tenosynovitis tends to occur in localized areas of high concentration—often on production lines, where repetition rates are high. Kuopojarvi *et al.* (1979) found a prevalence of 56% for tenosynovitis and related conditions in assembly line packers, compared with 14% in a control group of shop assistants. (The high prevalence in the control group suggests that the criteria used were probably quite wide.) Some of the assembly line workers were performing 25,000 motions per working day.

The highest concentration I have seen reported was in a tennis ball factory in Barnsley, Yorkshire, where 49 out of 50 women working in the ball-covering department were alleged to have sought medical advice for problems with their hands or wrists. (But we do not know how many of these women would have had tenosynovitis as such.) The women were employed on piece rates and each aimed to produce 1,200 balls per day (*The Guardian*, 9 July 1984).

Some specific ergonomic risk factors which have been reported are summarized in Table 4.3. The worst problems characteristically occur in jobs in which *repetitive motions* are combined with *forceful gripping actions*, with the *wrist in a deviated position* (i.e. outside the middle third of its range). There are a number of reasons why the deviated wrist should be especially undesirable:

(i) the strength of the grip is reduced (p. 266);
(ii) the muscles of the forearm may be subject to static loading;
(iii) in ulnar deviation of the wrist, the tendons of abductor pollicis longus and extensor pollicis brevis, will rub against the lateral surface of the radial styloid.

## Ganglion

The term *ganglion* (Greek: knot), or *ganglionic cyst*, is used to describe a round fluid-filled swelling, usually about the size of a grape (but sometimes the size of a plum), which commonly occurs on the back of the wrist or hand. Ganglia may communicate with the synovial sheaths of the extensor tendons—the point of insertion of extensor carpi radialis brevis (on the 3rd metacarpal) is a common location. Ganglia are rarely painful and cause no disability, but some people regard a high frequency of ganglia in a working group as a warning sign of possible ergonomic problems (Peter Buckle, personal communication).

## Carpal Tunnel Syndrome

The eight small carpal bones of the wrist form a trough or gutter, which is covered by a tough fibrous sheet (the flexor retinaculum) forming the carpal tunnel (Figure 4.9). The tendons of the finger flexor muscles pass (in their synovial sheaths)

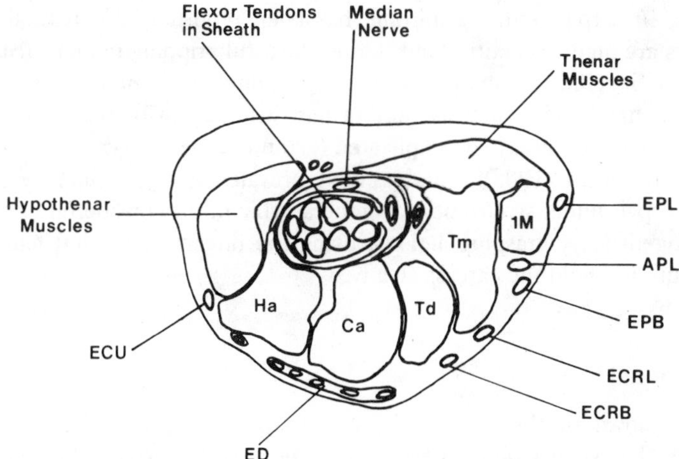


Figure 4.9 *Section across the wrist and base of thumb, showing the carpal tunnel and the extensor tendons. 1M = 1st metacarpal; Tm = trapezium; Td = trapezoid; Ca = capitate; Ha = hamate; EPL = extensor pollicis longus; APL = abductor pollicis longus; EPB = extensor pollicis brevis; ECRL = extensor carpi radialis longus; ECRB = extensor carpi radialis brevis; ED = extensor digitorum and extensor indicis (in sheath); ECU = extensor carpi ulnaris*

through the tunnel, along with the median nerve, which supplies sensation to the thumb, most of the palm, the index and middle fingers and part of the ring finger. (It also provides the motor supply to the thenar muscles at the base of the thumb and the two small lumbrical muscles of the index and middle fingers.)

*Carpal tunnel syndrome* is caused by an increase of fluid pressure within the tight compartment of the tunnel, which results in compression of the median nerve. The first symptom is generally a numbness and tingling in the distribution of the median nerve which wakes the patient during the night. Abnormal sensations may also spread to the remainder of the hand and up the forearm. In more advanced cases the pain may be severe and continue during the day; loss of sensation may make it difficult to pick up small objects and there may be weakness and wasting of the muscles. The condition occurs most commonly in the dominant hand but may also be bilateral.

Between 60% and 90% of patients are women—mainly aged between 40 and 60 years. Carpal tunnel syndrome may be secondary to other injuries (e.g. Colles fracture) or may be associated with a variety of systemic conditions, including various endocrine dysfunctions, rheumatoid arthritis and renal failure. These are present in up to 20–30% of cases. Oral contraceptives and premenstrual water retention may be contributory factors. About 25% of women experience the symptoms of carpal tunnel syndrome during pregnancy (Turner and Buckle, 1987). Tenosynovitis of the finger flexors (trigger finger), which is commonly work-related, may be regarded as a direct cause of carpal tunnel syndrome. In one series of carpal tunnel patients who were in employment, 10% had predisposing medical conditions; of the remainder, 79% were in jobs requiring repetitive hand and finger movements. Amongst housewives with no other employment, 51% gave knitting as their main hobby (Birkbeck and Beer, 1975). A significantly increased prevalence of carpal tunnel syndrome has been reported in industrial workers whose jobs are highly repetitive and require forceful gripping motions (Silverstein *et al.*, 1987). Specific ergonomic risk factors are summarized in Table 4.3. The use of vibrating hand tools such as buffers and grinders, with frequencies in the 10–60 Hz range, has also been implicated (Cannon *et al.*, 1981).

Falk and Aarnio (1983) found a high prevalence (approximately 50%) of left-side carpal tunnel syndrome in butchers, which they considered to be due to the prolonged heavy grasping actions, using the fingers of the left hand, which were required to hold the carcass steady.

## "Occupational Cramps"

The best-known of the "occupational cramps" or "craft palsies" is *writer's cramp*—also known as *scrivener's palsy* or *graphospasm*. Ramazzini (1713) described it thus:

> "Incessant driving of the pen over the paper causes intense fatigue of the hand and whole arm because of the continuous and almost tonic strain on the muscles and tendons, which in course of time results in failure of power in the right hand."

Writer's cramp seems to have been particularly common in nineteenth century England. As Samuel Solly, a London surgeon, put it (Solly, 1864):

> "... the greatest part of the middle classes of London get their bread by the use of the pen, either as the exponent of their own thoughts or the thoughts of others, or in recording the sums garnered, lost or spent in this great emporium of commerce—this vast Babylon."

The comparison with modern "epidemics" of "RSI" is interesting.

The symptoms of writer's cramp are variable and inconsistent. The first is a clumsiness of writing coupled with a sensation of fatigue in the hand—followed by spasm, pain, tremor, weakness, loss of muscular co-ordination and finally paralysis. Writing becomes impossible. In advanced cases other activities may also be affected.

Cramped and awkward writing postures and the writing technique itself are causative factors—a mode of writing which places an excessive loading on the small muscles of the hand is basically to blame. The cure is rest, and the acquisition of a "freer" writing style in which the hand rests lightly on the paper and the motion across the page comes from the shoulder rather than the wrist.

A number of other trades are subject to similar conditions, notably telegraphists (using the old Morse key), florists, cotton twisters and musicians—particularly pianists and violinists. Typists are also affected. The common factor is the demand for fast manipulative actions, not truly repetitive but generally demanding a high degree of skill and concentration. The occupational palsies are prescribed industrial diseases in the UK.

Writer's cramp has been known for centuries and its work-relatedness is in no doubt; but the underlying pathology is highly contentious. Many authorities consider the occupational cramps to be basically psychogenic. Hunter (1978), for example, describes them as a group of "psychoneuroses". Crisp and Moldofsky (1965) go so far as to regard writer's cramp as a psychosomatic manifestation of repressed anger.

These conditions doubtless have a psychogenic overlay. But to regard them solely as manifestations of neurotic disturbance seems to be based on a curious historical misconception. Late nineteenth century English writers such as Gowers (1888) described writer's cramp as an "occupational neurosis". But in those days the term "neurosis" had a very different meaning: it indicated a disorder of the central nervous system.

Sheehy and Marsden (1982) propose that writer's cramp is a focal dystonia—that is, a disorder of the neural control of the voluntary muscle. They found that people with occupational cramps were no more or less psychiatrically disturbed than the rest of us. They speculated that writer's cramp was a breakdown of the "motor program", stored in the brain, by which we generate our characteristic script—which remains identifiably ours whether we are writing in a small notepad or on a large blackboard. Cohen and Hallett (1988) found electromyographic evidence to support this contention.

Occupational cramps (and other upper limb problems) are common in musicians. Wind players may have similar problems with the muscles of the embouchure (lip), tongue, soft palate and throat. The player complains of a loss of speed and accuracy in the execution of difficult passages—he tends to fumble the "twiddly bits", which he could once play with ease, and not unnaturally he gets depressed. (Orchestral musicians who fall below standard soon end up out of work.) Equally naturally, he tends to tense up when approaching such passages—which makes the problem worse. But these are not adequate grounds for classing a condition as a "psychoneurosis". Fry (1986) considers these conditions to be the result of cumulative muscular damage resulting from the combination of static loading (due to the weights of the instruments and the awkward postures in which they are often played) and the repetitive motions of the fingering and bowing actions. The preventative measures he recommends are basically aimed at eliminating this unnecessary muscular tension. The methods which nineteenth century writers recommended for the rehabilitation of writer's cramp were based on the same principles.

## Some Other Conditions

### Bursitis, etc.

A bursa is a lubricated synovial sac, which provides a freely sliding surface at a point of friction. Bursae may be subcutaneous (where skin rubs against bone) or deep (where muscles or tendons rub against bone). Some bursae are normal anatomical structures; others are adventitious—that is, they develop in response to an abnormal pattern of loading.

Both anatomical and adventitious bursae may become inflamed and distended with fluid as a result of excessive loading (bursitis).

The best-known of these conditions is *housemaid's knee*, an inflammation of the pre-patellar bursa which lies in front of the upper part of the kneecap. There is another condition called *clergyman's knee* or *nun's bursitis*, which is an inflammation of the superficial infra-patellar bursa, just below the knee. The difference is said to be that housemaids sit back on their heels as they work, whereas clergymen kneel more upright—hence, loading the knee at slightly different points.

Other examples include *weaver's bottom* (or coachman's or lighterman's)—sub-ischial bursitis due to sitting on hard wooden benches; *student's elbow* (or miner's)—inflammation of the olecranon bursa from resting the elbow on desks; *hodman's shoulder*—the sub-acromial bursa from carrying hods of bricks; *Covent Garden hummy*—an adventitious bursa on the top of the head from carrying baskets of fruit and vegetables; *Billingsgate hump* or *humper's lump*—over the 7th cervical vertebra, in fish porters and timber porters, respectively; *dustman's shoulder*—over the shoulder and clavicle; and *tailor's ankle*—over the lateral malleolus, from sitting cross-legged (Hunter, 1978).

The terms *beat knee*, *beat elbow* and *beat hand* were originally used by miners. They may refer either to a bursitis (such as housemaid's knee, etc.) or to a cellulitis (inflammation of subcutaneous tissue due to infection, usually from a hair follicle). These are prescribed industrial diseases in the UK. Beat hand (which is always a cellulitis) occurs in workers who use picks, shovels, brooms, etc.

## Osteoarthrosis and Heberden's Nodes

Osteoarthrosis (which is more commonly but less accurately known as osteoarthritis) is a degenerative condition of the articular cartilages of the synovial joints, which is caused by "wear and tear". Any synovial joint may be involved, but the weight-bearing joints of the lower limb (hip, knee, etc.), the apophyseal joints of the spine and the interphalangeal joints of the fingers are the ones most commonly affected. Osteoarthrotic changes at one or more of these sites are more or less universal with age.

Characteristic patterns of localized osteoarthrosis occur in many trades where a particular joint (or joints) is exposed to unusual stress. These include the shoulders in bus drivers, and the ankles in footballers and dancers. Coal miners and dockers have a high prevalence of osteoarthrosis of the hips, knees and spine, which occurs at relatively early ages. In seamstresses and tailors the fingers of the working hand are affected—in tailors it is the index finger alone, whereas in seamstresses both the index and middle fingers are involved (Hunter, 1978). The difference is presumably due to differences in the sewing action.

Osteoarthrosis of the fingers is characterized by a deformity of the distal phalanges called Heberden's nodes. Lawrence (1961) studied weavers, spinners and other manual workers in a cotton mill in Bolton. He compared them with controls matched for age and sex, drawn from similar communities in the North of England, who had never worked in a cotton mill. The prevalence of Heberden's nodes was as follows:

| | *Men* | *Women* |
|---|---|---|
| Mill workers | 38% | 35% |
| Controls | 12% | 22% |

These differences point to an interaction between the occupational stresses of spinning and weaving in the mill and a sex-linked hereditary predisposition to generalized osteoarthrosis.

## Chapter Five

# Posture

"By reason of the frailty of our nature we cannot always stand upright."

*The Book of Common Prayer* (Collect for the Fourth Sunday after Epiphany)

"It follows that whenever occasion offers, we must advise men employed in the standing trades to interrupt when they can that too prolonged posture by sitting or walking or exercising the body in some way or other."

*De Morbis Artificum*, Bernardini Ramazzini (1713)

For any static unsupported posture to be maintained, two sets of equilibrium conditions must be satisfied:

(i) a vertical line, drawn through the centre of gravity of the body, must fall within the body's base of support;
(ii) the net torque (or moment) about each articulation of the body must be zero.

If condition (i) is not satisfied, the body will topple over, whereas if condition (ii) is not satisfied, it will collapse or fold up at its joints (and may then topple over as a consequence of condition (i) having been violated).

In a symmetrical upright standing position, the centre of gravity of the human body lies in the midline, somewhere near the second sacral segment. A vertical line drawn through the centre of gravity (sometimes called the *line of weight*) will meet the floor midway between the two feet, about halfway between the heels and the balls of the feet. That is, it will fall close to the centre of the person's *footbase*. The limits of this base of support effectively determine how far the person can lean or reach, in any particular direction, before he loses his balance. Suppose he leans or reaches forward. The centre of gravity of the upper part of his body (head, arms and trunk) will come to lie in front of each of the articulations of the lower part of the body. The weight of the superincumbent body parts acting about these points will exert a torque or moment: tending to flex the spine and hips, extend the knees and dorsiflex the ankles. In each case, the magnitude of the torque is given by the product of the weight of those parts of the body which are above the joint concerned and the horizontal distance between the centre of gravity of these parts and the centre of rotation of the joint (Figure 5.1). To calculate this torque, you need a table of data which tells you the weight of each body segment and anatomical locations of its centre of gravity (see Pheasant, 1986, for such a table).

For equilibrium to be maintained, muscles must exert an equal and opposite countertorque. In this case we might expect the back muscles (erector spinae—see Figure 5.6), the glutei, the hamstrings and the calf muscle to be active. It is not

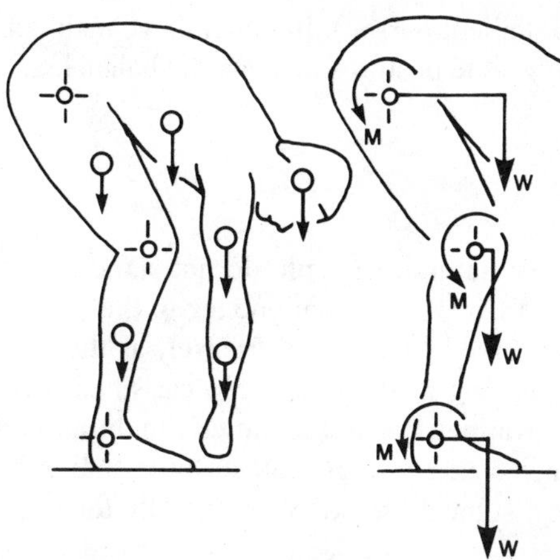


Figure 5.1 *Person leaning forward:* left, *showing the centres of rotation of the hip, knee and ankle joints and the centres of gravity of the body segments (head, trunk, etc.);* right, *showing the distance at which the weight (W) of the superincumbent parts acts in front of each joint and the muscular torque (M) required to counteract this weight*

quite this simple, however. At the limits of joint range, the countertorque may be supplied by tension in ligaments rather than muscles. In the posture shown, the spine is fully flexed—so the posterior ligaments may be taking over the role of the back muscles (Floyd and Silver, 1955).

## Balance

The stability of an inert object is dependent upon the size of the base of support and the height of the centre of gravity. Thus your most stable position is lying flat on your back (or front, as the case may be). The base of support is the whole posterior surface of the body, and the centre of gravity is only a short distance above it. It has been said that you're not really drunk if you can still lie flat on the floor without hanging on. Sitting is somewhat less stable than lying, so it requires a certain amount of effort and co-ordination, and standing is really quite precarious. We can improve our stability in the standing position by adopting a broad or long stance. We do it more or less automatically when standing on the rolling deck of a boat. It improves our stability in the direction of the long axis of the stance—but not in the direction perpendicular to this axis. So we fall over if the boat pitches unexpectedly.

The ballerina performing an *arabesque en pointe* has a footbase of about 2 square inches. The tightrope walker's footbase has negligible width (perhaps the thickness of the rope), although it may be quite long. The ballerina and the tightrope walker keep their balance by continually making minor corrections of position. (In a skilled performer these corrections may become so tiny that you do not see them.) We call this *dynamic balancing*, as against the *static balancing* of the inert object (or the recumbent person). In general, we maintain stable postures by static balancing, and unstable postures by dynamic balancing.

## Standing

When allowed to behave naturally, people do not stand still for long—they sway back and forward, shift their weight from one leg to the other, lean on things, and so on. Smith (1953) observed people unobtrusively as they stood at bus stops, on railway platforms and talking in the street. He classified their standing postures into symmetrical and asymmetrical and he timed the duration of each stance. The average duration of each stance was only 30 seconds, and 93% were held for less than a minute. Smith's subjects stood symmetrically for only about 20% of the time.

The fact that a symmetrical upright stance is sometimes called "the normal standing posture" is based on a misconception. The nineteenth century German anatomists used the term *die Normalstellung* to denote a hypothetical posture of perfect vertical alignment which could be maintained without muscular effort: a single vertical line passes through the centres of rotation of each of the major articulations of the body (hips, knees, ankles, etc.) and the moment about each joint is zero. The word "normal" is used here as a physicist might employ it—to mean perpendicular. But we do not actually stand this way, except perhaps transiently. In fact, it would be dangerously unstable to do so, since, when the line of weight falls through the ankle joint, it has just about reached the posterior unit of the footbase and we would be on the point of falling over backwards.

When "standing at ease" in a symmetrical posture, we sway slowly back and forward within our footbase. The line of weight of the superincumbent body parts probably passes in front of the ankle joints and behind the hip joints for most of the time—so electromyographic studies generally show activity in the calf muscles (particularly the soleus, which has a preponderance of slow fibres) and in the iliopsoas, which is acting as a hip flexor. The anterior tibial muscles are only occasionally found to be active (presumably when sway approaches the posterior limits of the footbase). Since moments about the knee are generally negligible, little activity is found in the thigh muscles—either the quadriceps or the hamstrings (Joseph, 1960; Basmajian and De Luca, 1985).

In postures which approximate to vertical alignment, the line of weight of the whole body and the line of weight of the superincumbent parts (at any level) will be approximately the same. In some textbook accounts of standing posture, they are treated as if they are one and the same thing. This suggests a fundamental misconception—although the magnitude of the resultant errors will be small (and will decrease as you approach the feet).

It is sometimes also said that the line of weight (presumably of the whole body) intersects the curve of the spine. This may well be true—but it is of no consequence. Most electromyographic studies show intermittent or low level activity in the lumbar part of erector spinae (Floyd and Silver, 1955; Joseph,

1960)—suggesting that the weight characteristically falls in front of the axis (of flexion) of the lumbar spine rather than behind it. But people vary in this respect. Asmussen and Klausen (1962) found the rectus abdominis muscle to be active in a sizeable minority of people (about 1 in 4) and the erector spinae in the majority—but they never found simultaneous activity in both muscle groups. Other studies have shown fairly consistent activity in the lower part of the internal oblique muscle—which is probably acting to safeguard the inguinal canal (Floyd and Silver, 1950). And the psoas muscle (which is active in relaxed standing to prevent hip extension) is also an extensor of the lumbar spine. So it presumably complements the postural activity of the erector spinae (and may well be the more important of the two).

Ligaments are probably of relatively minor importance in the maintenance of the upright standing position—except in the arches of the feet, the sacro-iliac joints and possibly the knee joints (which may be "locked" in some phases of postural sway).

The upright standing position is physiologically highly efficient. Balance is dynamic rather than static, and deviations from the ideal perpendicular position are small. Thus energy expenditure is not raised much above the resting level—and no muscle group is subject to consistent static loading at more than a very low level. So why do we feel "fatigue" in our legs and feet when we have to stand up for a long time? It is probably mainly to do with venous congestion and the pooling of fluid in the lower limbs (see below).

## The Spinal Curves

The child in the womb and the newborn infant have a single C-shaped antero-posterior spinal curve, which is convex to the rear. This *primary curve* is due to the shape of the bones—the vertebral bodies are slightly wedge-shaped. (If you pull the strings tight on the vertebrae from a boxed set of bones, you can reproduce the primary curve of infancy.) When the child begins, firstly to hold up its head and subsequently to sit up and to crawl around, *secondary curves* (concave to the rear) develop in the cervical and lumbar regions. The radii of these curves depend upon the tension in the muscles which stretch across the concavities. The secondary curve in the lumbar region is the *lumbar lordosis*. In the thoracic region the primary curve is retained throughout life—the *thoracic* (or *dorsal*) *kyphosis* (Figure 2.10—p. 55). As from middle age onwards the discs lose their turgor and the disc spaces get narrower, the lumbar curve flattens out and the thoracic kyphosis tends to increase. This may be accentuated, particularly in post-menopausal women, by osteoporotic crush fractures—giving the familiar "dowager's hump". In old age, the spine returns to the single primary curve of infancy.

The shape of the lumbar curve is mainly determined by the angulation of the sacrum and pelvis. In the standing position, the anterior superior iliac spines and the pubic symphysis are typically in the vertical plane, and the top of the sacrum is inclined forward at an angle of about 30° to the horizontal (see Figure 5.2). We may assume that motion in the sacro-iliac joint is negligible and that the sacral

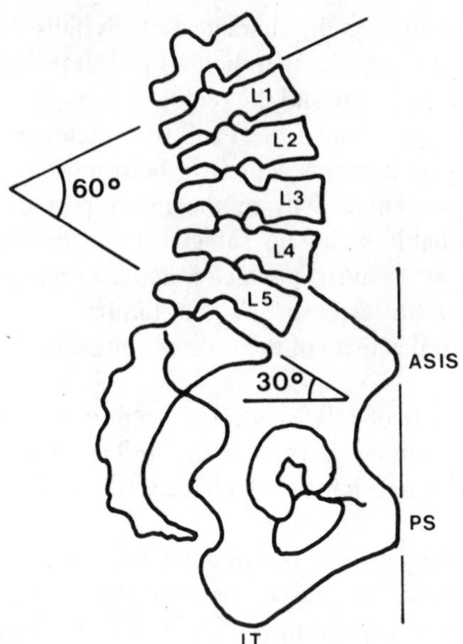


Figure 5.2 *Typical orientation of the lumbar spine and pelvis in the standing position. ASIS = anterior superior iliac spine; PS = pubic symphysis; IT = ischial tuberosity*

angle is dependent upon the position of the pelvis. Increasing the sacral angle by tilting the pelvis further forward requires a compensatory increase in the lumbar curve if the trunk is to remain vertical. Conversely, tilting the pelvis backward flattens the lumbar curve. In either case, the pelvis is rotating about the fulcrum of the hip joint—forward tilt is equivalent to hip flexion. *Note that a flattening of the lumbar curve is equivalent to flexion of the spine and an accentuation of the lumbar curve is equivalent to extension.*

The antero-posterior curve of the normal spine is extremely variable. Some people have a particularly flat back, whereas others have an unusually pronounced lumbar curve (hyperlordosis or sway back). Sometimes the thoracic curve is exaggerated (roundback) and sometimes both curves are exaggerated (kypholordosis). Some poeple have localized "reverse curves" where the smooth contour of the back is interrupted for a few vertebrae. These variants on the somewhat hypothetical norm may be found in both sexes at almost any age and in people of all body builds.

An accentuated lumbar curve with a forward pelvic tilt is often associated with slack gluteal and abdominal muscles and a *visceroptosis* (Greek: drooping guts). The sagging of the abdominal wall is not necessarily associated with obesity (although this is often the case) and it is quite common in people of slender build. A variety of adverse systemic effects have been attributed to visceroptosis, including disturbances of gastric emptying and reduced diaphragmatic excursion, leading to impaired respiratory capacity and venous return (Goldthwaite, 1934). I have not been able to find much empirical evidence for these assertions, but they make sense in principle.

A spine which is perfectly straight when viewed from behind is a rarity. Most backs have a slight lateral curve or *(physiological) scoliosis*. (The qualifying term "physiological" is sometimes employed to distinguish the benign normal variant from a much more sinister condition in which the spine twists into a corkscrew shape and finally collapses altogether.) The main part of the curve tends to be concave to the side of the preferred hand, and there may be compensatory opposite curves above it, below it, or both. Scoliosis is often associated with an inequality of leg length—the curve is typically convex to the side of the shorter leg. (It is more common for the left leg to be shorter than the right.)

Studies reviewed by Andersson and Pope (1984) indicate that scoliosis is only a risk factor for back pain when it is relatively pronounced and that unequal leg length carries a measurable increase in risk, but not a very great one.

There is a school of thought which regards an accentuated lumbar lordosis as a direct antecedent cause of low back pain (e.g. Caillet, 1968; Fahrni, 1975). The adherents of this school are found mainly in North America. They argue variously that hyperlordosis causes the posterior fibres of the annulus to get slack and soggy; that it increases the loading on the apophyseal joints; that it can result in a nipping of nerve roots in the intervertebral foramen; and so on. Clinicians of this school therefore regard the elimination of the lumbar lordosis to be a high priority in the management of back pain.

There is no epidemiological support for this theory whatsoever. Studies have repeatedly shown that people who have a pronounced lordosis in the standing position are no more or less likely to suffer from low back pain than people who do not. Neither Hult (1954) nor Rowe (1969, 1971) found any evidence for such an association. The lumbar lordosis may be defined quantitatively by the angle subtended by the surfaces of the L1 and S1 vertebrae, as seen in a lateral X-ray. This angle has been found to be approximately normally distributed in the population (and it has an average value of about 60°, as shown in Figure 5.2). The range of values is the same in people who suffer from back pain as it is in people with no history of the condition (Hansson *et al.*, 1985; Pope *et al.*, 1985).

Of itself, the form of the lumbar spine *in the upright standing position* is not, therefore, a predictor of an individual's susceptibility to back problems. Neither people with a hyperlordosis nor people with a flat black are at increased risk (compared with those who have a spine curve in the middle part of the normal range). Many clinicians (both orthodox and heterodox) would disagree with this conclusion—on the grounds that their clinical observations tell them otherwise. The question remains controversial. The majority of people with low back problems seem to be more comfortable when their lumbar spines are supported in a position of extension (although this need not necessarily be extreme). A minority may be more comfortable in flexion. This minority probably includes those whose problems are caused by certain specific underlying mechanical pathologies—such as spondylolisthesis and spinal stenosis (an abnormally narrow spinal canal). These findings have important implications for the ergonomics of seat design (see Chapter 11).

## Sitting

The action of sitting down (on a seat of average height but without a back rest) involves a flexion of the knees and a flexion of the trunk on the thighs (of about 90° in both cases). In most people the comfortable limit of hip joint flexion is only about 60°—beyond which the passive tension in the hamstring muscles increases quite rapidly. So the sitting action is completed by a backward rotation of the pelvis of 30° or more, as shown in Figure 5.3. The weight is taken on the ischial tuberosities, and the top of the sacrum (which in the standing position is inclined forward by about 30°) comes to lie more or less horizontally. If the overall line of the trunk is to remain vertical, there must be a compensatory flexion or flattening of the lumbar spine (by an amount equal to the backward rotation of the pelvis). According to Åkerblom (1948), the lumbar spine is flexed to its limit in the relaxed or "slumped" sitting position—*so the weight of the trunk is probably supported by the ligaments rather than the muscles.*

You can demonstrate the movements of the pelvis quite easily on yourself. Stand up and place one hand on one of your anterior superior iliac spines (the two palpable eminences on the front of the hip bone just below your waist) and the other on your pubic bones in the mid-line. You will find they lie in the same plane. Now sit down in a relaxed way. Your anterior superior iliac spine now lies well

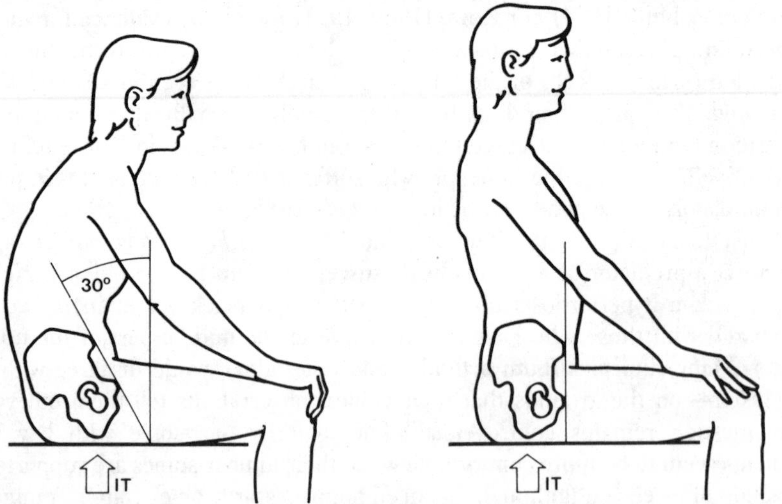


Figure 5.3 *The orientation of the pelvis in the sitting position:* left, *relaxed;* right, *upright. The ischial tuberosities (IT) act as a fulcrum. Reproduced from Pheasant (1986)*

behind your pubis. If you palpate the hamstring tendons (just behind the knee), you will find that they are slack. Sit up straight and they tighten.

To "sit up straight" requires a muscular effort to overcome the tension in the hamstrings. The pelvis rocks forward on the ischial tuberosities and the lordosis is restored. The principal muscle involved in this movement (of resisted hip flexion) is the iliopsoas. This has been confirmed electromyographically by Nachemson (1966). Muscular effort is required to support the weight of the trunk, now that the lumbar spine is in its mid-range of motion and the ligaments are no longer under tension. Other emg studies have shown that both the erector spinae and the trapezius are active in the upright sitting position but quiescent in relaxed sitting or when the weight of the trunk is supported on a suitably designed back rest (Lundervold, 1951).

The flexion of the lumbar spine which occurs during relaxed sitting involves a deformation of the intervertebral discs, which increases both the tension on the posterior part of the annulus and the pressure within the nucleus. Using a transducer mounted in a long hypodermic needle, Andersson *et al.* (1974) found that the intra-discal pressure during relaxed sitting is around 40% higher than it is when standing; but the pressure when sitting upright is halfway between the two. So by slouching we reduce the need for muscle effort but we increase the loading on the discs.

In the relaxed upright standing position, the pressure on the disc is more or less in equilibrium with the colloid osmotic pressure of the nucleus (p. 62)—so higher pressures will dehydrate the disc and lower pressures will rehydrate it.

People with back problems are often advised to "sit up straight"—in order to maintain the lordosis and reduce the loading on the disc. This seems to be bad advice if it requires static tension in the back muscles in order to do so.

Intra-discal pressure is determined by three factors: the weight of the superincumbent body parts acting directly downward, the deformation of the disc and the tension of the muscles of the back (which results in an approximately equal and opposite compression of the spine). In relatively upright positions, the weight of the body parts is probably the most important.

When you sit on a chair which has a back rest, your pelvis will tend to rotate backward until your back contacts the support, so the spine is again flexed. This may be exacerbated if the buttocks slide forward (as is often the case—p. 217). But the greater the inclination of the backrest, the greater the extent to which the weight of the superincumbent body parts is supported and the lower the disc pressure (this is a simple matter of cosines). A pad in the lumbar region helps restore the lordosis and reduce the disc pressure further. *In a seat with a well-designed backrest, therefore, the weight of the trunk is supported, the muscles are relaxed and the lordosis is maintained.* So the intra-discal pressure may be 30–40% less than it is in the standing position (Andersson *et al.*, 1974). Relative values for the intra-discal pressure in various sitting postures and during other activities are shown in Table 5.1. The lifting figures are for actions performed very slowly. (This is not surprising considering the way in which the pressure was measured.)

*Table 5.1 Intra-discal Pressure in Various Postures and Activities*

| Posture/activity | | Pressure % |
|---|---|---|
| Lifting 20 kg: | knees straight, back bent | 485 |
| | knees bent, back straight | 265 |
| Sitting in forward flexion | | 180 |
| Sitting and typing | | 175 |
| Standing with a 20° forward inclination of the trunk | | 170 |
| Writing at desk | | 145 |
| Sitting with relaxed slump | | 140 |
| Sitting up straight | | 125 |
| Office chair—resting on backrest | | 120 |
| Standing upright | | 100 |
| Lying on side | | 75 |
| Sitting reclined with lumbar support: | 20° backrest | 75 |
| | 40° backrest | 55 |
| Lying supine | | 35 |
| Lying supine with 30 kg traction | | 10 |

All figures quoted as percentages of pressure recorded during relaxed upright standing.
Data from Nachemson and Elfström (1979); Andersson and Örtengren (1974); Andersson *et al.* (1974).

## Lying

Lying prone results in a marked lordosis. Lying supine results in anything from a modest lordosis to a marked kyphosis, depending on the nature of the supporting surface. Disc pressure is much higher in sideways lying than supine—because the breadth of the shoulders and hips causes side bending of the lumbar spine.

Back pain sufferers are often advised to sleep on a hard bed. This may be based on a slight misconception. The mattress should be soft enough to conform to the surfacc of the body and distribute pressure evenly. (This is largely a matter of the design of the springs). But it must not sag—this is more a characteristic of the bed's underframe than of the mattress itself. Some back pain sufferers report that hard mattresses actually make their symptoms worse.

Lying with the legs raised and supported so that both the hips and knees are flexed to about 45° is widely believed to be a good position for resting. It is sometimes called the "semi-Fowler position". Raising the legs assists venous return. Back pain sufferers often seem to find this position helpful—but it is not clear why. Flexing the hips slackens off the iliopsoas muscle. This might reduce the compressive loading on the discs—but it will also tend to increase spinal flexion slightly, which in other circumstances we should regard as undesirable. It is of some interest that the legs-raised lying position is very similar to the posture which astronauts automatically adopt under conditions of weightlessness—in which all the articulations come to rest close to the mid-points of their ranges of motion. Neck pain with pronounced muscle spasm commonly comes on overnight. This is widely believed to be due to sleeping position. An arrangement of pillows which supports the head and neck in the symmetrical neutral position may be helpful. Feather pillows are better than sponge pillows in this respect.

## Squatting

To squat on your haunches so that the thighs meet the abdominal wall, you must fully flex your knees, your hips and your lumbar spine. Some people who think that lordosis is a bad thing therefore think that squatting is a good thing (Fahrni, 1975), and some have asserted that oriental and African people, who squat rather than sit, suffer less back pain than Western people who sit on chairs—supposedly with a lordosis. None of this is true. Western people do not sit with a lordosis unless they have unusually well-designed chairs or a great deal of self-discipline—and there are no valid cross-cultural data on the prevalence of back pain.

A squatting position is said to aid the easy emptying of the bowels (e.g. Hornibrook, 1934). I am not aware of any controlled experimental studies of this proposition—but I believe it to be true. The mechanism is unclear. Hornibrook says that the pressure of the thighs helps support the abdominal wall—particularly in the weak region of the inguinal canal. Perhaps full flexion of the spine also assists the action of the transversus abdominis muscle or raises the diaphragm (by releasing its crural attachments), thus allowing it to bear down more effectively.

Sitting cross-legged on the floor (as in the tailor's position or the lotus position used in hatha yoga) is a very stable posture: the centre of gravity is low and the support base is large. The full flexion of the knees slackens the hamstrings enough to allow a right angle between the thighs and trunk to be maintained without the backward rotation of the pelvis and the loss of lordosis. But the degree of outward rotation of the hips which is required, is beyond the capacity of many people.

## Working Posture

A good working posture is one which can be sustained with a minimum of static muscular effort—and in which it is possible to perform the task at hand more effectively and with least muscular effort. *In general, a varied working posture is better than a fixed working posture, but a working posture which is static and relaxed is better than one which is static and tense.*

### Systemic Effects of Sitting and Standing

Sedentary work is known or suspected to have an adverse effect on a number of bodily functions. In discussing these, it is not easy to distinguish the consequences of the sitting posture *per se* from those of the absence of physical activity. (In some respects they amount to much the same thing.)

A number of epidemiological studies have shown a significantly increased risk of cancer of the large bowel in sedentary workers. Garabant *et al.* (1984) found that the incidence of colon cancer in men in Los Angeles was 1.6 times greater in those who had sedentary jobs compared with men in physically active jobs. In some sedentary occupations (e.g. book keepers) the risk was 4 times higher than in certain manual jobs (e.g. longshoremen). The association has been confirmed by Vena *et al.* (1985), who reported a relative risk of 1.3 in a 19 year follow-up study of 1.1 million Swedish men. None of these studies found a significant association for cancer of the rectum. There is no evidence that sedentary workers have a different kind of diet from manual workers (Gerhardsson *et al.*, 1986). Physical activity has been found to increase the number of synchronized propulsive mass movements of the bowel and to decrease the number of random non-propulsive movements (Holdstock *et al.*, 1970). Thus the duration of the exposure of the bowel wall to potentially harmful substances in the stool may be less.

Given this physiological finding and the strength of the association between sedentary work and serious bowel disease, it seems likely that there will be a similar association for less severe disorders. It is widely believed that sedentary people tend to be constipated—and to suffer from various other varieties of gastrointestinal dysfunction. Slackness of the abdominal muscles may be another part of the story. Grandjean (1988) refers to this as "sedentary tummy", although the connection does not seem to have been studied on a statistical basis.

It is generally recognized that lung function (vital capacity, forced expiratory volume, etc.) is diminished in the sitting position. Wind and brass instrument players generally find that they play better standing up. But the long-term effects of sedentary work on lung function do not seem to have been investigated—and again the effects of sitting *per se* would be difficult to distinguish from the effects of inactivity.

Prolonged sitting in a cramped position has been known to cause deep vein thrombosis (the formation of blood clots in veins of the legs), which may result in pulmonary embolism (the transmission of parts of these clots to the lungs, which may prove fatal). Fatalities were recorded in people sleeping in deckchairs in air raid shelters during the London blitz (Simpson, 1940). The rung of the deckchair presses into the underside of the thigh. The number of cases decreased as bunks were provided. Deep vein thrombosis and pulmonary embolism are recognized risks of long distance air travel—where mild hypoxia and dehydration may be contributory factors (Homans, 1954; Symington and Stack, 1977; Cruikshank *et al.*, 1988).

Venous pressure in the lower limbs increases as a function of the vertical distance between the feet and the right atrium of the heart. The pressure will therefore be higher in the sitting position than it is when lying down, but lower than it is when standing. Increased hydrostatic pressure restricts the return of venous blood to the heart, leading to pooling of venous blood in the veins, oedema and an increase in the volume of the leg and foot. Muscular activity directly assists venous return (the *muscle pump*). Thus fluid congestion in the lower limbs is likely to be less during activities involving a mixture of standing and walking than during sitting or standing still.

Prolonged sitting (e.g. over an 8 hour working day) leads to an average increase in the volume of the leg and foot of between 2% and 5% (depending on the amount of leg movement which is allowed) and up to 7% in some individuals, compared with about 2% for standing and walking (Winkel, 1985). The swelling is greater at high ambient temperatures (owing to peripheral vasodilatation) and in chairs which compress the underside of the thigh (Pottier *et al.*, 1969). The swelling is associated with sensations of distension, heaviness and discomfort—and with a fall in tissue temperature, which leads some people to complain of cold feet (Winkel, 1985). Both swelling and discomfort are less if sedentary subjects are given additional tasks to perform which use the leg muscles, such as propelling the chair round on its casters at intervals during the working day. The provision of a footrest with a built-in treadle action has also been recommended (Stranden *et al.*, 1983).

It has been suggested that prolonged sitting may be a contributory factor in the aetiology of varicose veins (Alexander, 1972). Varicose veins are common in Western countries but rare in Asia and Africa, whereas the prevalences in blacks and whites in the USA are about the same. This suggests that the national differences are likely to be due to habitual patterns of activity rather than constitutional factors—notwithstanding the fact that varicose veins tend to run in families. But although the hypothetical association between varicose veins and sedentary lifestyle seems reasonable physiologically, it is not borne out by the epidemiology. Mekky *et al.* (1962) studied women working in cotton mills in England (Rochdale and Carlisle) and in Egypt. The prevalence of varicose veins in the English women was 32%, compared with only 6% for the Egyptians. In both countries, women who sat at work had the lowest prevalence and those who stood at work had the highest (18% and 57%, respectively, in the English mills). Those who mainly walked around and those whose work involved a mixture of standing, sitting and walking had intermediate prevalences (29% and 35%, respectively). Other significant factors included age, childbirth, body weight and corsetry.

The association between varicose veins and prolonged standing work was recognized by Ramazzini (1713):

> "The trades that require them to stand make the workers peculiarly liable to varicose veins . . . the blood that is returning from the arteries into the veins does not receive from the impulse of the arteries the force needed to make it rise perpendicularly . . . those who stand are less fatigued if they put their weight now on one foot now on the other."

It is a matter of common experience that prolonged standing leads to aching legs and feet. Buckle *et al.* (1986) studied a sample of women who worked in supermarkets and department stores. Over one-third of their subjects reported regular pain or discomfort in their feet. The prevalence increased with the percentage of the working day spent standing or walking. There was a marked threshold effect at 30%. In those spending more than 30% of the working day on their feet, the prevalence was 48%; in those spending less, the prevalence was only 7%. Work-related foot problems have not been much investigated—hard, unyielding flooring materials may be a contributory factor. Prolonged standing is also associated with a high prevalence of low back pain, although there is not necessarily a direct connection (p. 68).

The provision of suitable forms of seating, for people who habitually stand at work, is a basic ergonomic necessity. And there is much to be said in favour of workstations which permit the user to sit or stand at will. The controls and working surfaces should be set at levels which are suitable for a standing person, and a high stool should be provided to allow him to sit when he chooses. A footrest will be required, preferably separate from the stool itself. (Rungs, etc., attached to the legs of the stool are rarely satisfactory. They tend to immobilize the user's legs—often with the knees too much flexed for comfort.) Knee and thigh clearance is often a problem in sit–stand workstations. Most of the advantage of sitting is lost if the user cannot get his knees under the table; in fact, in terms of the loading on his spine, he would probably be better off to stand all the time.

## Common Postural Problems

The seven basic guidelines concerning working posture which are set out in Table 5.2 are quoted from Pheasant (1986). They were based extensively on the work of Corlett (1978).

*Table 5.2 Seven Basic Guidelines Concerning Working Posture*

1. Avoid forward inclination of the head and neck
2. Avoid forward inclination of the trunk
3. Avoid requiring the upper limbs to be used in a raised position
4. Avoid twisted or asymmetrical postures
5. Where possible, keep joints within the middle third of their range of motion
6. Provide an adequate backrest in all seats—and design the seat and workstation in such a way that the backrest can be used to full advantage
7. Where muscular forces must be exerted, the limbs should be in the position of greatest strength—unless by doing so one of the foregoing principles is violated

After Pheasant (1986); Corlett (1978).

*The posture of the head and neck* is principally determined by the visual demands of the task (see Chapter 10). The location of visual displays (and other sources of visual information which are necessary in the performance of the working task) is critical in this respect. Displays which are too low result in a forward inclination of the head and neck. If the display surface is horizontal or oblique (e.g. a desk top or keyboard), this will be achieved by tilting the head forward at the atlanto-occipital joint and flexing the cervical spine (particularly the upper part). The weight of the head is supported by tension in the cervical part of the erector spinae group and the upper fibres of the trapezius muscle. But if the display surface is vertical (requiring an horizontal line of sight, as in a VDU screen), a more complex postural change occurs (Figure 5.4). The sitting person will slouch in his seat, flexing the whole of his spine except the upper part of the cervical region. The thoracic spine is probably flexed (anatomy textbooks notwithstanding), giving the

Figure 5.4 *VDU user showing the "yuppie hump"—from an original in the author's collection*

Figure 5.5 *Stooped working posture at an assembly line which is very much too low—from an original kindly supplied by Peter Buckle*

characteristic "yuppie hump". This brings his eyes down to the correct level. The upper cervical spine is extended rather than flexed (to maintain the line of sight), drawing the back of the head down into the shoulders and poking the chin forward. (In the case of the VDU operator, the shoulders are often raised.) This position of the head is maintained by the sub-occipital muscles (semispinalis capitis, rectus capitus, etc., and possibly splenius) and also probably by the sternocleidomastoid. These muscles act in a shortened position, which exacerbates the problems which result.

*Stooped working postures* (in which the trunk is inclined forward) result from working surfaces which are too low and/or from the need to *reach* over obstacles (Figure 5.5). The latter may in turn be due to inadequate *clearance*—for example, lack of knee room in a seated workstation. Toe recesses in standing workstations are also important for this reason. The shoulders are commonly protracted. The

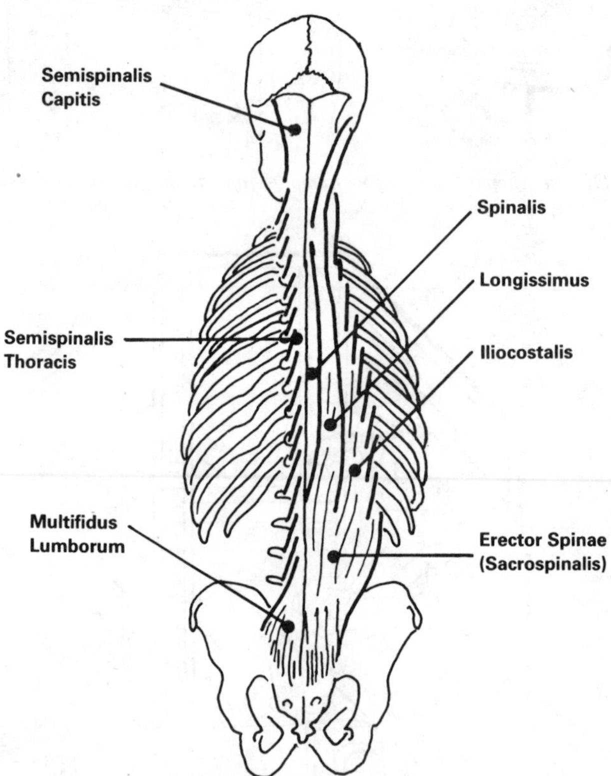


Figure 5.6 *The postvertebral muscles. Right-hand side shows superficial layer; left-hand side shows deep layer. The erector spinae group (sacrospinalis) arise by a common tendon of origin from the sacrum and iliac bone. Just below the level of the 12th rib it divides into three columns of muscle: iliocostalis, longissimus and spinalis. The multifidus and semispinalis muscles lie deep to these. Semispinalis is particularly well developed in the cervical region and multifidus in the lumbar region. Deeper still lie the tiny horizontally running rotatores. Multifidus, semispinalis and rotatores are collectively known as transversosopinalis*

forward inclination of the trunk usually involves a spinal flexion (but this need not necessarily be the case in a person with stretchable hamstrings who has a good range of flexion at the hip). The weight of the upper part of the body is supported by the post-vertebral muscles of the erector spinae group, especially the long fibres of the superficial layer—see Figure 5.6. These run parallel to the spine, so any tension in the muscles results in an equal and opposite compression on the discs. The muscles work at a mechanical disadvantage, so the loading involved is considerable (see p. 280). The deeper oblique muscles—particularly multifidus lumborum—probably serve to stabilize the spine within itself. Lifting actions performed in a stooped position are particularly hazardous (p. 279).

*Working with the arms in a raised position* (i.e. with the shoulders elevated and flexed, abducted, or both) imposes a static loading upon the muscles which raise the shoulder girdle (upper trapezius, levator scapulae, with serratus anterior acting to prevent these from drawing the scapula backward) and the muscles connecting the shoulder girdle to the upper limb (deltoid, supraspinatus and long head of biceps, etc.). Working with raised arms also increases the loading on the heart (p. 34). A rule of thumb, which is sometimes used, states that the elbows should not be raised above "mid-torso height", or that of the shoulder should not be

Figure 5.7 *Asymmetric posture due to a one-handed task on a work surface which is too high—from Roth (1861)*

abducted by more than 60° (note that these are maximum acceptable levels rather than optima).

*Twisted postures and asymmetric postures* which involve side bending increase the loading on the spine. Incorrect positioning of displays and controls is commonly to blame—such as "needing to have eyes in the back of your head". In some motor cars the pedals are off-set from the mid-line of the seat, thus imposing a constant slight twist on the spine. Some manual tasks (e.g. writing) are unavoidably one-handed—the resulting postural asymmetry will be greater if the work surface is too high (Figure 5.7).

The standing worker should at least have the option of taking his weight equally on both limbs, although he will not choose to do so all the time. The operation of pedals in the standing position is therefore undesirable. Frequent changes of footing are probably desirable rather than otherwise, since they lead to change of spinal loading. Victorian public houses commonly had a brass rail around the bar about 6 inches (150 mm) above the ground, allowing the drinker to rest each foot on it in turn.

## Postural Deformities

Postures which are initially adopted for occupational reasons may become habitual outside the working context and finally become irreversible—owing to the shortening and fibrous contractures of muscles and soft tissues. Ramazzini (1713) cites several examples, including tailors and cobblers, who:

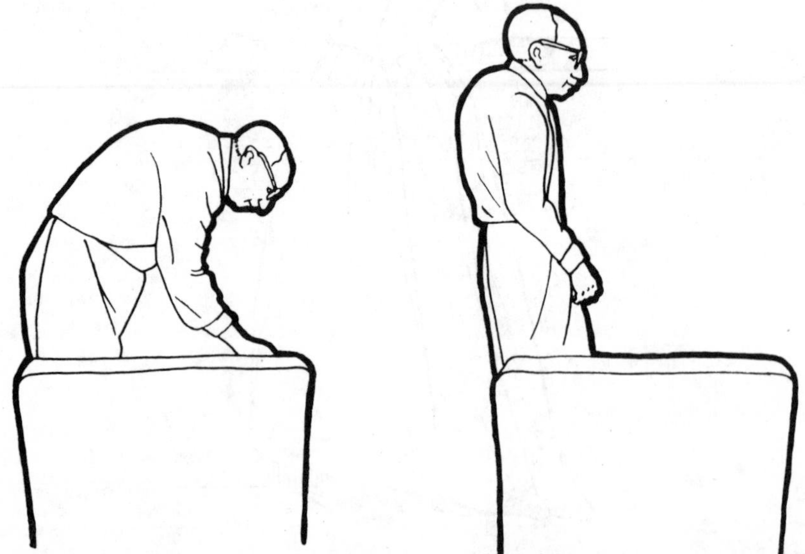

Figure 5.8 *The upholsterer:* left, *working position;* right, *standing position*

"Become bent, hump-backed and hold their heads down like people looking for something on the ground.... Since to do their work they are forced to stoop, the outermost vertebral ligaments are pulled apart and contract a callosity, so that it becomes impossible to return to the natural position."

Referring to feast days on which the guilds of cobblers and tailors march in procession through the city, he describes them as:

"... that troop of stooping, round-shouldered, humping men swaying from side to side; they look as if they had all been carefully selected for an exhibition of these infirmities."

Epidemiologically, the subject has been sadly neglected—but observation suggests that the most common work-related postural deformity is an accentuated dorsal kyphosis (round-back), coupled with a retraction of the head and a shortening of the cervical spine, and often with a sideways list so that one shoulder is markedly higher than the other (Figure 5.8). This occurs in turners (lathe operators), bricklayers, upholsterers, surgeons, dentists, sewing machinists and people who work at drawing boards. My impression is that tall people are worst affected. These deformities presumably take some years to develop and you rarely see them before middle age.

## Chapter Six
# Work Design

## The Systems Approach

In the early years of its development, the science of ergonomics borrowed much of its way of looking at things from two closely related disciplines called *cybernetics* (Wiener, 1954, 1961) and *general systems theory* (von Bertalanffy, 1968). In most respects, we could regard these two disciplines as one area of study with two different names. The exponents of these disciplines were principally concerned with the control mechanisms which govern the behaviour of complex dynamic systems. They very often sought to draw analogies between the complex systems designed by engineers and self-organizing systems such as living organisms. The ergonomic approach to work design is sometimes characterized as *systems-orientated.*

Figure 6.1 shows a simple *man–machine system.* The user feeds information into the machine via the *controls* and receives information from the machine via the *displays* (in this case the keyboard and screen, respectively). The displays and controls together make up the *man–machine interface.* The role of the "man in the loop" is principally one of *information processing.* In technologically less advanced systems, the human operator is also a source of mechanical power; and in technologically sophisticated systems he may surrender an increasing number of his decision-making functions to the machine.

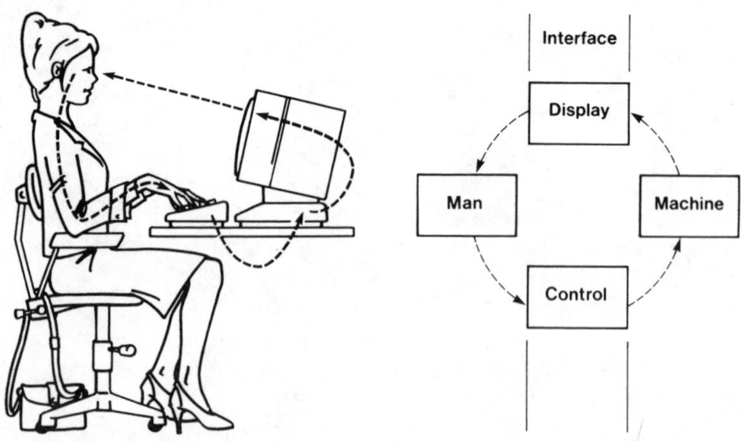


Figure 6.1 *A simple man–machine system. Reproduced from Pheasant (1986)*

By analogy, any source of sensory information used in the working task may be regarded as functionally equivalent to a "display" and any tool or device may be regarded as a "control". So our simple system model, of "the man in the loop", may be applied to a wide range of working tasks. Consider, for example, the tasks of driving a car, using a pneumatic drill or cooking a meal.

But the simple system does not exist in isolation. The "systems approach" to work design encourages us to take a broad view of the human being in his (or her) physical working environment and social (or organizational) context. A working system in which a number of human beings interact with a number of machines (in pursuit of a common goal) is sometimes referred to as a *socio-technical system.*

In principle, this process of "enlarging the system boundaries" is limitless. It is sometimes useful to think in terms of a systems hierarchy of the kind shown in Table 6.1. The systems under consideration at each level of the hierarchy are *sub-systems* of those at the next level up.

*Table 6.1 The Systems Hierarchy*

| Level | System type |
|---|---|
| 9 | The global economy; the biosphere |
| 8 | Socio-economic systems; ecosystems |
| 7 | Communities; urban systems; industries |
| 6 | Socio-technical systems; working organizations |
| 5 | Man–machine systems |
| 4 | The human being |
| 3 | Organ systems |
| 2 | Cellular systems |
| 1 | Sub-cellular systems |

By way of illustration, it might be informative to apply the somewhat abstract concept of the systems hierarchy to a practical ergonomic problem, such as RSI in data entry workers. Each level in the hierarchy may tell us something interesting and useful about the problem—from the biochemical changes occurring in the damaged muscle cell to the role of information technology in the global economy—but none on its own tells us the whole story. You could regard this as a *holistic* approach to ergonomics.

The strength of the systems approach is that it helps to prevent us from concentrating our attention too exclusively on isolated features of a complex real world problem to the exclusion of other features which might be equally important. Its weakness is that it may divert our attention from those limited aspects of the overall problem which are under our direct control. And as we direct our attention to the higher levels of the systems hierarchy, we tend to find ourselves dealing less with questions of fact (which are properly within the domain of empirical science) and more with questions of value (which are properly within the domain of ethics or politics).

## Task Demands and Working Capacity

Human beings vary in their physical and mental characteristics, and thus in their capacity and inclination for different types of work. Much of the applied science of ergonomics is concerned with the measurement of this variability, and with matching the demands of the task to the capacities of the working population or to the users of a particular product, system or environment (Figure 6.2).

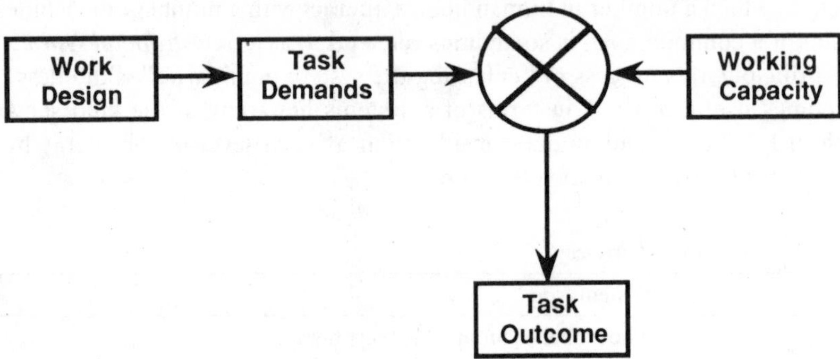


Figure 6.2 *Task demands and working capacities*

### Statistical Aspects

The variability of most measurable bodily characteristics may be described (to a variable degree of accuracy) by the symmetrical bell-shaped curve of the *normal distribution*. The use of the word "normal" in this context is a bit confusing. It does not refer to "normal people" or to "normal limits" for the variable concerned. You could perhaps say that it means the typical distribution for data of this kind.

Figure 6.3 shows the distribution of stature (i.e. standing height) in adult British men. The quantity plotted vertically is the *relative frequency* with which you would expect to encounter individuals of a given stature (in a sample of men drawn at random from this population). The curve is symmetrical about its highest point (i.e. the stature you encounter most frequently). Fifty per cent of the population lie to the left of this point and 50% lie to the right. We therefore call this the 50th percentile stature. In distributions which are symmetrical about the mid-point, this is the same thing as the arithmetic average or *mean*. Somewhere in the left-hand "tail" of the distribution is a point at which we could say: "5% of people are shorter than this". We call this the 5th percentile. An equal distance to the right of the mid-point is a point at which we could say: "5% of people are taller than this". We call this the 95th percentile. In general, *n*% of the population are smaller than the *n*th percentile.

It follows from the shape of the normal curve that the percentiles are close together in the mid-range of the distribution and widely separated in its tails (because people of average size or shape are relatively common, but extremely

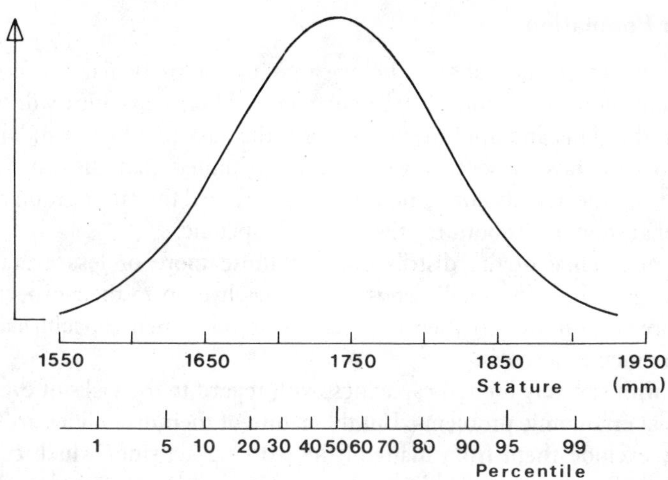


Figure 6.3 *The distribution of stature in British men. Reproduced from Pheasant (1986)*

large or extremely small people are relatively rare).

Mathematically, the curve of the normal distribution can be completely described by two *parameters*—the mean and the standard deviation. The mean locates the centre of the distribution on the horizontal axis. The standard deviation is a statistical measure of dispersion—that is, of the extent to which individual data points are scattered about the mean. If the mean and standard deviation are known, any percentile can be calculated. For a more detailed discussion of the properties of the normal distribution and its importance in applied anthropometrics, see Pheasant (1986, 1991).

The symmetrical normal curve describes the variation of linear bodily dimensions (such as stature and limb lengths) fairly accurately in most population samples. (The exceptions to this rule tend to occur in samples drawn from working populations which have been selected for height in the first place—such as guardsmen, policemen, etc.) But the distribution of body weight (and correlated dimensions such as bust, waist or hip circumference) is in most cases *positively skewed*—that is, the distribution curve is asymmetrical, with the right-hand tail being longer than the left. In other words, there are an excess of heavy people in the population sample and a dearth of light people, obesity being more common than emaciation (at least in our society). Strength measurements usually show a pronounced asymmetry in their distribution; and this would probably also be true for other measures of physical performance, such as aerobic power (although the data sets available are not yet sufficiently extensive for us to be certain). In part, these asymmetries probably result from the interaction of hereditary and environmental factors in the determination of the characteristic concerned, but sampling bias is probably also involved. Thus you may well see a marked positive skew in the exam results of a population of students, because those individuals with limited capacities in this particular respect tend to be excluded from the sample.

## The User Population

Consider the range of people who wish to use a particular *product* (system, environment, etc.) to perform a particular *task*. Their capacities will vary: from those with the skills and aptitudes to perform this task to a high standard without much effort to those whose capacities are so limited that they are effectively excluded from the activity concerned. We might regard the latter group as *disabled*, but note that the term is both relative and task-specific.

In the mid-range of the distribution are those more or less averagely unfit people we regard as "normal", whose capacities live up to their expectations, or who are prepared to modify their expectations to meet their capacities and do not suffer too much as a result.

People with severely limited capacities, with regard to the tasks of everyday life, pose special ergonomic problems. But the nature of their disabilities are such as to effectively exclude them from many of the working activities which concern the ergonomist. This process of *disablement* is often due, however, to unresolved problems in the design of the working and living environment—as in the case, for example, of the mobility problems of the wheelchair user. And the exclusion of the less able individual from the user population is very often more a matter of economic or operational expediency than physical inevitability (see Chapter 16).

In practice, it is the user in the lower part of the normal range who tends to be of particular concern to the ergonomist: the individual who is more than averagely unfit, but not so unfit as to be excluded from the population of people who use a particular product, and who therefore experiences a greater than average degree of difficulty in doing so. People who fall into this category may be sensitive to minor deficiencies in product design which go unnoticed by more able individuals.

We could call the hypothetical least able member of a user population the *limiting user*. In general, we should expect a design which is satisfactory for the limiting user to be satisfactory for the more able majority (although there are exceptions). Compare this with the *fallacy of the average person* (p. 15).

## The FPJ–FJP Model

Consider a working situation in which the demands of the task and the capacities of the workers may be described by overlapping distributions, as shown in Figure 6.4. At a particular workplace, for example, there may be heavy and light jobs—and the demands of a particular job may vary according to circumstances. The members of the workforce will also vary in strength, fitness, etc., and the capacities of an individual worker will vary according to how fatigued he is. When the demands of the task exceed the capacity of the worker, he is *overloaded*, and thus potentially at risk, as indicated by the shaded area of the diagram. Brown (1982) has used a model of this kind in discussing road accidents—but it would be equally applicable to lifting and handling injuries, RSI, stress, etc.

To deal with the problem, we have to minimize the shaded area—either by shifting the distribution of capacities to the right or by shifting the demands

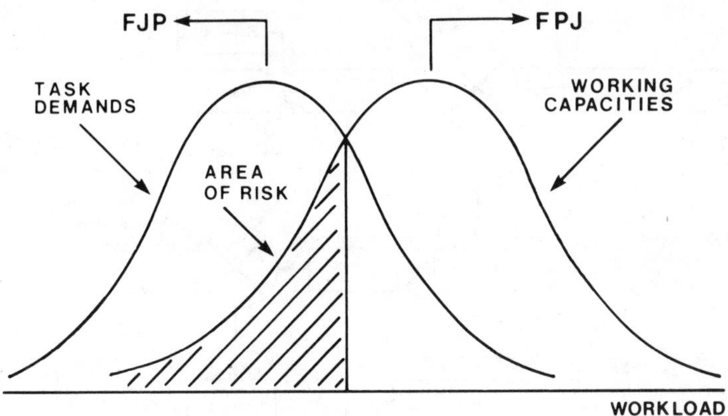


Figure 6.4 *The FPJ–FJP model*

distribution to the left (or, at least, by truncating the overlapping tails of the distributions).

Measures aimed at increasing the capacities of the working population could be characterized as *fitting the person to the job* (FPJ): this includes most conventional forms of selection and training, as well as any other form of therapeutic or prophylactic intervention aimed at increasing the individual's *fitness for work*. Ergonomic improvements in work design (aimed at eliminating the heaviest, most dangerous tasks, etc.) may be characterized as *fitting the job to the person* (FJP). In principle, the two approaches should be seen as complementary, and elements of both would be required for an effective programme of preventative medicine in the workplace (Table 6.2). In the days when "men" were matched to jobs, the late Professor Alec Rodger used to speak of "the FJM–FMJ framework".

An engineer defines mechanical stress as load per unit area. Similarly, the relationship between the demands of the task and the working capacity of the individual may be held to define the mental or physical stress to which he or she is exposed in the working situation (Figure 6.5). The analogy is more exact for physical work: as when we compare the energy expenditure of a task with the individual's aerobic power (p. 31), or the weight of a load with his lifting strength (p. 287). Psychological stress is more complicated, since excessively high and

*Table 6.2 FPJ–FJP*

| Fitting the person to the job (FPJ) | Fitting the job to the person (FJP) |
|---|---|
| Selection and screening | Ergonomics—work design |
| Skills training | Safety engineering |
| Safety training | Environmental control |
| Fitness training | Organizational change |
| Health education | |
| Back care education | |
| Stress management | |

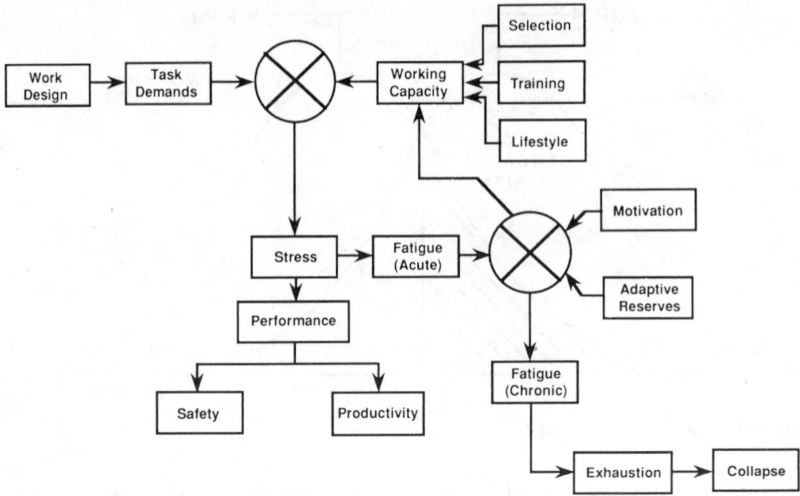


Figure 6.5 *Task demands, working capacities and the adaptation process*

excessively low levels of mental stimulation may both be subjectively stressful. When mental or physical stress levels are high, the well-motivated person may draw upon his adaptive resources in order to maintain a high level of performance despite the effects of fatigue. In doing so he may run up an *adaptation debt* (the term was suggested by Jackie Nicholls). This may lead to more severe problems in the long term—as in musculoskeletal conditions due to cumulative overuse, stress-related psychoneurotic disorders, and so on. Conversely, chronic physical underload leads in the long term to a reduction in fitness and in working capacity.

## Anthropometrics

The branch of ergonomics which deals with body measurements is called *anthropometrics*. I have dealt with this aspect of the subject at length in my earlier books (Pheasant 1986, 1991). The main stages we might reasonably expect to go through, in applying the principles of anthropometrics to a design problem, are summarized in Figure 6.6.

At the outset, it is necessary to define the user population. For adult populations, the principal factors to take into account will be sex, nationality, age and occupation—probably in that order. (If children are involved, age moves up to first place.) A detailed discussion of these issues, together with an extensive compilation of anthropometric data for different populations, will be found in Pheasant (1986).

Quite early in our hypothetical design project it will be useful to establish the criteria which define a satisfactory match between the product and its user. Design criteria are by their nature hierarchical: there are high-level criteria which deal with generalities like comfort and safety, and low-level criteria which are much

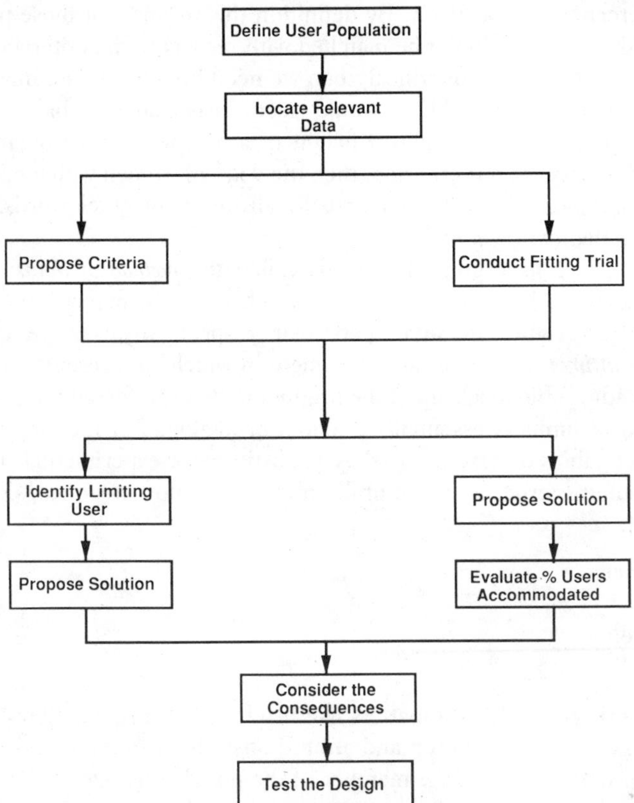


Figure 6.6 *The stages of a typical design project involving applied anthropometrics—after Pheasant (1991)*

more specific. Take the case of a chair. We presumably want it to be comfortable and to give adequate support. In order to achieve these overall objectives, a variety of subordinate criteria, dealing with seat dimensions and so on, must be fulfilled. If the seat is too high (for example), it will press on the underside of the user's thighs, causing discomfort and worse; so the seat height should not exceed the popliteal height of a short person (the popliteal region being the back of the knee—see p. 216). This low-level criterion leads directly to a design recommendation—provided, of course, that we have an appropriate figure for the popliteal height of "a short person".

How short is "short"? It necessarily follows, from the shape of the normal distribution, that it will become increasingly difficult to cater for increasingly extreme members of the user population. That is, it becomes less cost-effective, if you treat the constraint which it imposes upon your design as a cost, and the number of potential users you satisfy as a benefit. In practice, therefore, we will usually need to impose arbitrarily chosen cut-off points in either tail of the distribution. These are called the *design limits*. The ones usually chosen are the 5th and 95th percentile, respectively. By definition this means that those people who fall outside these limits will not be matched with respect to the criteria concerned. If these mismatches are at all critical, then we need broader design limits.

In the case of seat height, it would seem reasonable to base our design recommendation upon the popliteal height of a 5th percentile woman. If she is satisfied with respect to our criterion, then the 95% of women with longer legs will also be accommodated (as well as virtually all men). In other words, we could regard her as the *limiting user.*

The process we have gone through is called the *method of limits*: it involves searching for those boundary conditions which make an object "too big", "too small" or "just right" in some particular respect. Instead, we could have conducted a *fitting trial*: that is, an experiment in which a representative sample of users set an adjustable mock-up of the product to their preferred range of heights. The method of limits is essentially a model or analogue of the empirical fitting trial, in which tables of data are used as a substitute for experimental subjects.

Anthropometric criteria fall naturally into four main categories, dealing with questions of:

- clearance;
- reach;
- posture;
- strength.

*Clearance* criteria deals with matters like headroom, leg room, elbow room, and so on. Access problems between and around obstacles also fall into this category. The limiting user will be a *large* member of the population—generally one who is 95th percentile or more in the relevent aspect. *Reach* criteria include those concerned with the location of controls or the storage of materials, and with a variety of situations where it is necessary to reach over an obstacle to perform a task. The limiting user will be a *small* member of the population—usually 5th percentile. *Postural* criteria include those concerned with the location of displays and controls at the heights of working surfaces. In these cases it may not be possible to identify a limiting user, and a different approach is required.

## Working Height

The optimal working height for a particular activity is the one which minimizes the overall degree of effort required for the performance of that activity—particularly with regard to the static muscle loading which the working posture requires but also taking into account things like the visual demands of the task.

Heavy tasks, particularly those involving the application of downward forces, should be performed at well below elbow height. People speak of being able to "get their weight on top of the work". This is a slight misconception. It is actually more to do with moment arms. If the working level is low enough, you can get your shoulders and elbows directly above the workpiece—and thus apply a vertical force along the axis of an almost straight limb. But if the working surface is too low, an unacceptable stooped posture will result.

Tasks involving fine visual discrimination and hand/eye co-ordination sometimes need to be performed at close to eye level—say at a little below shoulder height. A suitably padded elbow rest or wrist suport may help to stabilize the hands and reduce the loading on the shoulder muscles—provided that it does not hamper the user's movements by immobilizing his forearms.

The following working heights are widely recommended:

- for manipulative tasks involving a moderate degree of both force and precision—from 50 mm to 100 mm below elbow height;
- for lighter and more delicate tasks (including writing)—from 50 mm to 100 mm above elbow height (generally with wrists supported);
- for heavier tasks (particularly if they involve downward pressure on the workpiece)—from 100 mm to 250 mm below elbow height;
- for lifting and handling tasks—from knuckle height to a little above elbow height;
- for two-handed pushing and pulling actions—a little below elbow height;
- for hand-operated controls—between elbow and shoulder height. (Controls with associated visual displays should be in the upper part of this region.)

Some relevant anthropometric data are given in Table 6.3. Note that the measurements concerned are taken on an unshod person standing upright with his arms hanging loosely by his sides, as shown in Figure 6.7. For practical purposes it will generally be necessary to add an appropriate correction for the person's shoes (e.g. 25 mm for men; 45 mm for women).

A workbench which is too high for a short user is no better or worse than one which is too low for a tall user. In other words, there is no limiting user. In practice, we have two possible options: either we must build an adjustable workbench (so that each user can set it to his own preferred height) or we must settle for a suitable compromise which accommodates the greatest possible number of people and minimizes the inconvenience for the remainder. Suppose our workbench was for "tasks involving a moderate degree of both force and precision", as specified above. If the range of adjustment extended from 100 mm below the elbow height of a 95th percentile user to 50 mm below the elbow height of a 5th percentile user, then 90% of the user population would be accommodated with respect to the criterion. The best compromise, for a fixed workbench, would be 75 mm below the elbow height of an average (50th percentile) user. This follows from the shape of the normal distribution. To decide whether this compromise was acceptable, we would need to consider how serious the postural problems which it created for the 5th and 95th percentile would actually be. Kitchen worktops are an interesting case in point (p. 325). For a more detailed discussion of these questions, see Pheasant (1986).

Many ergonomists have misgivings about adjustable workstations, mainly because users often "don't bother to adjust them". People with strong healthy backs are commonly insensitive to the spinal loading caused by a working surface

*Table 6.3 Anthropometric Data for British Adults in the Standing Position*

| | Men (percentile) | | | Women (percentile) | | |
|---|---|---|---|---|---|---|
| | 5 | 50 | 95 | 5 | 50 | 95 |
| 1. Stature | 1,625 | 1,740 | 1,855 | 1,505 | 1,610 | 1,710 |
| 2. Eye height | 1,515 | 1,630 | 1,745 | 1,405 | 1,505 | 1,610 |
| 3. Shoulder height | 1,315 | 1,425 | 1,535 | 1,215 | 1,310 | 1,405 |
| 4. Elbow height | 1,005 | 1,090 | 1,180 | 930 | 1,005 | 1,085 |
| 5. Hip height | 840 | 920 | 1,000 | 740 | 810 | 885 |
| 6. Knuckle height | 690 | 755 | 825 | 660 | 720 | 780 |
| 7. Fingertip height | 590 | 655 | 720 | 560 | 625 | 685 |

All dimensions in mm.
Source: Pheasant (1986).

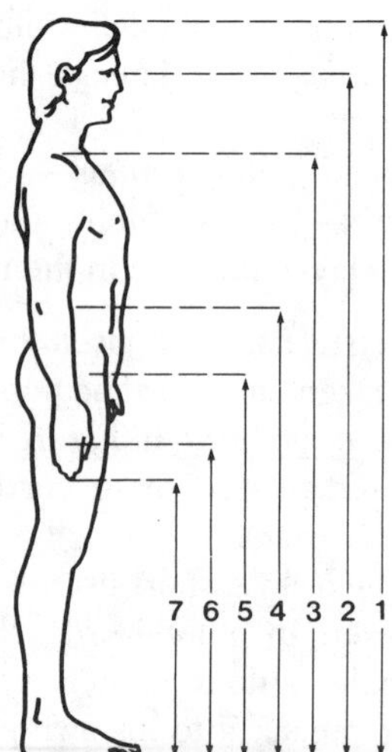


Figure 6.7 *Anthropometric dimensions of the standing person. Reproduced from Pheasant (1986)*

which is too low—and indeed even when their back begins to ache they may fail to make the connection. People with back problems are, of course, acutely aware of the need to stoop. Curiously enough, everybody seems to notice a work surface that is too high. This may be due to the fact that upper limb muscles fatigue more rapidly than back muscles.

## Force Application

Problems concerning acceptable levels of force in the operation of controls occur widely in ergonomics. As a rough rule of thumb:

- up to 60% of an individual's maximum strength is acceptable for occasional exertions;

- up to 30% maximum strength is accepted for frequent exertions;
- sustained exertions should be eliminated wherever possible.

Ideally, the operating resistance of a control should be comfortably within the capacity of the weaker members of the user population (5th percentile or less, depending on the circumstances). But setting the resistance this low may cause problems for a stronger or heavier person. For example, if the resistance of the accelerator pedal of a car is too low, a heavy-footed driver may have to exert a continuous muscular effort to prevent himself from depressing it when he does not intend to. This may cause anterior tibial pain—particularly in people who drive long distances on motorways.

## Working Efficiency

An *efficient* course of action is one which achieves its ends with a minimum of wasted effort. The work of the operators at the chemical plant, which we referred to in the prologue—who had to run up and down the stairs to operate a valve and monitor a display—involved a degree of effort which was disproportionate to the ends it achieved. It was inefficient therefore—and hazardous as a result.

In general, people seek to minimize their effort—at least up to a point—and will actively resist attempts to persuade them to do otherwise. It follows, therefore, that if you want people to do things in a certain way (for reasons of safety, quality of service to customers, etc.), you must design the working system so that the *correct* way of doing things is also the *easy* way of doing things. In many respects this is the cardinal principle of applied ergonomics.

In evolutionary terms, the minimization of effort makes sense. If you don't know where your next meal is coming from, you should minimize the energy consumed in finding it. But in an advanced society, with abundant food supplies, the minimization of effort may be maladaptive. You could argue, for example, that the workers at the chemical plant were getting valuable aerobic exercise—and indeed

*Table 6.4 The Physical Characteristics of Work*

**BAD**

- Any single activity performed continuously over an excessively long period—especially if it involves a fixed working posture
- Explosive efforts of near-maximal intensity
- Static muscle loading
- Repetitive motions—particularly if forceful, jerky, awkward, etc., or if they involve small muscle groups

**GOOD**

- Any change of activity
- Dynamic work of moderate intensity involving large muscle groups
- Gently stretching muscles—especially those which have been subject to static loading
- An appropriate balance between work and rest

some people advocate stair-climbing programmes at work, in the interests of physical conditioning.

What kinds of effort are good for you; what kinds are bad for you? This is not an easy question. The position (at least as I understand it) is summarized in Table 6.4. In practice, most real manual activities have both a static and a dynamic component—that is, there is commonly a steady background of static loading due to the nature of the working posture, superimposed over which there are peak dynamic loadings of varying intensities. Psychological stress may add to the static component.

## The Rationalization of Workspace Layout

The four principles set out in Table 6.5 for the elimination of wasted effort by the rationalization of workspace layout were originally formulated by the late Ernest J. McCormick (1970). You could regard these principles as a formalization of common sense. They are applicable to a wide range of ergonomic problems, such as the arrangement of displays and controls on a console; the materials and machines in a working area; the functions and services in a public building; or even the information in a database.

*Table 6.5 Principles of Rational Workspace Layout*

- **Importance Principle**—the most important items should be in the most advantageous or accessible locations
- **Frequency-of-Use Principle**—the most frequently used items should be in the most advantageous or accessible locations
- **Function Principle**—items concerned with closely related functions or actions should be grouped together
- **Sequence-of-Use Principle**—items which are commonly used in sequence should be grouped together and laid out in a way which is compatible with that sequence

After McCormick (1970).

## Physiological Efficiency

To the physiologist (who borrows his definition from the mechanical engineer):

$$\text{efficiency} = \frac{\text{work output}}{\text{energy expenditure}} \times 100\%$$

The physiological efficiencies of some common working activities are given in Table 6.6. To a great extent, these differences are due to the static components of the tasks concerned (which may be regarded as wasted effort). The nature of the resistance encountered is also a factor: for example, much of the energy consumed in shovelling is dissipated in driving the shovel into the material rather than actually moving it. It may also be that the size of the muscle groups involved also make a difference.

The physiological definition of efficiency is limited in its applicability to the

*Table 6.6 Physiological Efficiency of Various Activities*

| | Efficiency (%) |
|---|---|
| Walking on slight incline | 30 |
| Walking on the level | 27 |
| Cycling | 25 |
| Walking up and down stairs | 23 |
| Hand cranking | 21 |
| Climbing and descending ladder | 19 |
| Hammering | 15 |
| Lifting weights | 9 |
| Shovelling in normal posture | 7 |
| Screwdriving | 5 |
| Shovelling in stooped posture | 3 |

Data from Grandjean (1988).

fairly narrow range of activity in which it is possible both to determine the mechanical work performed and to make a steady state measurement of oxygen consumption. The determination of mechanical work output is simple for some tasks and difficult for others. To calculate the work performed in level walking, for example, we need to know the kinetic and potential energies of each segment of the body at each instant of the walking cycle (see Winter, 1979; Inman *et al.*, 1981, for a discussion of these questions).

For many practical purposes, a comparative measure of the physiological cost (i.e. oxygen consumption) of performing a task under different conditions is all we require. Strictly speaking, we should refer to this as a measure of *physiological economy* rather than efficiency as such. The distinction is more than just a semantic one. Consider the case of two runners who have different rates of oxygen consumption at a particular speed. This does not of itself imply that they have a different physiological efficiency—since the runner with the lower energy expenditure may have a better technique such that he performs less mechanical work. He thus runs more economically (Cavanagh and Kram, 1985).

The work of Bedale (1924) on the physiological cost of load carriage is a classic in this area. Bedale measured the oxygen consumption of a woman carrying a range of loads (from 20 lb to 60 lb) in eight different ways. The most economical method was to distribute the weight symmetrically on a yoke across the shoulders; the least economical was to carry it on one hip. The differences could be due to three factors: postural compensation, interference with normal gait and fixation of the chest wall. Postural compensation is thought to be the most important. The greater the displacement of the centre of gravity of the upper part of the body, the greater the static work component of the task.

## Motion Economy

The principles of motion economy, as formulated by Barnes (1963) and summarized in Table 6.7, deal with the optimization of "working efficiency"—as

*Table 6.7 The Principles of Motion Economy*

1. The two hands should start and finish their motion at the same time.
2. The two hands should not be idle at the same time, except during rest pauses.
3. Motions of the upper limbs should be opposite and symmetrical (about the mid-line of the body).
4. Use the smallest body movement compatible with effective task performance.
5. Momentum should be employed to assist movement—but minimized if muscle action will be required to overcome it.
6. Smooth curved motions of the hands are preferable to straight-line motions involving sudden sharp changes in direction.
7. Ballistic (free-swinging) movements are faster, easier and more accurate than movements under continuous muscular control.
8. Work should be arranged to permit an easy, natural rhythm.
9. Eye fixations should be as few and as close together as possible.

Adapted from Barnes (1963).

measured in terms of the *motions* which are called for and the *time* it takes to perform them. They are widely accepted amongst work study practitioners—and Barnes cites considerable case historical evidence for their validity. They pose a number of interesting questions from the ergonomic standpoint.

The first three principles deal with the co-ordinated use of the hands. The even distribution of workload between the two hands ensures that neither is disproportionately loaded; and symmetry and balance of movement will reduce both the loading on the spine and the need to stabilize the body. The advantages of symmetrical and opposite limb movements are a matter of common experience; it is perhaps to do with the central connections of the nerves which supply the muscles. (Try describing a clockwise circle with one hand and an anticlockwise with the other—then try moving both hands in the same direction.) Barnes argues that finger movements are faster and more efficient than (and therefore preferable to) movements of the hand at the wrist, which in turn are faster than forearm movements, and so on (Principle 4). Taken together, however, the first four principles could easily have become a prescription for the sort of fast repetitive movements, leading to the overloading of small muscle groups, that are most undesirable in ergonomic terms.

The subsequent principles are less contentious. The exploitation of gravity and momentum (Principle 5), where it can be achieved, is a sound strategy for the reduction of effort. Indeed, it is the underlying principle of the "kinetic" school of lifting and handling training (p. 285). The easiest hand movements to perform (Principle 6) tend to be those which approximate to the arc swept out by the forearm as the flexed limb rotates about the shoulder. McCormick (1970) cites studies showing that hand movements are fastest and most accurate when made outward rather than inward obliquely across the body at an angle of 45–60° to the straight ahead action—which corresponds well to the "natural" movement of the forearm. A "free-swinging" movement (Principle 7) will in general be one which follows the natural pattern of limb rotation and is aided by momentum. Although we all think we know what we mean by a "natural rhythm", the concept is not easy to pin down; but you could probably argue that the eighth principle is really a summary of the seven which precede it.

The physiologically efficient natural activity of walking actually has many of the characteristics of economical motion which Barnes identified. The limbs act alternately in a continuous, symmetrical, reciprocating pattern of smooth curvilinear movements (see Figure 6.8). The swinging limb moves freely about the hip; the thigh is accelerated first and the knee is straightened by momentum transfer rather than muscle action—at least at moderate speed.

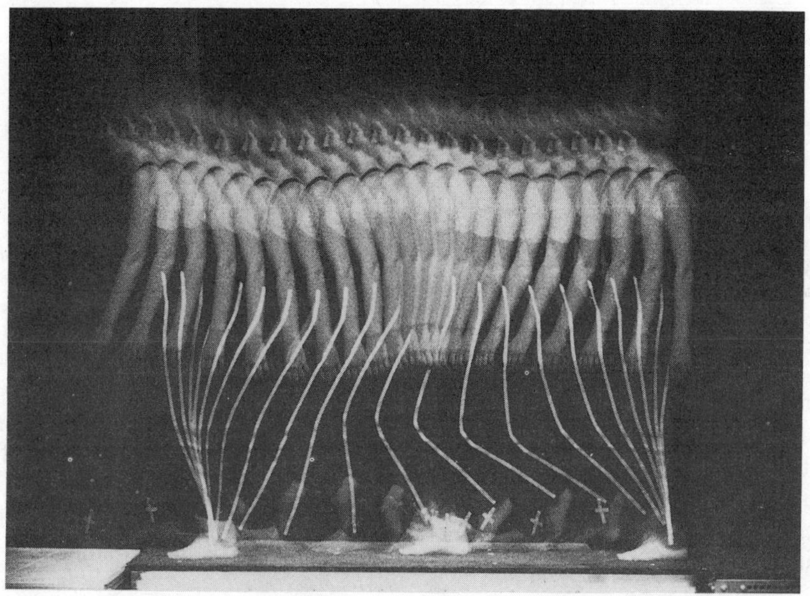

Figure 6.8 *The walking action—chronocyclograph kindly supplied by Professor D. W. Grieve*

Many of the muscles employed in the walking cycle go through a phase of active stretch before performing positive work, and part of the kinetic energy which the body possesses as the foot strikes the ground is stored by the deformation of elastic tissue and released in the subsequent propulsive phase.

The rate of energy expenditure in both walking and running increases as a curvilinear function of speed. Energy expenditure *per unit distance travelled* is plotted as a function of speed, for both walking and running, in Figure 6.9. Both curves show minimum energy conditions, which are presumably determined by the interaction of the various factors described above. When left to their own devices, people tend to adopt a speed close to the one which minimizes the energy required to cover a given distance (Inman *et al.*, 1981). Furthermore, we "break into a run" at about the speed where the two curves cross and it becomes physiologically more economical to do so, and attempts to maintain the walking pattern at higher speeds (as in race walking) may result in an overloading of the tibialis anterior muscle (D. W. Grieve, personal communication).

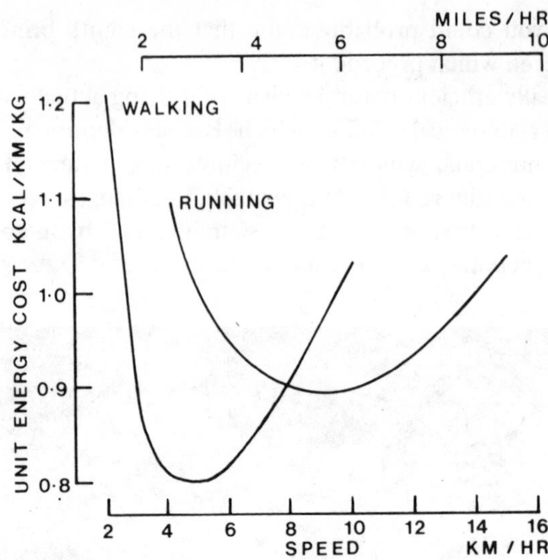


Figure 6.9 *Unit energy expenditure, plotted as a function of speed, in walking and running—based on data from Inman* et al. *(1981) and Böje (1944)*

The likelihood that people would gravitate towards minimum energy conditions when allowed to perform repetitive actions at their own rate was recognized in the last century by the French physiologist Marey (1895). We have surprisingly few data in this area. Some examples given by Drillis (1963) are shown in Table 6.8. In general, we should expect the optimum tempo to be inversely proportional to the weight of load or resistance encountered—just as you tend to slow down when walking uphill. Freivalds (1986a) has noted that this is indeed the case for shovelling.

*Table 6.8 Optimum Tempi for Working Activities*

| | |
|---|---|
| Filing | 60–78 strokes/min |
| Chiselling | 60 strikes/min |
| Cranking | 35 revolutions/min |
| Shovelling | 14–17 tosses/min |

## Some Ergonomic Aspects of Efficiency

Efficient movement is a learned skill. Person (1956) studied engineering apprentices learning the use of the hacksaw. Electromyographic records from the flexors and extensors of the elbow were compared with measurements of the forces applied to the workpiece. Novices showed uneven and erratic force records, with extensive co-contraction of the opposing muscle groups (i.e. there were periods when flexors and extensors were both active at the same time). As the apprentice became more skilled, the co-contraction disappeared and the force records became smoother. In a rhythmic action like sawing, co-contraction represents a hidden form of static work—and, hence, a loss of efficiency. Conditions like tenosynovitis are said to affect novices more frequently than experienced workers (p. 90)—perhaps the novices have less efficient patterns of movement and are thus subject to a greater degree of overload.

Figure 6.10 shows the lifting strengths of a group of ten untrained men and women. Part of the variation in strength was due to differences in technique. The stronger subjects were able to exert greater forces, partly because they were skilful enough to adopt more advantageous postures, in which the internal loading was less for a given external force.

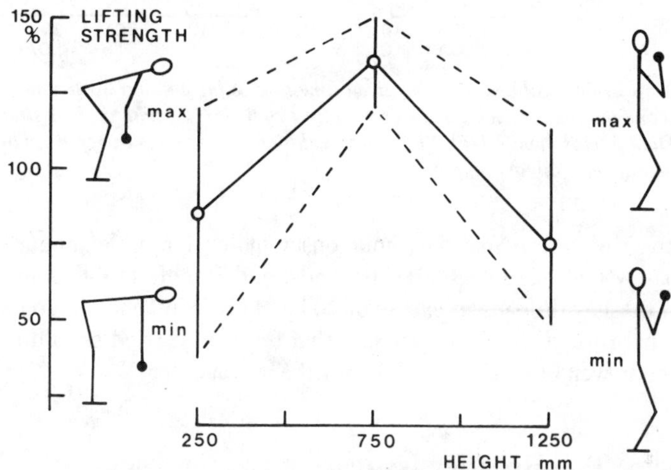


Figure 6.10 *Variation in lifting strength (means and ranges) at three heights above the ground—showing the postures adopted in the highest and lowest position by the strongest and weakest subjects (sample of 5 men and 5 women). Note the distance of the load from the body in each case. Data from Grieve and Pheasant (1982)*

In general we should expect working techniques which minimize internal loading to maximize physiological efficiency. But there are exceptions. For example, a lifting action in which the knees are bent and the back remains straight imposes less loading on the spine than one in which the back is bent and the knees remain straight. But it is physiologically less economical because more energy is wasted in lifting and lowering the weight of the body (Figure 6.11—see also p. 311). The energy consumed in raising and lowering the weight of the body will be more or less independent of load. This component of the overall energy expenditure may be regarded as wasted effort; thus physiological economy will be optimized with relatively heavy loads.

According to Lehmann (1962), the overall quantity of energy consumed in carrying a given quantity of material from one location to another is minimized with loads of 50–60 kg. But loads of this magnitude may result in loadings on the spine which are well into the risky range (p. 281). Since carrying tasks have a high

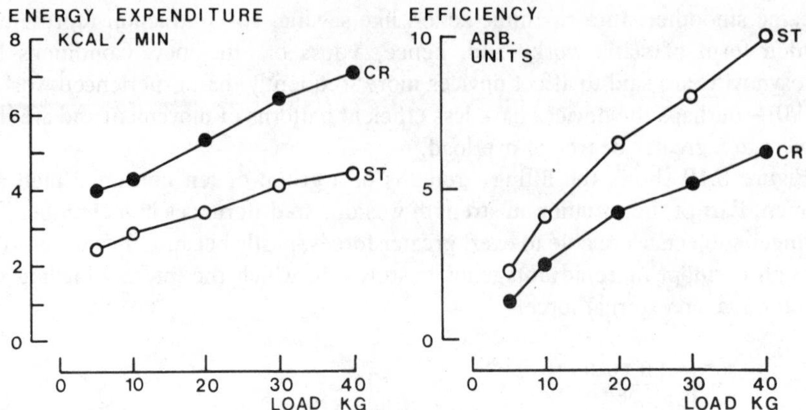


Figure 6.11 *Physiological cost of a repetitive lifting task: using a* stoop lift *technique (ST) with the back bent and the knees straight; and a* crouch lift *technique (CR) with the back straight and the knees bent. Data from Brown (1976). "Efficiency" calculated by dividing energy expenditure by load weight and setting the optimum condition to 10*

static component, there may be rapid onset of local muscle fatigue and disproportionately high cardiovascular demands; and Randle (1988) found a sharp increase in these with loads in excess of 20 kg. This is in general agreement with the traditional rule of thumb which says that the load carried should not exceed one-third body weight—or about 25 kg for the average man.

## Job Design: The Broader Issues

What is a good job? The answer which you give to this question will depend on what you think people want from their working lives. One view (to which many people who consider themselves well informed would subscribe) is typified in the words of Henry Ford (1922):

> "The average worker wants a job in which he does not have to put much physical effort. Above all, he wants a job in which he does not have to think."

Henry Ford founded an empire on this belief. It is doubtless true of some people and equally untrue of others. But the extent to which it is true (in the sense of being empirically verifiable) is less important than the fact that it will inevitably tend to act as a self-fulfilling prophecy. If you follow Frederick Taylor, Henry Ford and many modern industrialists in believing that the blue collar worker is motivated solely by the desire to earn as much as possible for as little work as possible, then you will necessarily create the sort of working environment (both physical and cultural) in which your belief will be confirmed by their behaviour. By Ford's own admission, the jobs in his plants engaged a very narrow range of human faculties (p. 13). But this is what he had set out to achieve. Ford's greatest achievement was to subdivide the assembly of a motor car (which had hitherto been built by skilled artisans) into a sequence of discrete, repetitive, short-cycle, machine-paced operations, which were within the capacity of just about anybody with a pair of hands (or less, as the case may be). By doing so, he achieved the economies of scale necessary to transform the motor car into an object of mass consumption.

Job simplification—the fragmentation of the production process—creates working tasks which can be performed by fragments of people. Just a few isolated attributes of a human being are required to serve the purposes of the machine—the rest of the person is left to atrophy. Thus is work dehumanized. Fragmentary tasks, devoid of intrinsic interest, are by no means unique to the assembly line. Clerical work is often equally lacking in meaning—from the penpushing of an earlier era to the keypunching of the electronic office. The data entry operator is just as much an appendage of the machine as her sisters on the production line. The individual worker is thus alienated both from the corporate goals of the organization and from the better parts of himself—at least insomuch as these might be manifest in his working life. The better parts are not required. You are not paid to think. It is not work for the whole person.

Alienation (in this sense of the word) has little to do with the ownership of the production process (as the classical Marxist argues). It is as much a characteristic of the state-owned bureaucracy as it is of the autocratically run family firm—in many cases, more so. Given the working conditions which prevail, it is inevitable that the relationship between labour and management should become one of the opposing factions engaged in a process of mutual exploitation. Bargaining is thus carried out in an atmosphere of recrimination and mistrust—and both sides maintain the status quo by coertion. Thus is the class struggle of nineteenth century Marxism perpetuated in the consumer age.

So far we seem to have learned something about what constitutes a bad job, but little about what constitutes a good job.

The antithesis to Ford's bleak and uncompromisingly pessimistic view of human nature is to be found in the ideas of the humanistic psychologists. Abraham Maslow (1954) proposed that human beings are motivated by a hierarchy of needs. In general, low-level needs must be satisfied (at least in some measure) before higher-level needs become pressing matters for concern. At the lowest level in the hierarchy are physiological drives like hunger and the need for sleep. At successively higher levels are needs for safety (security, stability, order, etc.); needs for love and social affiliation; and needs concerned with self-esteem (competence, achievement, recognition, etc.). Somewhere near the top of the hierarchy are needs concerned with aesthetic and intellectual experience. And at the very summit is the need for self-actualization—to do or to become that in which one's full individual potential is realized.

You could say that the lower levels in Maslow's hierarchy are concerned with the avoidance of suffering and the provision of *comfort*, and the upper levels are concerned with *personal growth*. On the basis of data gathered in semi-structured interviews with engineers and accountants in Pittsburgh, USA, Herzberg and co-workers (Herzberg *et al.*, 1959; Herzberg, 1968) drew a parallel distinction between those things which people associated with bad feeling about their work and those they associated with good feelings. The former include working conditions, relationships and the way the organization was run; the latter include achievement, recognition, the acceptance of responsibility and the content of the job itself. Interestingly enough, pay does not seem to fall exclusively into either category—Herzberg's subjects mentioned pay as a source of good feelings and bad feelings almost equally often.

Not all subsequent studies have pointed to quite so clear a distinction—possibly because not everybody shares the needs and aspirations of a Pittsburgh accountant. It may also be that Herzberg's engineers and accountants generally had reasonably meaningful jobs. If you were to ask an assembly line worker what he dislikes about his job, he might mention the noise and the dirt—but he would be much more likely to say that it was repetitive and boring. These limitations aside, studies like those of Herzberg point us to some important conclusions.

Good pay, a safe and comfortable working environment and good human relations in the workplace do not in themselves lead to a positive feeling of satisfaction with the quality of working life—although their absence is likely to be a source of dissatisfaction. They may be necessary but they are not sufficient. (Some dirty, dangerous and ill-paid trades may even have psychological rewards which, in subjective terms, outweigh their disadvantages—mining is a case in point.) Human beings are not motivated solely by the stick and the carrot—given the opportunity, they will seek out situations which are conducive to personal growth. For work to be psychologically rewarding, it must be seen to have meaning.

This is a much more optimistic view of human nature than that of Frederick Taylor or Henry Ford. What are its implications for job design? Table 6.9 is a

*Table 6.9 Eklund's Checklist: Characteristics of a Good Job*

| |
|---|
| Variation; a job which consists of different subtasks |
| Overview of the entirety of the production process |
| Freedom to move around physically |
| Long cycle time |
| Self-paced work |
| Influence on the choice of working methods and their order |
| Influence on production quantity and quality |
| Planning and problem solving |
| Control and adjustment of the results |
| Few temporal deadlines (time pressure) |
| Few temporal constraints (time binding) |
| Continuous development of skills |
| Freedom of action |
| Responsibility and authority |
| Participation |
| Work demands in parity with ability |
| Positive work management climate |
| Group organization |
| Social support and interaction with colleagues |

From Eklund (1988)—reproduced with permission.

checklist of "the characteristics of a good job", compiled by Jorgen Eklund, which he has kindly permitted me to quote in its entirety. The checklist was originally intended as a tool to be used in the redesign of blue collar jobs in the manufacturing industry, and in particular for the evaluation of alternatives to the traditional type of production line. But most of the items on the checklist have a more general applicability. (Try applying these criteria to your own job—or perhaps that of your secretary.) This approach to job design is diametrically opposed to the principles of Taylor and Ford in almost all respects.

The basic prerequisite for an improvement in the quality of working life, either for the blue collar industrial worker or for his clerical counterpart, is an increase in job content. In practice, this is generally achieved either by *job rotation* or by some form of *job enlargement*.

*Job rotation* is the less radical approach. The jobs themselves remain unchanged, but the worker gets more variety by moving from one job to another. This is easy and cheap to put into practice because it does not involve any significant changes in production methods; but from the worker's point of view the benefits will depend upon how different from each other the separate jobs involved actually are. In the worst case, he may simply move from one repetitive, short-cycle, machine-paced task to another, none of which engage his mental faculties to any appreciable extent—and since the various jobs on an assembly line may well impose very similar patterns of musculoskeletal loading, the physiological benefits may also be minimal.

*Job enlargement* may be "horizontal" or "vertical". In the former case, the job content is increased by giving the worker a greater number of tasks to perform, all of which involve similar levels of responsibility. In its simplest form (sometimes known as *job extension*) the worker may perform more operations on one product before passing it down the line—thus increasing the cycle time and getting more variety of working posture (perhaps). Vertical job enlargement is also known as *job enrichment*. It involves the line worker (or his equivalent) taking over some of the responsibilities which were previously allocated to his foreman or supervisor or to a quality control inspector, such as testing or checking the product, etc. Job enrichment programmes may be combined with attempts to increase the line worker's participation in decisions which are relevant to his activity, or the setting up of autonomous self-regulating working groups who were allocated collective responsibility for a sector of the production process which they may organize as they see fit.

A large number of studies of the results of such experiments have been reported in the literature. Industrial organizations are complex cultural entities, and experimental changes in their working methods do not always lend themselves to rigorous analysis. Summarizing the available research evidence, Warr and Wall (1973) conclude that in the vast majority of cases "the introduction of greater variety and discretion into jobs is welcomed by employees" and furthermore that "jobs which offer variety and require the individual to exercise discretion over his work activities lead to enhanced well-being and mental health".

Robotization and other forms of advanced manufacturing technology are obviously going to have a considerable impact on the nature of industrial work. Will robots take over all the boring repetitive dehumanizing tasks—freeing the human worker for better things? If human beings continue to be employed (as they generally have in the past) merely to plug those gaps in the production process that cannot yet be automated, then the jobs of tomorrow will be no more meaningful than those of today. Working at the pace of a robot is no different from working at the pace of any other machine.

Perhaps, instead of asking what people want from their working lives, we should ask what they can reasonably expect. This has the effect of removing the question from the domain of empirical science to that of ethics and politics—which is where, in many respects, the debate properly belongs. The quality of working life debate is more about values than about facts. In the free market economies of the Western world the purpose of a company is to make a profit. In the command economies of the Eastern bloc it is to make those things which the state orders to be made (although this is changing at the time of writing). In neither case do we have any special reason to expect that the condition of mind of the employee will be high on the agenda of concerns, except insofar as it affects his working efficiency. Perhaps it does—at least sometimes. The human relations school of management was seemingly based on the belief that if you talk nicely to somebody he might not notice that he is being exploited.

Production line jobs commonly have a high labour turnover and a high sickness absence rate. When labour is scarce, this poses a problem for the company; when labour is plentiful, it does not. In general, therefore, company interest in the quality of working life waxes and wanes with the state of the labour market (Eklund, 1988). We have no cause to be surprised by this—nor, indeed, to be particularly shocked.

The people who work for the company can, however, reasonably expect that they will not be damaged in the process, and the company has a moral obligation to avoid damaging its workers. The sorts of physiological damage that are the principal concern of this book are easy enough to define. But it does not stop there. People with good jobs live longer. Palmore (1969) found that job satisfaction was actually a better predictor of longevity than things like smoking habits or certain indices of cardiac function. Does this simply reflect the well-known social class difference in mortality (Townsend *et al.*, 1988)? Or is there a more direct connection? What of the more subtle kinds of damage that the dehumanization of work may inflict—the overwhelming sense of futility which works like a slow poison in the soul?

## Chapter Seven

# Stress, Fatigue and the Working Environment

"If you can keep your head when all about you
Are losing theirs and blaming it on you"

"If," from *Rewards and Fairies*, by Rudyard Kipling

". . . it takes all the running you can do to keep in one place. If you want to get somewhere else, you must run at least twice as fast as that!"

The Red Queen in *Through the Looking Glass*, by Lewis Carroll

A leading article in *Newsweek* magazine recently described stress as "the dirty little secret of the office age" (Miller, 1988). The deleterious effects of stress are both widespread and diverse, to the extent that many people would regard stress as the principal threat to human well-being in the advanced industrialized societies. A partial list of conditions which are widely considered to be stress-related is given in Table 7.1. The list includes both "symptomatic conditions" (some of which are

*Table 7.1 Stress-related Conditions*

**Headache:** tension headache, migraine
**Musculoskeletal disorders:** pain and dysfunction at many sites but particularly the back and neck
**Cardiovascular disorders:** symptomatic conditions, arrhythmias, hypertension, coronary heart disease (angina pectoris, myocardial infarction, cardiac arrest), stroke
**Gastrointestinal disorders:** dyspepsia, peptic ulcer, constipation, colitis, "irritable bowel syndrome"
**Amenorrhoea**, dysmenorrhoea, premenstrual tension, sexual dysfunction (impotence, frigidity)
**Allergies**, asthma, skin rash, "sick building syndrome"
**Suppression of immune function:** increased susceptibility to infections, cancer (?)
**Sleep disorders:** insomnia, nightmares, "non-restorative sleep"
**Chronic fatigue effects:** lethargy, tremor, dizziness, faintness, palpitations, "effort syndromes", "post-viral syndrome", "sick building syndrome"
**Disorders of mood state:** impaired concentration, emotional lability, suppressed anger, anxiety, depression, lack of self-esteem, apathy, "nervous breakdown", burn out
**Behavioural effects:** smoking, escapist drinking, drug abuse, aggression and antisocial acts, accident- and error-proneness, risk-taking, errors of judgement
**Organizational effects:** absenteeism, labour turnover, accidents and errors, labour relations problems, the SNAFU syndrome

vague) and clinical entities of a more specific nature (some of which are life-threatening). I have not attemped to differentiate between the two categories. Many of the conditions in the list have a multiple aetiology, in which stress is one contributory factor amongst many. The relative importance of stress, in any individual case, will depend upon the presence or absence of other risk factors.

Bodily states which have a major psychogenic component in their aetiology are sometimes described as *psychosomatic*. But the description is unhelpful, since the onset or progression of just about any condition may be influenced in one way or another by the individual's perceptions of his or her circumstances. And since stress has a widespread effect on the individual's psychological and physiological adaptive mechanisms and resistance to disease, any list of stress-related conditions must necessarily be incomplete.

## The Semantics of Stress

The use of the word "stress" is inconsistent, and no amount of academic debate is likely to change this. In everyday speech we can say "He's been under a lot of stress" or "He's in a state of stress". In the former case, stress is a cause; in the latter, it is an effect. The same inconsistency occurs in the scientific literature. Some researchers define stress in terms of the demands of the environment—and investigate the changes in behaviour which result from stresses like noise, lack of sleep, etc. (e.g. Poulton, 1970; Broadbent, 1971). Others regard stress as the organism's response to environmental demands (e.g. Selye, 1956). This causes some confusion. For example, to interpret the statement "a certain amount of stress is good for you" you have to know which usage is employed.

Many authorities now prefer to frame their definition of stress in transactional terms, regarding stress as a psychological condition which arises when there is a perception of imbalance between the demands placed upon an individual and his or her capacities to meet these demands (e.g. Lazarus, 1976; Cox, 1978, 1987). This has the advantage of emphasizing that it is the individual's appraisal of his situation that determines his experience and behaviour. But it has the awkward consequence of relegating "stress" to the status of an intervening variable which we cannot observe directly. In this scheme of things, stress is seen as a cognitive precursor of the *stress response*—which has subjective, behavioural and physiological components. If we see a man tearing his hair, or measure his adrenalin levels, we are observing "the stress response" rather than stress as such. A simple five-stage model of stress, which avoids this difficulty, is shown in Figure 7.1.

For many practical purposes, these distinctions are more subtle than we require; and in the discussion which follows I shall sometimes follow everyday usage in employing terms like "work stress", which refers to *both* the pressures imposed upon the individual by his or her working life *and* the individual's response to these pressures. In doing so, I hope that I shall make the issues involved clearer rather than otherwise.

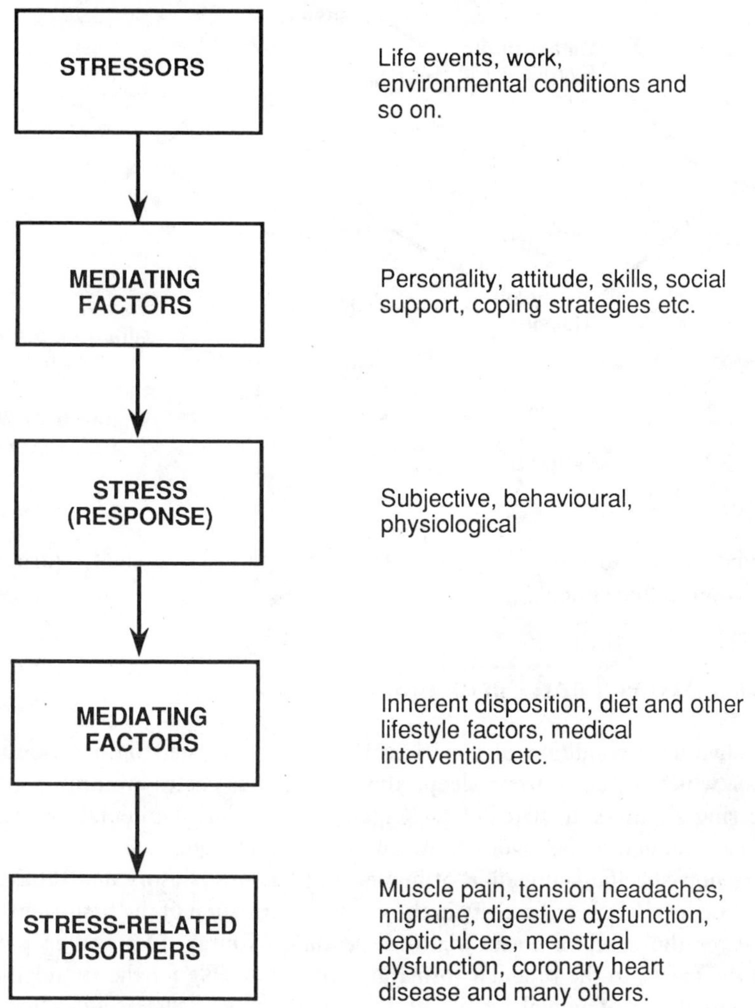


Figure 7.1 *A simple model of stress*

## A Taxonomy of Stress

The stresses of life may be divided into two categories: isolated catastrophic events and ongoing everyday hassles. The first kind knock you flat; the second kind grind you down. The acute catastrophic category includes natural and man-made disasters and major life events: changes in our circumstances which tax our powers of adaptation—bereavement, marriage, divorce, redundancy, etc. (Figure 7.2).

Work falls into the second category, along with disharmonious domestic circumstances, commuting and the various pressures of city life. Work stress may

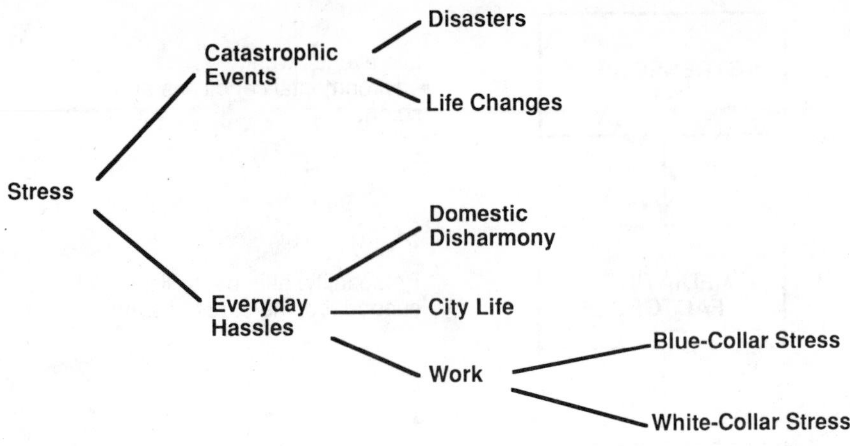


Figure 7.2 *A taxonomy of stress*

be subdivided into blue collar and white collar forms. We shall return to this distinction in due course.

## Stress, Arousal and Performance

An individual's condition of consciousness may be located on a continuum of *arousal* which extends from sleep, through various states of drowsiness and increasing alertness, to states of excitement. The underlying neural mechanisms, which are thought to be involved, are summarized in Figure 7.3.

The non-specific summation of inputs from various sensory modalities determines the level of overall activity in the reticular formation of the brainstem, which stimulates the cerebral cortex via an ascending pathway. Descending pathways from the cortex to the reticular formation are responsible for the stimulating (i.e. arousing) effects of mental activity, and for the process of *autoarousal*, whereby the well-motivated person may "will himself" to stay alert when fatigued, etc. Reciprocal connections between the reticular formation and the spinal centres which control muscle function result in an increase in muscle tension in states of high arousal. Conversely, we may "sit up straight and pay attention" or wake ourselves up by going for a walk. Sensory information from the gut seems to have an inhibitory effect on the reticular formation—thus we may feel sleepy after a meal. All of these effects are superimposed over the normal daily cycle of sleep and wakefulness (p. 166). Arousal may also be directly affected by drugs.

In general, it is variation and change in mental input that is most likely to reach consciousness. Constant or repetitive stimuli (unless they are very intense) tend to be "filtered out". (This process is called *habituation.*) Hence, you do not "notice" the sound of a ticking clock unless it stops. Thus, when the environment is bland or unvarying, we tend to drift off into drowsiness and sleep.

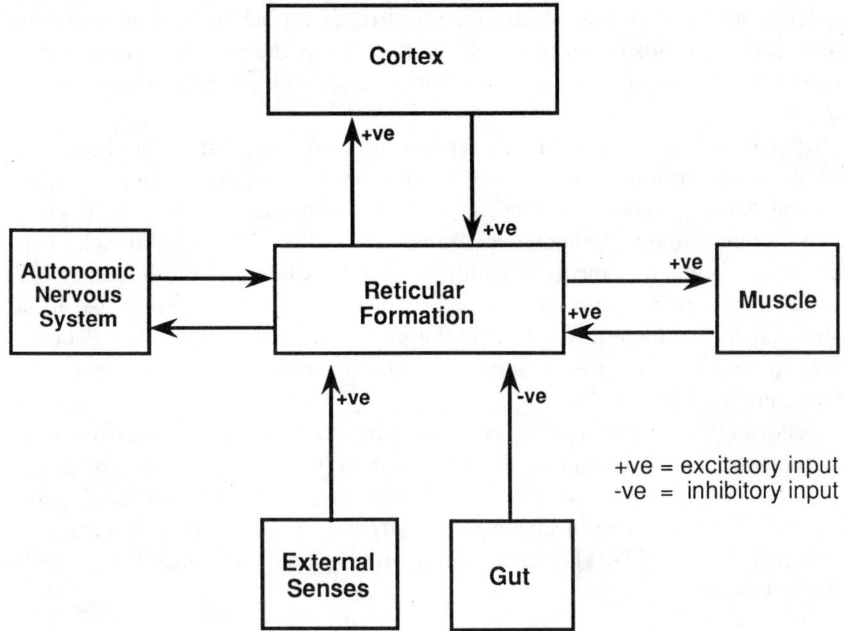


Figure 7.3 *The neurophysiology of arousal*

Neither extremely low levels nor extremely high levels of arousal are compatible with the effective performance of practical tasks. The relationship between arousal and performance is described by an inverted U-shaped curve, as shown in Figure 7.4. This relationship is sometimes known as the *Yerkes–Dodson law*, after the two psychologists who first described it (Yerkes and Dodson, 1908). At extremely low arousal levels (as when we are on the point of falling asleep or just waking up) the central nervous system is unresponsive—sensory messages may not get through, motor control is poor, attention drifts and we slip in and out of consciousness. At

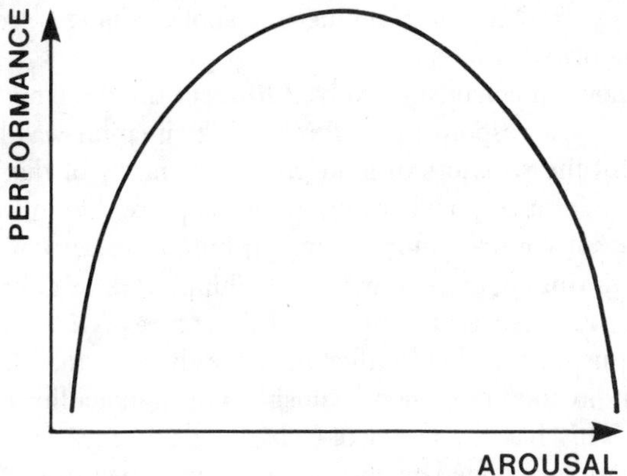


Figure 7.4 *Arousal and performance—the inverted-U relationship*

moderate levels of arousal we are alert and ready for action, and performance is optimal. Beyond this point, performance declines as the nervous system becomes over-responsive—our behaviour becomes disorganized and finally we may "freeze" altogether.

A person's level of arousal will depend, *inter alia*, upon the complexity of the task he is performing and upon the environmental condition under which he is working. Environmental stressors like noise act to increase arousal, whereas lack of sleep reduces arousal. In some experiments, the effect of noise and lack of sleep have been found to cancel each other out (Broadbent, 1971; Poulton, 1971). Incentives (i.e. rewards for success and punishments for failure) increase arousal. They may have a beneficial effect on the performance of simple tasks (where the inherent level of arousal is presumably low) but may have a detrimental effect on more complex ones.

Welford (1973) regards stress as arising when there is a departure from optimal environmental conditions or levels of stimulation (with respect to the task at hand) *which the individual is unable to correct.* Both overload and underload may be stressful. Thus the relationship between stress and arousal may be regarded as U-shaped (Levi, 1972). This conveniently predicts that performance will fall off as a linear function of stress.

## The Physiology of Stress

Stress has a profound effect on most of the organ systems of the body. The best-known features of the physiological stress response are summarized in Figure 7.5. These involve the sympathetic portion of the autonomic nervous system and the endocrine secretions of the adrenal medulla and the pituitary/adrenocortical axis. The adrenal medulla secretes the catecholamines adrenalin and noradrenalin. Acting with the sympathetic nervous system, these mediate the famous *fright, fight and flight* response of Cannon (1929), which prepares the body for physical action (increased heart rate, blood pressure and sweat rate; reduced salivation and gut motility; dilation of the pupils, etc.). The catecholamines and the hormones of the adrenal cortex initiate far-reaching metabolic changes resulting in the mobilization of energy stores.

These physiological mechanisms evolved to meet the demands of an environment in which the best response to a threatening situation was likely to involve physical action. But the stressors of modern life are rarely of this kind. Thus the body is made ready for a response which never happens. The physiological stress response is *adaptive* if you face mortal combat; but *maladaptive* if you are waiting for the telephone to ring. (Compare with the minimization of effort—p. 127.)

The adrenalin response is accentuated by *hyperventilation*—rapid or erratic shallow breathing using the chest rather than the diaphragm, which reduces the amount of carbon dioxide in the blood, causing a drop in acidity and disturbances of ionic balance. This has the effect of "charging up" the sympathetic nervous system. Excessive or prolonged hyperventilation may result in abnormalities of cardiac function, as well as a variety of neurological symptoms—disturbances of

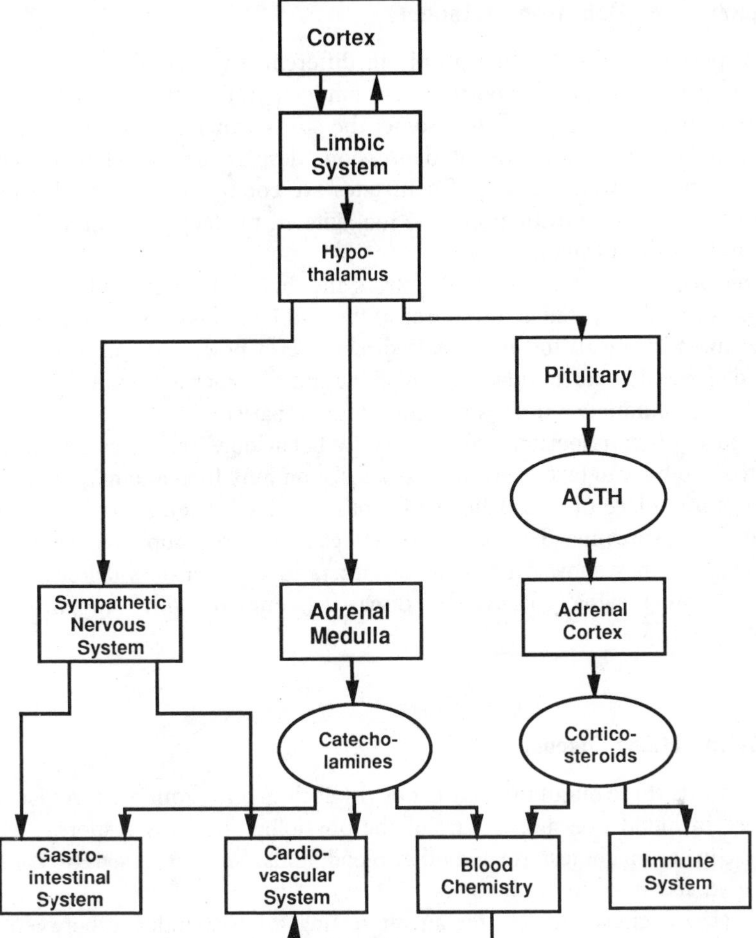


Figure 7.5 *Some features of the physiological stress response*

consciousness, dizziness, disturbances of vision, numbness and tingling sensations, and so on. As the blood becomes more alkaline, lactic acid is produced to compensate, resulting in weakness and fatigue (Perera, 1988).

Stress (mediated by the adrenocortical hormones) suppresses the function of the immune system. This increases our susceptibility to infectious diseases. Thus we are more prone to colds and 'flu, etc., when we are feeling "run down". Kasl *et al.* (1979), for example, found that West Point military cadets who suffered badly from academic pressures (particularly those who were well-motivated but unsuccessful) were more susceptible to the Epstein–Barr virus, which causes glandular fever. It is thought that immune suppression in response to stress may play a part in the onset or progression of some forms of cancer (Fox, 1978).

## Subjective and Behavioural Aspects

The experience of stress hits people in different ways. As Cox (1985) puts it: "... there is no single diagnostic stress emotion; rather there is a mixture and variety of negative feelings". For some, the stress experience is one of agitation and anxiety; for others, one of depression, despair associated with a loss of self-esteem, and finally apathy. Such states are commonly associated with sleep disturbances. The individual may be too agitated to sleep at night and may feel tired and listless during the day.

Some people become angry under stress and their inner rage spills over into acts of aggression. If directed at the source of the problem, this may be an effective way of dealing with it; all too often it is directed at a powerless substitute. Stress-related aggression surrounds us on all sides and becomes a stressor in itself, thus threatening to initiate a downward spiral into barbarism.

Stress is often associated with unhealthy behaviour. The stressed-out person may feel too busy to take exercise, may subsist on junk food and may turn to drink or drugs for solace or as a palliative for other stress-related symptoms. But such palliative measures may have knock-on effects; and, preoccupied as he is with his problems, he may show what seems to others to be a reckless disregard for the "early warning signs" of organic disease. (Compare with "cognitive tunnel vision"—p. 155.)

## Stress and Heart Disease

Stress affects three out of the four major risk factors for coronary heart disease. An increase in blood pressure is part of the physiological stress response, as is an increase in serum cholesterol and other blood lipids. Smokers' cigarette consumptions tend also to go up under stress.

Cox (1978) cites evidence for an interesting interdependence between stress and the effects of a high-fat diet. Populations who live on very restricted diets have a low prevalence of heart disease despite extremely high stress levels, whereas populations who have a diet which is high in animal fats, but lead tranquil, non-competitive lives, have a low prevalence of heart disease despite high blood cholesterol.

According to Friedman and Rosenman (1974), there is a strong connection between coronary heart disease and a constellation of behavioural traits which they refer to as the Type A personality. The Type A individual is intensely competitive, and habitually works against self-imposed deadlines. He thus deliberately places himself in stressful situations which make multiple demands on his attention, and is unwilling (or unable) to slow down. Friedman and Rosenman go so far as to say that in people who do *not* exhibit Type A behaviour, coronary heart disease is almost unknown before the age of 70 (diet, smoking and lack of exercise notwithstanding); and they categorically regard Type A behaviour as self-inflicted.

Friedman and Rosenman's conception of the Type A personality has been widely accepted; but it is not altogether borne out by subsequent research. There is mounting evidence, for example, that the increased risk of heart disease is associated less with the Type A person's self-imposed deadlines than with the feelings of hostility and repressed anger which may result when he is thwarted (Wood, 1986).

Russek and Zohman (1958) found that 91% of a sample of coronary patients (aged 25–40) reported a period of prolonged stress preceding their heart attacks, compared with 20% of controls. The stress was mainly to do with their work. Forty-six per cent of the coronary patients had been working 60 hours or more a week and 25% had been holding down two jobs. Buell and Breslow (1960) found that light industrial workers under 45 years of age who put in more than 48 hours a week are around twice as likely to have fatal heart attacks. Shiftwork is also a significant risk factor (p. 172).

In both prospective and case control studies, Karasek *et al.* (1981) found strong associations between a self-reported measure of symptomatic coronary heart disease and two particular job characteristics: high job demands and low decision latitude (i.e. a low degree of individual discretion concerning working practices). Both were significant predictors and the combination carried a particularly high risk. The same factors were also significant predictors of a composite measure of the psychological response to stress—based on reports of sleep problems, physical symptoms and depressed mood (Karasek, 1979). To what extent, therefore, is it reasonable to regard the deleterious consequences of "Type A behaviour" as self-determined?

## Stress and the Musculoskeletal System

Musculoskeletal pain and dysfunction commonly has a significant psychogenic component in its aetiology. Let us summarize the physiological mechanisms which might mediate this connection.

(i) Psychological stress is accompanied by a *generalized muscle tension*—which could, for example, trigger pain in previously sensitized structures (p. 53).

(ii) It is possible also that certain psychological states may be associated with characteristic patterns of *localized muscle tension.* The sternomastoid seems somehow to be involved in the expression of effort (you see this in dancers, runners, etc.). When startled, we hunch the shoulders and poke out the chin (sternomastoid and trapezius). Responses of this kind are part of the body language of stress. We stand differently when depressed, elated, etc. William James (1890) recognized this; and every actor knows it. Neck pain is alleged to be associated with repressed anger; low back pain with anxiety.

(iii) People concentrating on a demanding task may fidget less—thus allowing areas of abnormal tension to be maintained (p. 213).

(iv) The mechanisms of local muscle fatigue have a strong neuropsychological component (p. 47).

(v) People under stress may prolong work beyond the normal limit, take shorter rest pauses, etc.—allowing chronic fatigue states to develop (p. 163).

(vi) People under stress are accident-prone (pp. 155, 179). They may blunder around, move awkwardly, overexert themselves.

(vii) Sleep disturbance is known to play a part in the onset of some forms of muscle pain (p. 54).

(viii) Hyperventilation increases blood lactate (p. 144); conversely, deep relaxation decreases lactate below the ordinary resting level (Benson, 1975).

## Work Stress

Why is work stressful? This question has no simple answer. The stressful aspects of working life are too diverse and too idiosyncratic. The overworked and emotionally drained nurse, the bloody-minded production line worker and the burnt-out yuppie are equally suffering from the effects of stress at work.

Some important and commonly recognized sources of stress at work are summarized in Table 7.2. But in some ways a catalogue of this kind misses the point. The concept of work stress embraces all those negative aspects of working life which we experience as sources of annoyance and frustration. In the last chapter we looked at some of the characteristics of a good job (Table 6.9). A bad job—as defined by the absence of these positive features—is necessarily a stressful job.

A distinction may be drawn between *blue collar stress* and *white collar stress.* These have features in common; but in practical terms, the differences between them are more important than the similarities. Blue collar stress results from monotony, alienation and the feeling of being an extension of the machine. White collar stress results from having too much to do and too little time to do it in. New technology has brought blue collar stress into the office environment. Both varieties of stress become more pressing to the extent that the individual lacks autonomy and a sense of purpose in the tasks he is called upon to perform. Boredom and monotony are not altogether the same thing. A quite complex task can be boring if it lacks

*Table 7.2 Sources of Work Stress*

- *Task-related factors:* a task which is beyond the individual's capacity (or for which he has not been properly trained), information overload (too much to do in too little time or multiple demands on attention), information underload (monotony and boredom), machine pacing, deficiencies of equipment design (e.g. low turn-round times and delays caused by the breakdown of computer systems), anything which stops you from getting on with the subjectively important parts of your work, etc.
- *Interpersonal factors:* physical overcrowding, responsibility for the welfare of others, conflicts with superiors and subordinates, abuse and harassment, role ambiguity (not knowing what is expected of you), role conflict (being called upon to meet conflicting requirements by different people), the general sense of being "just a cog in the wheel"—that is, of not being valued or respected as an individual in an impersonal organization, etc.
- *Environmental factors:* noise, heat, lighting, dirt, squalor and general nastiness, etc.
- *Personal threat:* to physical safety, economic security or self-esteem.

meaning and relevance for the individual concerned. When work has no meaning, it is mere drudgery—an endless torment.

## Noise

Noise is unwanted sound. Typical noise levels encountered in a variety of situations are summarized in Figure 7.6. Noise is considered to be a potential threat to the hearing at levels in excess of 85 dB or 90 dB (for a review of criteria, see Pheasant 1987; for guidance concerning current UK regulations, see HSE, 1990). But noise may be a source of stress or may interfere with communication (and thus be a safety hazard) at much lower levels. The individual's subjective response to noise is determined more by its nature and context than its intensity. Thus intermittent noise and noise with information content (e.g. speech) are much more annoying than continuous unstructured noise of a similar intensity (e.g. machine noise), which tends to be habituated; and all forms of noise are more stressful when you are trying to concentrate on something else. Noise in certain high frequencies tends to be more stressful than low-frequency noise of equivalent

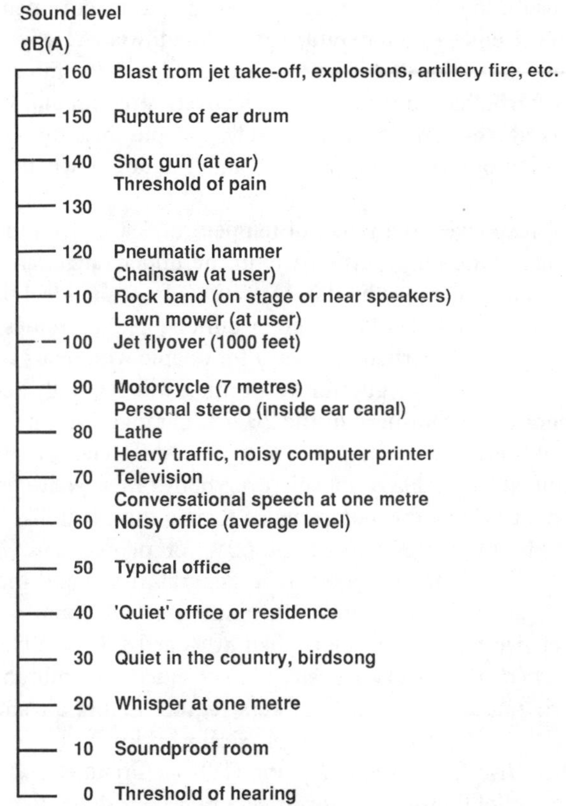


Figure 7.6 *Some typical noise levels. Reproduced from Pheasant (1987)*

intensity. (This is probably the case even when you allow for the varying sensitivity of the ear.) Noise which is construed as evidence for someone else's inconsiderateness is the most stressful of all. Music is noise when you don't want it, and if you don't like it. For a further discussion of noise see Pheasant (1987) and Grandjean (1988).

## Temperature and Humidity

Extremes of temperature may be physically hazardous. High temperatures (particularly if combined with a heavy physical workload, high humidity or low air speed) may lead to dehydration, exhaustion and collapse, abnormalities of cardiac function and a variety of other problems. These are discussed at length by Leithead and Lind (1964). Methods of environmental measurements and exposure criteria for hot working conditions are the subject of a series of international standards (ISO 7726; ISO 7243; ISO 7933) which are summarized in Pheasant (1987).

Sub-zero temperatures (particularly if combined with high wind speeds) can cause frostbite in exposed flesh (surprisingly common in agricultural workers), and prolonged exposure to cold may cause hypothermia in inactive people.

Between the extremes of temperature and humidity is a relatively narrow range of environmental conditions which are considered to be comfortable. Relatively minor deviations from comfort may be subjectively stressful and lead to impaired performance (and thus safety); conversely, people become less tolerant of sub-optimal conditions (particularly heat and humidity) when they are under stress.

Some typical responses to a range of temperatures are given in Table 7.3 (but these are a rough guide only, particularly in the upper range, and should not be used for determining tolerance limits). Deviations from the ideal temperature (for any particular task) are tolerated less well by unfit people. Extremes of temperature (in either direction) are a particular hazard for people with heart conditions.

Sedentary people wearing light indoor clothing will generally be most comfortable at a temperature somewhere in the 20–24 °C range, with a relative humidity of 40–50%. But there are considerable individual differences in this respect, and no single temperature is likely to suit everybody. In a well-controlled indoor environment, the best possible compromise temperature is unlikely to be considered exactly right by more than about 60% of people, and in practice the proportion will often be as low as 40%. In general, drafts and thermal gradients (e.g. between head and feet) should be minimized. A room will start to feel "stuffy" at a relative humidity of more than 80% at 18 °C or 60% at 24 °C. Low relative humidity (<30%) leads to desiccation of mucous membranes, etc. (p. 160) and electrostatic nuisance. Manual workers require higher air speeds and lower temperatures (19–15 °C, depending on physical workload).

For a further discussion see Pheasant (1987), Grandjean (1988), etc. The definitive treatment of thermal comfort is generally considered to be that of Fanger (1973), whose research forms the basis of an International Standard (ISO 7730);

*Table 7.3 Some Typical Responses to Temperature*

| Temperature °F | °C | |
|---|---|---|
| 110 | 43 | Just tolerable for brief periods |
| 90 | 32 | Upper limit of reasonable tolerance |
| 80 | 26 | Extremely fatiguing to work in. Performance deteriorates badly and people complain a lot |
| 78 | 25 | Optimal for bathing, showering. Sleep is disturbed |
| 75 | 24 | People feel warm, lethargic and sleepy. Optimal for unclothed people |
| 72 | 22 | Most comfortable year-round indoor temperature for sedentary people |
| 70 | 21 | Optimum for performance of mental work |
| 64 | 18 | Physically inactive people begin to shiver. Active people are comfortable |
| 60 | 16 | Manual dexterity impaired (stiffness and numbness of fingers) |
| 50 | 10 | Lower limit of reasonable tolerance |
| 32 | 0 | Risk of frost-bite to exposed flesh |

Note: Figures quoted are on the "new effective temperature" scale: that is, a thermal environment which is subjectively equivalent to the air temperature as given, with 50% of relative humidity, low air flow and minimal radiant heat.
Reproduced from Pheasant (1987).

but some of his conclusions have been subsequently challenged (Auliciems, 1989).

## Stress Management

Stress management has become an industry in its own right. Some possible approaches to stress management are summarized in Table 7.4. The majority of these are too well known to require further discussion here (see, for example, Cox, 1978; Quick and Quick, 1984; Cooper *et al.*, 1988). The techniques on the list may conveniently be divided into those which are based on fitting the job to the person (FJP) and those which are based on fitting the person to the job (FPJ). The latter are in the majority. By implication, most of these techniques seem to be more relevant to white collar stress than to blue collar stress.

Many of the most widely taught techniques for stress management are aimed at initiating a process of *attitude change* and *behavioural modification.* Some of the examples in the literature seem a bit far-fetched: for example, Friedman and Rosenman (1974) suggest that Type A people can learn to overcome "the hurry sickness" by doing things like reading Proust (because he takes so long to get to the point) and deliberately choosing restaurants where the service is slow and inefficient. To a Type A person (like me) this would be a subtle form of torture.

The most striking feature of the art and science of stress management—as it seems to be conceived at present—is its overwhelming emphasis on fitting the person to the job rather than the converse. Much of stress management is a spin-off from the commercial self-improvement culture. Few of the commonly

*Table 7.4 Approaches to Stress Management*

**STRESS AVOIDANCE**
- Work design/ergonomics, environmental control, organizational change

*Comments: FJP approaches involving corporate or collective action*

- Time management: prioritization, planning, delegation, effective decision making, avoidance of time wasting activities (travel, inefficiently run meetings, etc.)
- Assertiveness training, negotiating for change, quitting your job

*Comments: FPJ approaches leading to FJP on a self-help basis*

**COGNITIVE REAPPRAISAL**
- Psychotherapy and counselling: searching for the underlying causes of autodestructive behaviour; seeing the possibility for change; reassessing life priorities; taking control
- Positive thinking
- Religion and its analogues

*Comments: Approaches which stress the need for personal change. These may be regarded as forms of FPJ—but in some cases lead to FJP changes*

**RELAXATION**
- Relaxation techniques, meditation, yoga, visualization, autogenics, biofeedback, massage
- Sport and leisure

*Comments: Aimed at reversing the downward slide into exhaustion (equivalent to FPJ)*

**SUPPORT**
- Homelife, personal relationships, formal and informal social groups
- Self-help groups, victim support groups, workaholics anonymous
- Practical help for people under stress (e.g. crèches for working mothers)

*Comments: Act by reducing the overall psychological burden of the stressed individual (equivalent to FPJ)*

**LIFESTYLE**
- Giving up smoking
- Sensible drinking
- Diet
- Exercise

*Comments: Act by increasing the individual's adaptive capacity and fitness for work; and by eliminating contributory risk factors in the aetiology of stress-related disorders (equivalent to FPJ)*

taught techniques are aimed at eliminating the problem at source—for example, by getting to the social and organizational roots of "Type A" behaviour. Perhaps this is because the more radical approach—which we could regard as that of ergonomics—is likely to involve corporate or collective action, and stress management is characteristically seen in self-help terms.

Stress is a management issue. Much administrative work is simply unnecessary. High stress levels are often caused by the unstructured and inefficient use of working time: lack of prioritization and forward planning, and so on. Business trips, luncheon appointments and badly run meetings may not produce results commensurate with the time required and debilitating effects which ensue. (Think of the meetings which you have attended over the past month.) Vacillation and errors of judgement by superiors—themselves the result of high stress levels—may impose needlessly tight deadlines on subordinates.

Stress is built into the authoritarian and hierarchical organizational structures of corporate working life. People in subordinate positions find their stress levels increased by the capricious decisions of their superiors, which they are not in a position to question (or so it seems). Or they may work 60 hour weeks in order to conform to the prevailing social norms—on the grounds that "if I don't make the breakfast meeting, somebody else will". Their superiors take the view that "if you don't like the heat you should stay out of the kitchen". At each level in the hierarchy, the individuals concerned are willing to sacrifice their subordinates in order to meet goals set by their superiors (with whom they seek to identify). The person in authority may deliberately foster a climate of insecurity amongst his subordinates (with regard to promotional prospects, etc.) in order to keep them "on their toes". (When did your boss last compliment you on a job well done?) Thus people dissipate a disproportionate amount of effort in playing politics or in being "competitive". This is as much a feature of publicly funded institutions like universities as it is of the private sector. As the late Professor Alec Rodger was fond of saying, "there are no daggers longer than academic daggers". From the standpoint of the working organization, effort devoted to playing politics is largely wasted—except insofar as actions undertaken by the individual, in pursuit of his selfish ends, chance to have a corporate pay-off.

As the stress levels mount, it all seems increasingly futile—until you reach burn-out. For the individual who is caught up in this process, relaxation techniques or a reappraisal of his circumstances may be a life saver. But high stress levels will remain endemic in any working organization which is unwilling to initiate an overhaul of the management practices which lead to this futile waste of human potential.

## Mental Workload

Human beings have a finite capacity for processing information. The cognitive processes involved are summarized in Figure 7.7. Psychologists differ in the way they look at these matters—so we should not treat the details of such a model too seriously. When the capacity of the system is exceeded, performance falls increasingly short of the ideal, as shown in Figure 7.8(A). In situations of *overload*, where demands exceed capacity, the person may adopt one of a number of strategies for *off-loading*.

In many everyday contexts it is possible to trade-off speed against accuracy (Figure 7.8B). Typists do this, as do musicians if they are allowed to play a piece at their own speed. The point of trade-off which an individual chooses in any particular case will depend upon his appraisal of the rewards and penalties involved. Choosing accuracy over speed will cause a backlog of unprocessed information to develop. This acts as a *buffer* store which permits irregularly spaced input to be smoothed out and a steady workload to be created. The buffer may be external (the papers on the desk, the patients in the waiting room) or internal (the

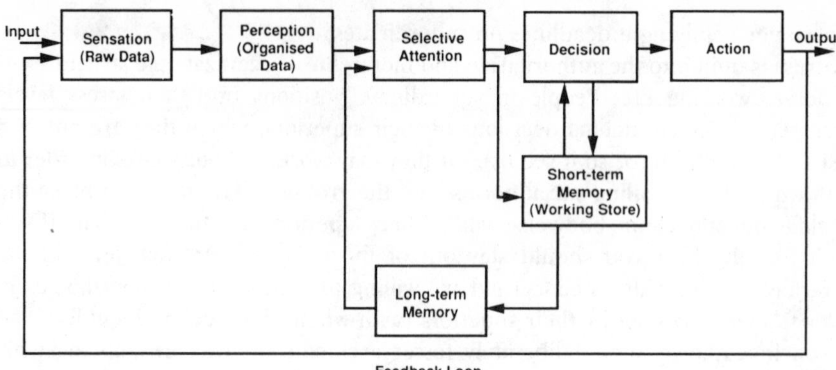


Figure 7.7 *The human being as an information processing system*

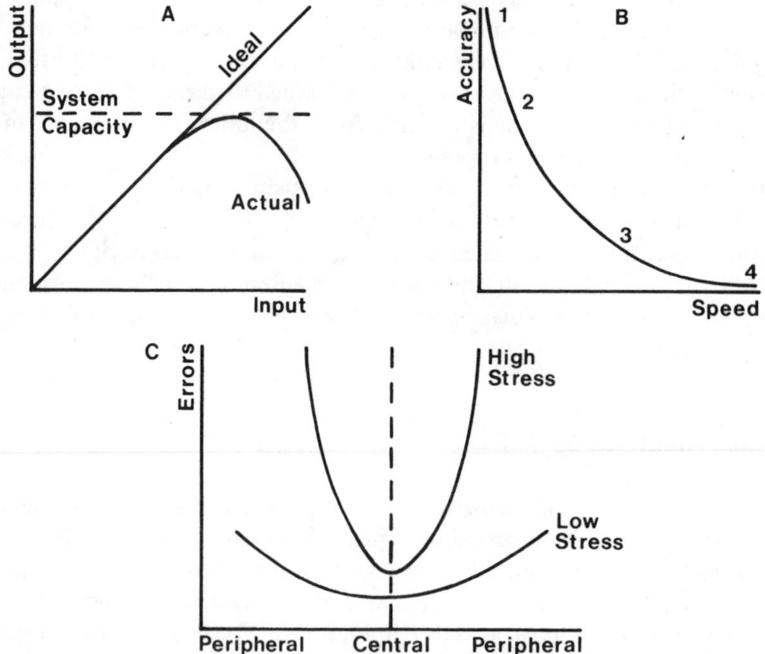


Figure 7.8 *Performance under stress: (A) information overload; (B) the speed/error trade-off. 1, limit of human reliability; 2, quick and dirty; 3, slow but sure; 4, random responses. (C) the cognitive tunnel vision effect*

things you "have in mind"). If the task is not truly self-paced, critical items may wait in the buffer beyond the time when they need to be dealt with. This may result in errors.

In some tasks, an internal buffer may be set up by the cognitive strategy which psychologists call *preview*. Eye fixation studies reveal that skilled touch typists read about one second ahead of the letter they are typing (Shaffer, 1973). Musicians who are good at sight reading do something similar. The gap narrows as the passage becomes more complex—and when it closes completely, an error occurs.

In other tasks the short-term memory buffer is filled with information that has already been presented. Hence, when taking notes in a lecture, you start by writing down one fact—but as you are doing so, the lecturer presents another. So you place the second in the buffer whilst dealing with the first. But since you do not have the mental capacity to spare for continually "rehearsing" it (although you may well attempt to do so), it commonly fades before you write it down.

In complex situations which make multiple demands on our attention, we tend to concentrate on certain high-priority sources of information and to neglect the remainder (Figure 7.8C). Experiments have shown, for example, that under high workload conditions, signals presented in the peripheral parts of the visual field tend to be missed, whereas the performance of centrally located visual tasks remains relatively unimpaired (Poulton, 1970). This convergence of attention is sometimes called *cognitive tunnel vision.* It was first recognized in the classic "Cambridge Cockpit" experiments on simulated flying, which were conducted in the late 1940s (see Welford, 1968, for an account of these). Towards the end of a demanding two-hour "flight", pilots became increasingly likely to miss peripheral signals like the fuel indicator.

The effect is by no means limited to visual tasks: Welford (1973) observed that it "has obvious echoes in everyday life where, under stress, side-issues and niceties tend to be ignored". (Ignoring the "niceties" tends to stress other people—who then stress you in return, and so on.)

Under stress, we tend to take refuge in the familiar and predictable—that is, we revert to *stereotyped patterns of behaviour*. This represents an attempt to off-load by simplifying an environment which is too complex. (A similar process of simplification accounts for the attraction of authoritarian leaders, gurus, cults, etc.—it's all much easier than thinking.) Some psychologists like to think of well-practised patterns of behaviour as being stored in the long-term memory as templates or *schemata.* These tend to surface when the person is abstracted or when his

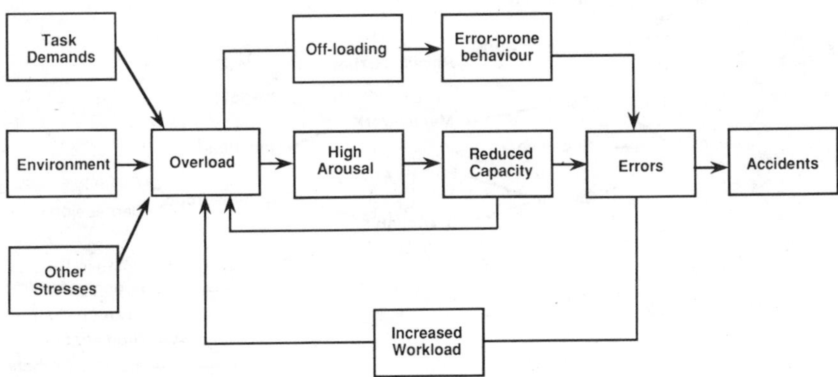


Figure 7.9 *Error-prone behaviour*

attention is diverted—for example, under stress. The execution of a pre-programmed routine may not be appropriate to the demands of a novel situation. Accidents may result from the failure to break into habitual patterns of behaviour when the situation demands it.

There are a number of ways, therefore, in which *off-loading* may result in behaviour which is *error-prone* in one respect or another. The errors which ensue may further increase the workload, thus initiating a self-perpetuating cycle of increasing stress, as shown in Figure 7.9. In Chapter 9 we shall consider some of the consequences of this process.

## Fatigue

The word "fatigue" is applied to a wide diversity of conditions. A provisional classification is given in Figure 7.10. The categories are by no means exclusive and may co-exist in various combinations. The underlying common feature of these conditions is a dimunition in the capacity and/or inclination for work.

General fatigue is the more difficult concept. The condition merges into a number of others which are equally difficult to define precisely—stress, boredom, depression, etc. The taxonomy of these conditions is in its infancy. Like stress, fatigue has subjective, behavioural and physiological aspects. The effects of local and general fatigue may be combined—for example, an individual's overall level of arousal would doubtless affect his performance on a test of muscular endurance. The after-effects of work are superimposed over the daily cycle of sleep and wakefulness—the downswings of which are sometimes called *nychthemeral* or *circadian* fatigue.

The generalized state of bodily exhaustion which results from prolonged heavy work (e.g. running a marathon) is the consequence of a depletion of the body's energy reserves. Relatively few everyday tasks are sufficiently physically taxing to

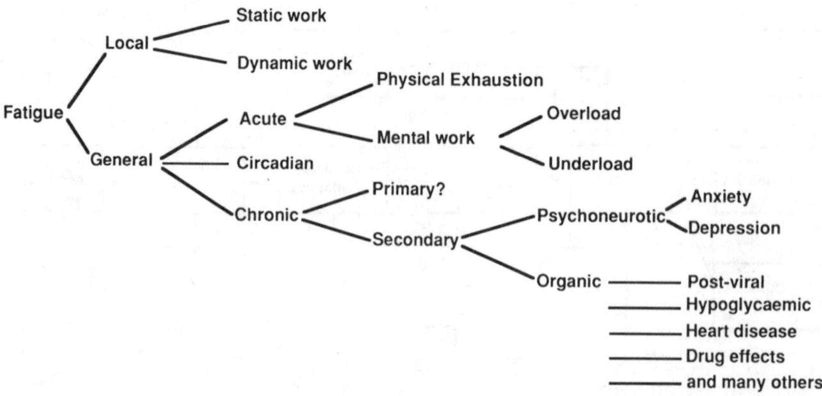


Figure 7.10 *A taxonomy of fatigue*

result in this kind of fatigue. The fatigue which results from most occupational tasks (or equally from a day at home with the children) is due to a more subtle psychophysiological process, although the subjective experiences seem similar.

Mental fatigue may result from both underload and overload; from a task which makes excessive demands or from one which is dull and uninteresting. Are there two separate varieties of fatigue—one associated with overarousal and the other with underarousal? The physiological evidence suggests that there are not. Grandjean (1988) cites a number of studies using the critical flicker fusion frequency (CFFF) test—a measurement of the minimum frequency at which a flickering light is perceived as continuous. A low CFFF is generally held to indicate low arousal. Grandjean concluded that a marked reduction in CFFF may result from the continuous performance both of demanding tasks (mental arithmetic, telephonist, air-traffic control, etc.) and of boring monotonous ones (long distance driving, inspecting bottles, etc.), whereas tasks involving intermediate levels of mental stimulation and comparative freedom of action (e.g. office work) show little or no such effect. In both high- and low-workload conditions the lowering of CFFF commenced after an hour or more of continuous work and was associated with subjective reports of weariness, etc. Similarly, Frankenhauser *et al.* (1971) found that extremely demanding and extremely monotonous mental tasks both resulted in marked increases of adrenalin secretion, whereas a task of intermediate complexity (reading a newspaper) did not.

The inference to be drawn is that overload and underload both increase stress—but if prolonged, they both reduce arousal. Grandjean's observation that a demanding task, which ought to increase arousal, in fact reduces it, is something of a paradox; and it stands as an awkward question mark over the inverted-U model.

But the performance decrements which occur in states of fatigue are generally task-specific in the way that the inverted-U model would predict. Military and naval watchkeeping tasks and industrial inspection tasks require an observer to detect faint, infrequent and unpredictable signals. These are called *vigilance tasks.* During World War II, radar operators reported 50% of all U-boat sightings during their first half-hour on watch; this figure fell to 23%, 16% and 10% in the successive half-hour periods which followed (Grandjean, 1988). This *vigilance decrement* is presumably due to the progressive fall in arousal. It generally becomes apparent after about half an hour on the task. Factors which increase arousal levels help to maintain performance levels (Welford, 1968).

There is evidence that the convergence of attention of the cognitive tunnel vision effect, which occurs as a consequence of task overload, results specifically from increased arousal, since it is increased by noise but not by lack of sleep (Poulton, 1971). Indeed, it may well be that under conditions of extreme sleep deprivation, the reverse occurs. Anecdotal evidence suggests that there is a divergence of attention, with undue regard being given to the periphery. This results in a break-up of the peripheral field and a sense of disorientation. The individual finds himself unable to focus attention on anything in particular.

In repetitive self-paced tasks (such as are commonly found in industry), the first sign of fatigue is typically an increase in the average length of time it takes to complete a cycle of activity. Careful scrutiny of the distribution of cycle times often reveals that this is due not so much to an overall slowing of performance as to the appearance within the distribution of a greater proportion of abnormally slow cycles (Murrell, 1969). This irregularity of timing is believed to be due to the appearance of short gaps in performance associated with lapses of attention called *blocks* (Bills, 1931). You could think of these as the central or cognitive equivalent of blinks (which also increase with fatigue).

Mental fatigue is characterized by a progressive failure of the normal processes of selective attention whereby most of the irrelevant sensory input which constantly bombards us is filtered out before reaching conscious awareness. The fatigued person may thus show various lapses in behaviour. Things like noise, heat and bodily discomfort become more intrusive. Welford (1968) speaks of the "disorganization of performance". The hierarchical structure by which a skilled activity is built up from its component parts has broken down. Thus the fatigued person may perform the right actions in the wrong order—put his tea bag in the toaster and so on.

## Blood Sugar

An adequate supply of blood sugar (glucose) is essential to cerebral function. (Unlike the muscles and most other tissues, the brain is unable to metabolize any other substance.) This underlies the subjective component of advanced states of physical fatigue. As the blood sugar level falls (despite the mobilizing effects of the adrenal hormones), the brain endeavours to shut down muscle activity so as to conserve the remaining sugar supplies for the maintenance of cerebral function. Thus work is perceived as more arduous and it takes a greater effort of will to continue. If blood sugar falls below a certain critical level, symptoms of impaired cerebral function ensue—disorientation, lack of co-ordination and a progressive waning of consciousness.

Blood sugar also has a more general role in mediating the connections between stress, fatigue and performance. Most of us would recognize the subjective symptoms of hypoglycaemia (low blood sugar)—the tired, irritable feeling that comes on when we need our lunch. Haggard and Greenberg (1935) studied the effects of blood sugar levels on shop floor workers in a tennis shoe factory. (The workers were engaged in relatively light manual tasks.) Throughout the morning period, output was substantially lower (by about 17%) in those workers who had not eaten any breakfast, but after lunch the difference evened itself out. The decline in performance, between the first and last hours of the morning work period, was less in those workers who ate both breakfast and a mid-morning snack than in those who worked through on breakfast alone. Thus "fatigue" in these workers would seem to have been very much a matter of hypoglycaemia.

Low blood sugar levels may well be implicated in the errors and unsafe behaviour which lead to accidents. Brooke *et al.* (1973a, b) observed that foundry workers who ate little breakfast had more accidents in the morning than in the afternoon and more accidents before the mid-morning break than after it.

Foundry workers who were given a high-energy drink of glucose and salts before starting work had fewer accidents than men who were given a drink containing salts alone. Cox (1978) has also noted that blood sugar levels in motorway crash victims are commonly abnormally low.

A light lunch improves working performance in the afternoon. But experience teaches us that there is an optimum weight of lunch, beyond which the performance of mental work in the early afternoon deteriorates. Hutchinson (1954) confirmed this for a typing task—but noted that the deterioration did not occur in well-motivated subjects (which underlines the importance of autoarousal effects). People who habitually have a light lunch suffer more from the effects of a heavy lunch; and people who habitually have a heavy lunch benefit more from a light lunch (Craig and Richardson, 1989). The effects of lunch are superimposed over those of the daily cycle of sleep and wakefulness (p. 166).

For maximum working efficiency (and therefore safety) it may make sense, therefore, to spread your daily calorie intake (i.e. energy requirements) over three meals plus two snacks—breakfast, second breakfast/elevenses, luncheon, afternoon tea, dinner/supper—rather than eating more calories on fewer occasions. This has the additional benefit of reducing the size of lunch and thus the somnolence which follows it.

## Chronic Fatigue

There was a time when "chronic fatigue" was a respectable medical diagnosis. The term has largely fallen into disuse—but in many respects it deserves to be revived. Feelings of abnormal tiredness, lethargy, etc., are widely recognized as secondary symptomatic manifestations of a range of systemic conditions—virus infections, heart disease, endocrine dysfunction, etc. Can we also recognize a condition of *primary chronic fatigue*?

It seems reasonable to argue that the after-effects of work could be cumulative—for example, if recovery was incomplete at night and on rest days, etc. Thus you may "wake up feeling tired". "Chronic fatigue" would seem to be as good a name as any other for the state which ensues: we might regard it as the cumulative non-specific response to prolonged stress. Grandjean (1988) uses the term in this way, and gives a list of symptoms for the condition which are essentially stress-related: headaches, giddiness, palpitations, cold sweats, loss of appetite, dyspepsia, diarrhoea, constipation, etc. Sleep disturbance is characteristically part of the picture—and many insomniacs show signs of daytime hyperarousal. Nixon (1982) believes that chronic hyperarousal associated with progressive states of exhaustion is a common precursor of heart attacks. (He describes this process by reference to a modification of the Yerkes and Dodson inverted-U curve, which he calls the "human function curve".) In Japan, sudden death (from heart attacks, strokes, etc.) due to overwork is called *karoshi.*

The lethargy and loss of drive encountered in states of chronic fatigue merge indistinguishably into the profound sense of apathy, despair and the futility of existence which characterize the group of psychoneurotic conditions which are known today as depression, and in the past were called melancholia. Depressive illness is also associated with sleep disturbance, loss of appetite, dyspepsia, constipation, etc. It tends to run in families and may be associated (in some cases at least) with metabolic abnormalities involving the neurotransmitter serotonin. It is tempting, therefore, to regard some forms of clinical depression as a progression of chronic fatigue which occurs in constitutionally vulnerable individuals.

From antiquity, both dyspepsia and melancholia have been regarded as occupational risks for academics. Hence, Ramazzini writes:

> "Almost every student who devotes himself seriously to the pursuit of learning complains of weakness of the stomach. For while the brain is digesting what is supplied by the passion for knowledge and the hunger for learning, the stomach cannot properly digest its own supply of foods. . . . As a rule, then, learned men are liable to fits of melancholia, and all the more when this temperament has been allotted to them originally. We know from observation that those who are genuinely devoted to learning are lean, wan, lead coloured and want to lead the life of hermits."

Abnormal tiredness and lethargy may occur in conjunction with musculoskeletal symptoms, as in fibromyalgia (p. 53), RSI (p. 79) and the condition known as myalgic encephalitis or post-viral fatigue syndrome—to the extent that some patients suffering from these conditions may be regarded as clinically depressed. In some cases of myalgic encephalitis there is evidence of chronic infection with an enterovirus of the Coxsackie group (Yousef *et al.*, 1988). Muscle pain sometimes starts with a virus infection, and there is evidence that people suffering from virus infections may be more susceptible to work-related musculoskeletal injuries (Hagberg, 1986). But many people think that myalgic encephalitis is predominantly psychological (and/or does not exist) and it is said to occur predominantly in overworked "high achievers"—hence the term "yuppie 'flu". So the condition may be basically due to a combination of chronic fatigue and a stress-related immune suppression.

Similar problems occur in athletes as a result of overtraining. Endurance athletes have been found to be more vulnerable to upper respiratory tract infections—possibly because of the effects of the adrenocortical hormones on the immune system (Lewicki *et al.*, 1987; Reilly and Rothwell, 1988).

## Sick Building Syndrome

The condition referred to as "sick building syndrome" occurs predominantly in people who work in tightly sealed, artificially ventilated and air-conditioned buildings—usually office workers (many of whom work with VDUs).

The symptoms of the condition are summarized in Table 7.5. Sick building syndrome *per se* is said to exist when these symptoms occur more frequently than might reasonably be expected in a particular building (Finnegan *et al.*, 1984). Up to 80% of the occupants of a building may be affected (Bardana *et al.*, 1988). Women are affected more frequently than men, and clerical and secretarial workers more frequently than managers (Burge *et al.*, 1987). Prevalences tend to be higher in open plan offices.

*Table 7.5* *Sick Building Syndrome* (Symptoms Investigated by Burge *et al.* (1987))

| | NV | AC | ALL |
|---|---|---|---|
| Lethargy, tiredness | 50 | 62 | 57 |
| Blocked nose | 40 | 52 | 47 |
| Dry throat | 36 | 51 | 46 |
| Headache | 39 | 46 | 43 |
| Itchy eyes | 22 | 31 | 28 |
| Dry eyes | 18 | 31 | 27 |
| 'Flu-like symptoms (aching muscles, etc.) | 15 | 27 | 23 |
| Runny nose | 19 | 25 | 23 |
| Tight chest/difficulty in breathing | 6 | 11 | 9 |

NV = naturally ventilated office; AC = air-conditioned offices.
Symptoms reported elsewhere include: *sinus congestion*, *non-productive cough*, *wheeze*, *nausea*, *skin irritation*, *abnormal taste*.

A number of possible causes have been proposed. Microbiological contamination of humidifiers and pipework is a favoured explanation—but no specific agents have been identified. An allergic mechanism may be involved. High temperatures and low humidity dry out the mucous membranes—which in turn may increase susceptibility to infection. Most modern office buildings are overheated in winter and the humidity is often too low. In the interests of energy conservation, up to 90% of the air in an air-conditioning system may be re-circulated. This may lead to an accumulation of a variety of mildly toxic contaminants: tobacco smoke, solvents (from photocopier machines, correcting fluid, etc.), and substances which evaporate from building materials, furniture and fittings. Air ionization may be part of the picture. The presence of electrical machines, plastics and other materials which can take up a charge (particularly carpets made of artificial fibres) may positively ionize the air in offices. Negative ionization has been shown to improve both subjective well-being and the performance of psychomotor tasks (Hawkins and Barker, 1978; Hawkins, 1981).

It seems unlikely, however, that the so-called sick building syndrome is solely due to either pathogens or air-quality. The symptoms of "sick building syndrome" overlap with those of "eyestrain" (p. 210). (In addition to being overheated, many modern offices are also over-illuminated.) The epidemiology of the condition strongly suggests that it might be in some measure stress-related. In analysing its causes, therefore, we must look at the whole of the working system.

## Working Hours

In the early years of the Industrial Revolution, a factory hand in one of England's dark Satanic mills might well have worked a 14 hour day, 6 days a week. The first calls for legislation to limit the length of the working day were due to the enlightened self-interest of a minority of humanitarian manufacturers—who found themselves at a commercial disadvantage compared with their competitors. The opponents of the *Ten Hours Movement* argued that the manufacturer made his profit in the last hour of work, and that a reduction in the working day would drive him out of business.

At present, the nominal working week in manufacturing industry is typically around 38 hours, and perhaps about 35 hours in offices. But it may be considerably extended by overtime. In the early 1980s, despite the recession in manufacturing, some 27% of manual workers in the UK were in fact working more than 45 hours per week, and 20% were working more than 48 hours. The 8 hour working day in industry is not yet a reality. The amount of overtime worked in other European countries is considerably less. Overtime is likely to remain popular with management because it increases flexibility of manning and holds down certain labour overheads (national insurance, pension contributions, etc.); and it is equally popular with labour because it increases real wages.

The early nineteenth century belief that excessively long working hours are necessary for maximum productivity periodically reasserts itself in times of national crisis. During World War I, a shortage of munitions led to working weeks of 100 hours or more in armament factories. But it rapidly became clear that attempts to increase output by lengthening either the working day or the working week were likely to be self-defeating. The longer working shifts were found to yield lower *hourly* output—and in some cases the magnitude of decrement was sufficiently great to result in an overall reduction in *daily* output. The effect was most marked in heavy manual jobs, but it was still present in lighter ones (Vernon, 1921). The situation repeated itself in 1940. Working hours in armaments factories were again increased—partly to satisfy the patriotic impulses of the workers. Production rose temporarily and then, to the alarm and amazement of the authorities, fell sharply. The fall was associated with a rise in sickness absence (Edholm, 1967).

Similar effects may be observed under normal peacetime conditions. Excessive overtime characteristically results in both a reduction in hourly output and an increase in the sickness rate (Grandjean, 1988). The inference to be drawn is that, faced with excessive working hours, people will pace themselves to last out the shift and will periodically "go sick" to recuperate from cumulative states of fatigue. In work which is machine-based, the effects on hourly output will be less, but we should expect the effects on sickness absence to be greater.

Whilst the nineteenth century mill owner worked far shorter hours than his factory hands, the reverse is probably true for management and labour today. Recent surveys have shown, for example, that senior executives in American companies work a 56 hour week on average; and that in the UK 54% of senior managers said they worked more than 10 hours a day, 71% said they took work home and 49% said they worked on the train (*Daily Telegraph*, 6 May 1987; 25 April 1988). Working hours of this length must presumably result in a performance decrement in the decision-making function—although the subject would be difficult to study empirically.

## Rest Pauses: The Actile Period

Rest schedules for heavy manual work may be determined according to physiological principles (p. 33). But the problem is less straightforward for light manual or mental tasks (in which the physiology of fatigue is not well understood). In order to overcome these difficulties, Murrell (1969, 1976) introduced the concept of the *actile period*—which may be defined as the length of time during which an individual is able to maintain uninterrupted performance of a particular task at an optimal level. Its duration varies with the individual, the task and the circumstances. When it is exceeded, a "fatigue decrement" in performance occurs, the nature of which depends on the task (see above).

In laboratory studies of simulated industrial tasks, Murrell showed that irregularities in performance typically showed up after about one hour of continuous work. The optimal time to take a break is somewhere near the end of the actile period—that is, just before the fatigue decrement occurs. If an adequate pause is taken at this point, a fresh period of optimal performance will follow. *But left to their own devices, people generally prolong their work well beyond the actile period*—presumably until subjective sensations of fatigue become intolerable. A rest pause at this late stage will be less valuable, giving only a brief restoration of performance.

Bhatia and Murrell (1969) tested this theory in a series of shop floor experiments, using a group of girls who made account book covers at a factory in Birmingham. Two schedules were tested. In the first, the morning and afternoon periods were each broken up by three pauses of 10 minutes each (at 50 minute intervals). In the second, there were two pauses of 15 minutes (at 65 minute intervals). Both schedules showed an increase in productivity (over previous working practices where rest pauses were taken at irregular intervals)—but the first showed a greater increase and was greatly preferred by the workforce.

In many universities, a ten-minute rest pause between lectures is considered to be good practice. (Perhaps this implies that university lectures and repetitive industrial work have similar psychological characteristics.) *It may be that a period of somewhere between 45 minutes and 1 hour represents a maximum time span of human attention in a fairly wide range of activities.* For vigilance tasks, however, the actile period is probably as short as half an hour; and for enjoyable and absorbing tasks it may be longer.

Creative, self-motivated people may choose to work uninterruptedly for long hours at a stretch. Computer programmers are very often a case in point. Does this imply that the actile period for mentally absorbing and psychologically fulfilling work is that much longer? The actile period might, for example, be prolonged by an auto-arousal effect. Self-imposed schedules are probably a poor indicator, however. Left to our own devices, we often prolong the working period, past the point of fatigue, in order to reach some self-imposed goal or tidy cutting-off point. Back pain patients in particular need to become aware of this tendency and to take active measures to correct it.

## Micropauses

Micropauses are rest pauses of very short duration (2–60 seconds). Their effects have not been extensively investigated but the idea seems very promising. In one study, data entry operators took micropauses of 10–15 seconds every 10 minutes. At the end of the day, they reported less than half as much fatigue in their neck and shoulder muscles as they did on an ordinary working day (Ehnström, 1981).

## Active Rest

In general it clearly makes sense to get up and walk around during a rest pause: to have a change of scene, get a breath of fresh air, chat with other people, and so on. Activities of a more formalized kind may also be beneficial. These are popular in Scandinavia, where they are called *pausgymnastik.* In some cases it may be possible to devise a special programme of exercises for stretching those muscles which have been subject to static loading during the working activity (p. 248).

Laporte (1966) notes that the concept of "active rest" dates back to research done in the Soviet Union and Germany in the 1930s, where it was applied mainly in the context of heavy manual tasks. The German research was done on chain gang workers. Recent experimental studies have confirmed that mental tasks or light physical activity during rest pauses aids recovery from muscular fatigue (p. 48). Active rest has also been found to be beneficial in mental work. Laporte (1966) studied the effects of a "gymnastic pause" in which post office workers did exercises to music. Active rest gave a greater improvement in mental performance and arousal (as measured by the critical flicker fusion test) than ordinary passive rest.

## Chapter Eight
# Shiftwork

"It is but lost labour that ye haste to rise up early, and so late take rest, and eat the bread of carefulness: for so He giveth His beloved sleep."

Psalm CXXVII (Book of Common Prayer version)

There has always been a need for some people to work whilst the rest of us sleep—shepherds, sentinels, nurses, and so on. In ancient Rome, heavy traffic was limited to the night hours, so as to minimize interference with the daytime activities of the citizens. Industrialization and the increasing complexity of urban life have led to a steady proliferation of night-time working.

Society requires some services to be provided on a 24 hour basis, and some industrial processes need to operate continuously for technical reasons. But in many cases the need for working at abnormal times is principally economic. A substantial proportion of a company's working capital may be tied up in plant and equipment. Lengthening the working day by dividing it into shifts may be seen as an advantageous way of utilizing these rapidly depreciating capital assets.

Round-the-clock working may be organized in different ways. In manufacturing and service industries, the most common is to divide the 24 hour period into three shifts of equal length. In Britain and Europe these typically run from 6 am to 2 pm (early shift), 2 pm to 8 pm (late shift) and 8 pm to 6 am (night shift)—or an hour (or occasionally two hours) earlier in each case. Workers typically rotate between shifts once per week, and less commonly a minority may be recruited onto "permanent nights", usually on a voluntary basis. In the USA, shifts typically start at 8 am, 4 pm and midnight, rotations are more likely to occur at longer intervals (e.g. a fortnight), and permanent nightwork is more common.

Some people, like office cleaners and market tradesmen, regularly do their work early in the morning. Some factories and late-opening retail stores offer a "twilight shift" (e.g. 5–9 pm) for women who work whilst their husbands look after the children. In some industries two-shift systems are encountered and the plant shuts down at night.

It has been estimated that between 15% and 30% of industrial workers in the developed countries are engaged in shift work of some kind (ILO, 1978) and that 8–15% of the economically active population work nights (Carpentier and Cazamian, 1977). Until the late 1970s at least, the percentage was growing steadily. In all probability it still is—although with changes in the distribution of employment between the manufacturing and service sectors of the economy, it may be that different sorts of people are involved.

## Circadian Rhythms

Many physiological processes show daily cyclic fluctuations. These are called *circadian rhythms* (Latin: *circa*, about; *dies*, day).

The best-known (and most widely studied because it is the easiest to measure) is the body temperature rhythm, which shows a daily fluctuation of about 0.5 °C either side of a mean value of about 37 °C (Figure 8.1). The cycle hits its low point at about 4 am. It begins to climb again at about 6 am (commonly before the person wakes) and climbs rapidly until noon and more slowly thereafter. Peak temperatures are reached anywhere between noon and late evening, but most commonly between 6 pm and 9 pm. From 10 pm onwards, temperature begins to fall rapidly. There are similar cyclic changes in cardiac, respiratory and renal functions, blood pressure, various endocrine secretions, and so on—although these may have their peaks and troughs at slightly different times (Minors and Waterhouse, 1985).

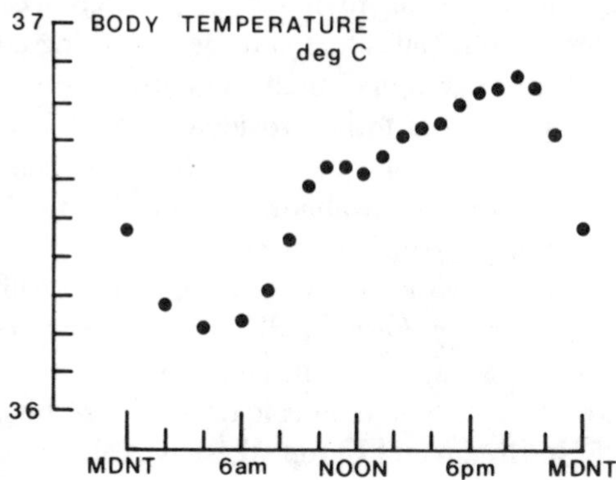


Figure 8.1 *The body temperature rhythm—after Colquhoun* et al. *(1986)*

Circadian rhythms are partly driven by internal "body clocks" (the nature of which are obscure) and partly synchronized to the external world by cues known as *zeitgebers* (German: *zeit*, time; *geber*, giver). The effects of these, on any particular physiological cycle, may, of course, be masked by activity, food intake, and so on. Curiously enough, the internal body clock, when left to its own devices, seems to run a little slow. People totally isolated from environmental time cues develop a body temperature rhythm with a period of $25 \pm 0.5$ hours (Wever, 1985).

The body temperature rhythm (and the other physiological cycles of which we may regard it as the representative) are intimately coupled to the cyclic changes in arousal which underlie our normal daily pattern of sleep and wakefulness. Thus there is a more pronounced tendency towards sleepiness at certain times of day, whether or not we have slept—but more so if we have not. We are likely to be at our sleepiest in the early hours of the morning (2–7 am) and to a lesser extent in the afternoon (2–5 pm). These are periods of vulnerability in the circadian cycle, during which we are most likely to "nod off" unintentionally (Figure 8.2) and during which drivers are most likely to be "asleep at the wheel" (Figure 8.3). At

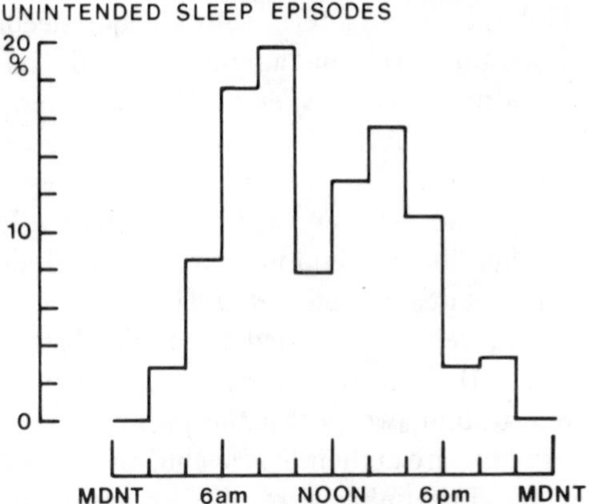


Figure 8.2 *24 hour temporal distribution of unintended sleep episodes (% of observed episodes occurring in each 2 hour period). Data from various studies collected by Mitler* et al. *(1988)*

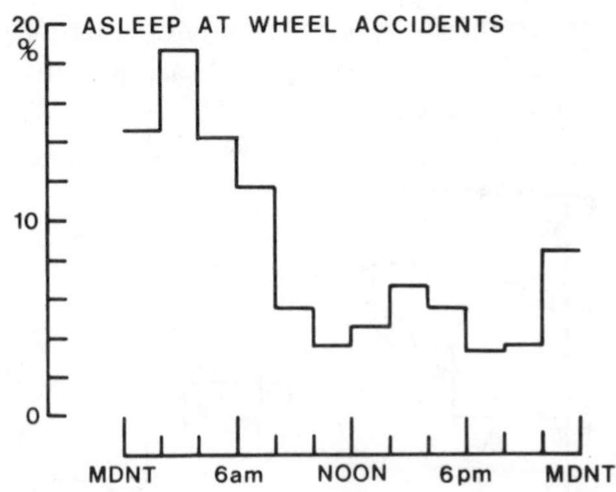


Figure 8.3 *24 hour temporal distribution of accidents attributed to drivers being asleep at the wheel (% cases in each 2 hour period). Data from various sources collected by Mitler* et al. *(1988)*

these times, *microsleeps* (involuntary and largely unavoidable episodes of sleep, lasting a few seconds only) may result in inattention, forgetfulness and other performance lapses (Mitler *et al.*, 1988).

Unlike the body temperature cycle, therefore, the sleepiness cycle is bimodal—that is, it has two peaks. The earlier (and larger) of these coincides with the trough of the body temperature cycle. The latter is often referred to as the *post-lunch dip* (Colquhoun, 1971). Experiments have shown that mental performance is generally impaired in the early afternoon, despite the fact that body temperature continues to rise. This is not just the result of post-prandial somnolence, since it may occur at about the same time in people who have not yet had their lunch (Blake, 1971). It is possible therefore that the post-lunch dip, occurring as it does at the mid-point of the circadian cycle, is a manifestation of a superimposed 12 hour *ultradian* rhythm, the nature of which is not yet understood. (An ultradian rhythm is one with a cycle time of less than 1 day.) Some people, isolated from environmental time cues, adopt a biphasic sleep pattern, with a major sleeping tendency at the low-point of their temperature cycle and a minor one at the mid-point (Zulley and Campbell, 1985). Furthermore, naps taken in the early afternoon (the traditional siesta time in many countries) have been shown to have a particularly beneficial restorative effect (Naitoh, 1981).

Circadian rhythms have an extensive and far-reaching effect on bodily functions. It is well known that "babies tend to arrive in the middle of the night" (i.e. midnight to 6 am). So does death. Mitler *et al.* (1988) have compiled data for the times of death for all causes, as recorded in studies based on a total of over 437,000 cases (Figure 8.4). Like the events indicative of sleepiness discussed above, these clustered around two peaks: the greater occurring in the 4–6 am period, when the tides of life are at their lowest ebb, and the lesser in the afternoon (2–4 pm). Since people are unlikely to die of eating lunch (at least, not immediately), the latter peak is presumably evidence that the post-lunch dip is truly an independent circadian effect. Heart attacks (fatal and otherwise) also tend to cluster around two peak periods—but the pattern is delayed by a few hours, showing its major peak in the 6–10 am period and its minor peak between 6 pm and 8 pm. Twenty-four-hour ecg tapes of heart patients also show greater frequencies of arrhythmias during these periods.

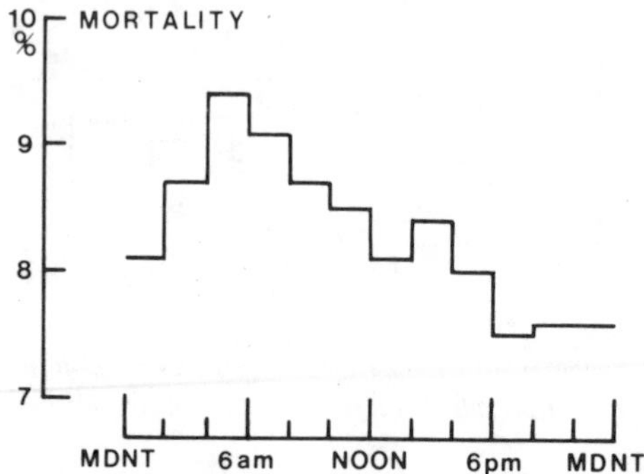


Figure 8.4 *24 hour temporal distribution of mortality from all causes (% deaths occurring in each 2 hour period). Data from various sources collected by Mitler* et al. *(1988)*

People with chronic back problems often tend to find that the pain is worse at certain times of day. Baxter (1987) found that the worst time of day is usually either the morning (35% of cases) or the evening (29%), rather than the afternoon (12%) or night-time (12%) or no particular time at all (12%). Back pain tends, therefore, to peak at one or other of the times of day when people are at their least sleepy. Men were more likely to have a morning peak; women were more likely to have an evening peak, with the pain becoming steadily worse through the day. People with a morning peak sometimes had a secondary lesser peak in the evening, and vice versa.

Following the early work of Kleitmann (1939, 1963), an extensive body of experimental research has confirmed that performance in a variety of work tasks follows the body temperature rhythm fairly closely—generally with a superimposed post-lunch trough (e.g. Colquhoun, 1971; Folkard and Monk, 1985a). Time-of-day effects are most marked in vigilance tasks and in others which are prolonged, simple and repetitive and therefore unarousing. Folkard and Monk 1985a) note, however, that complex tasks involving a high memory load may be an exception to the principle and tend to show peak performance much earlier in the day. (This may be a manifestation of the inverted-U effect, or it may be due to something altogether more complex.)

The most extensive field of study of round-the-clock human performance recorded to date remains that of Bjerner *et al.* (1955), who analysed some 75,000 meter reading errors recorded over a 20 year period in a Swedish gas works, where shifts comenced at 6 am, 2 pm and 10 pm and rotated once per week. The errors were distributed according to the classic two-peak pattern of the sleepiness cycle (Figure 8.5)—so the circadian effects presumably outweigh any "fatigue" decrements which occurred with time during each working shift. Some notable accidents resulting from errors committed during the 2–6 am period of vulnerability will be discussed in the next chapter.

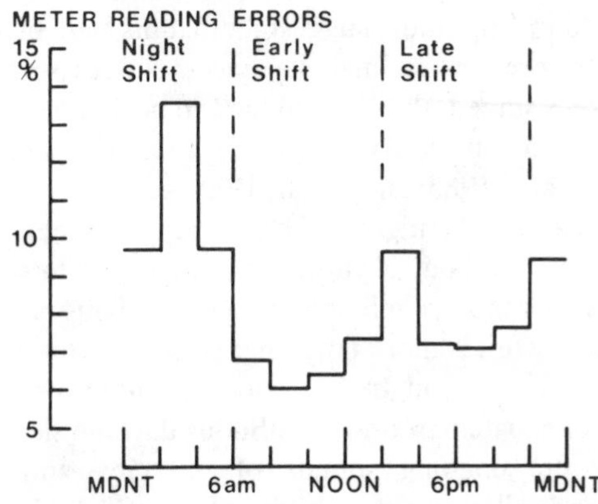


Figure 8.5 *24 hour temporal distribution of meter reading errors in a gas works. Data of Bjerner* et al. *(1955)*

Some people are "better in the morning" than others. Some of us are larks; other are owls. People who are *morning types* (on the basis of their reported mood states) have an earlier peak to their body temperature rhythm (by up to 3 hours) than *evening types.* Introverts are more likely to be morning types than extraverts. Morningness tends to increase slightly with old age, and women tend to be better in the morning than men (Kerkhoff, 1985).

## The Effects of Shiftwork on Health and Well-being

It is reasonable to assume that most people dislike irregular and antisocial working hours. But different individuals may balance the pros and cons differently. A minority find that shift work has tangible advantages—particularly time off when other people are working. The nightshift may also have a certain *esprit de corps*—which is supported by the various unofficial (and advantageous) working practices which can go on when the eyes of management are averted. Another minority find themselves unable to adapt to the demands of shiftwork (physiological and/or social) and seek an alternative form of employment within a relatively short time. The majority put up with it more or less stoically—at least for a time. Grandjean (1988) estimates that about two-thirds of shiftworkers suffer some form of ill-health, and about a quarter are eventually forced to give up shiftwork as a result.

## Sleep Disturbance

The night shift worker is a man at odds with his own body rhythms. With each successive working night, the body temperature cycle diminishes in amplitude and the peak of the cycle moves forward around the clock by about an hour. It may take a fortnight or more for the night worker's body rhythms to get in phase with the inverted pattern of his sleeping and waking life—and even then his rhythms tend to remain diminished in amplitude, suggesting that his state of adaptation remains incomplete. Furthermore, his normal daytime cycle reasserts itself as soon as daytime activities are resumed, thus he will have to begin the process of adaptation (such as it is) all over again after weeks or days off (Carpentier and Cazamian, 1977; Monk and Folkard, 1983; Akerstedt, 1985; etc.).

It follows, therefore, that the nightworker, on a typical weekly shift rotation, will be trying to work when his body rhythms are telling him to sleep—and trying to sleep when his body rhythms are telling him to work. Thus neither work nor sleep is entirely satisfactory: work is more fatiguing; sleep less restorative.

To make matters worse, the nightshift worker is out of time with the rhythm of life which continues unabated around him. So his daytime sleep, fitful at best, will be interrupted by the ongoing sounds of the city—and by the inevitable roundsmen, telephone calls and noisy children from which his long-suffering wife must do her best to protect him.

The quantity and the quality of his sleep both suffer. He may well sleep 8 hours or more on the nights preceding a late shift when he does not have to get up to go to work in the morning. But on the nights preceding the 6 o'clock start of the early shift, his sleep will almost certainly be curtailed by an hour or so. On the day preceding the first night duty of a series, he may well try to get his head down for an hour or so in the afternoon. But on the days following a night shift, he will probably get in 4–6 hours' sleep at most (e.g. Rutenfranz *et al.*, 1976).

Our typical nightshift worker is therefore building up a progressive sleep debt after each successive period of shortened and interrupted sleep—which is paid off by sleeping longer (during the night-time) at weekends. The deleterious effects of sleep loss on performance (and subjective well-being) at work will therefore be added to those of a partially adjusted circadian cycle.

According to various studies cited by Carpentier and Cazamian (1977), some

50–66% of night workers (including those on rotating shifts and those who work permanent nights) say that they have problems sleeping, as against 5–11% for comparable groups of day workers. Furthermore, 84–97% of former shiftworkers report having suffered from sleep problems in the days when they were working shifts—suggesting that the quality of sleep is one of the principal factors which determine whether an individual is able to tolerate shiftwork or not. Koller *et al.* (1978) found that former shiftworkers continue to report a significantly higher frequency of sleep disturbances (generally associated with noise) long after they have given up working shifts—suggesting that some kind of permanent sensitization may occur.

The normal sleeper passes fairly rapidly through three identifiable stages of increasing depth before settling down into a sleep pattern in which two further stages alternate on a 90 minute cycle. These are called Stage IV or deep sleep, and paradoxical or rapid eye movement (REM) sleep. In people who sleep by day, it is the REM sleep which is principally curtailed. REM sleep is associated with dreaming. Neuroses may be experimentally induced by selective REM sleep deprivation. Mental activity tends to increase subsequent REM sleep, whereas physical activity increases deep sleep. REM is thought to be the mentally restorative phase of sleep, whereas deep sleep is thought to be physically restorative. Some forms of muscle pain may be associated with disturbances of deep sleep (p. 54). Carpentier and Cazamian (1977) suggest therefore that the labourer is likely to stand up to shiftwork better than the worker whose job involves principally mental activity, since his shortened and interrupted daytime sleep will be adequate for recovery from physical exhaustion and fatigue, but not for recovery from mental fatigue.

## Psychological and Physical Health

Cross-sectional studies in which populations of shiftworkers are compared with dayworkers are subject to a number of sources of bias—the most important of which is a *healthy worker effect.* (That is, those individuals who suffer most from the effects of shiftwork will transfer to the day working population.) To overcome these difficulties it is necessary either to study the effects of shiftwork longitudinally or to treat former shiftworkers as a separate group. When these methodological precautions are taken, the deleterious effects of shiftwork emerge more clearly.

Bohle and Tilley (1989) have shown a clear and significant *increase* in self-reported psychological symptoms (such as depression, loss of self-esteem, difficulty in concentrating, etc.) in a group of nurses, 15 months after commencing shiftwork. Morning types were more susceptible, but a measure of neuroticism which had predictive value for symptoms reported before the commencement of shiftwork had no predictive value at the 15 month follow-up. Conversely, Akerstedt and Torsvall (1978) showed an equally clear *decrease* in sleep disturbances, mood disorders, gastrointestinal complaints and sickness absence in a group of steelworkers who were transferred from shiftwork to daywork because of a change in production requirements. And Frese and Semmer (1986) found a significant excess in psychological and gastrointestinal symptoms in both shiftworkers and former shiftworkers (who had left for reasons of ill-health) compared with dayworkers and shiftworkers who had left for other reasons. (The former shiftworkers who had left because of ill-health also had significant excesses of both heart and lung diseases.)

The poorly adapted nightworker suffers from a potentially progressive state of chronic fatigue, which may be manifest in episodes of irritability, loss of drive, depression, and so on—together with loss of appetite, dyspepsia, constipation and other disturbances of gastrointestinal function (Carpentier and Cazamian, 1977; Grandjean, 1988). As long ago as 1921, Vernon noticed an excess of gastric problems in shift workers in the armaments industry. These are presumably exacerbated by their irregular mealtimes. The nightworker may consume a heavyish meal in the works canteen half-way through his shift—as much for something to do to break up the night as because he is really hungry. At 2 am, the non-adapted body rhythms of his digestive functions may be at a low ebb. He may find himself sitting down to a daytime meal with his family, when he is not long out of bed and has little appetite—which is conducive neither to good digestion nor to domestic harmony.

We might suppose that symptoms of this kind are the precursor of more serious conditions such as peptic ulcer. Cross-sectional studies comparing shiftworkers with dayworkers show overlapping ranges of prevalence values—both for gastrointestinal symptoms in general and for ulcers in particular. Figures for former shiftworkers provide much clearer evidence for an association (Aanonsen, 1964; Carpentier and Cazamian, 1977; Rutenfranz *et al.*, 1981, 1985).

Taylor and Pocock (1972) found a slight increase in mortality over a 12 year period in shift workers compared with dayworkers, and a greater increase in former shiftworkers. But the differences did not reach statistical significance (except for cancer deaths).

More recently, evidence has emerged for a strong association between shiftwork and coronary heart disease. In a 15 year follow-up study of blue collar workers in a paper mill, Knutsson *et al.* (1989) found that 11–15 years of shift work gave a relative risk for coronary heart disease of 2.2, and 16–20 years of shiftwork gave a relative risk of 2.8 (compared with dayworking controls). The association was independent of both age and smoking. The relative risk fell sharply after 20 years—owing presumably to a healthy worker effect.

## Individual Differences

Tolerance of shiftwork decreases with *age.* The sleep disturbances and digestive dysfunctions of the shiftworker may not cause him serious problems until he reaches his fifties—perhaps after he has been working shifts for 30 years. Perhaps as few as 40% of men who work shifts continue to do so throughout their working lives (Carpentier and Cazamian, 1977)—although many of those leaving shiftwork may have been promoted into positions of greater seniority rather than changing because of ill-health. Seniority promotions from shiftwork to daywork are (sensibly) built into many organizational structures.

There is some suggestion that rotating shiftwork might be associated with menstrual dysrhythmias in female shiftworkers. But sex differences in the ability to adapt to shiftwork are unlikely to be marked—except insomuch as they result from domestic pressures.

Extroverts and evening types tend to adapt to shiftwork—both physiologically and psychologically—better than introverts and morning types (see Monk and Folkard, 1985, for a further discussion).

### Social Aspects

The quality of life and social well-being of the shiftworker will depend on his social circumstances and aspirations, and on the extent to which his family are able and willing to adapt their lives to meet his unusual needs. The shiftworking family man generally finds that he sees his children less often than he would wish—particularly in the early evening period, which is central to the integration of family life. He may dislike leaving his wife alone at night and during the social hours of the evening, and she in turn may feel lonely and isolated at these times. He may find that he "gets under her feet" during the daytime—and she may find her activities curtailed by the need to adapt to his habits and in particular to keep the household quiet when he is sleeping.

The value which the individual attaches to leisure time at different hours of the day or night will modify the effect which shiftwork has on his quality of life. The golfer or the birdwatcher will value free time in the morning much more than the man-about-town. Conversely, the shiftworker may find it difficult to take an active role in social groups which demand a regular commitment of time (sports clubs, adult education, etc.) and he may come to feel himself isolated from the mainstreams of social life.

Both shiftworkers and those whose work is regularly finished early in the day (i.e. roundsmen) have a notorious propensity for moonlighting. It is difficult to get an accurate estimate for how common this is, since people are naturally reticent on the subject. Adlauer (1988) asserts that "unofficial and more or less clandestine statistics show that 33% of shiftworkers have another job 'on the side'". If this is true, it seems very sad.

## The Design of Shiftwork

How should shiftwork be organized so as to minimize its potentially adverse effects? There are two schools of thought: the *rapid rotation theory* and the *slow rotation theory*. Adherents of the *slow rotation theory* argue that since physiological adaptation to nightwork may take very many days, the worker should change his shift as rarely as possible, so that the maximum degree of adaptation may occur. Thus permanent nightwork is said to be preferable to rotating shift systems, and fortnightly or (even better) monthly rotations are preferable to weekly ones. But people are likely to lose whatever adaptation they achieve on rest days, etc.—so in practice the benefits of slow rotation may be negligible.

Adherents of the *rapid rotation theory* therefore argue that since adaptation to nightwork is never likely to be more than partial, it would be better if it did not occur at all—since it is the constant process of de-adaptation and re-adaptation that is likely to be harmful. Workers should not, therefore, be required to work more than two or three nights in a row.

Adherents of both theories agree that weekly rotations are amongst the worst of possible options. At present, the majority of ergonomists support the rapid rotation theory (e.g. Murrell, 1969; Monk and Folkard, 1983; Grandjean, 1988), but a small minority favour slow rotation (e.g. Akerstedt, 1988). The issue is by no means closed.

Any change in shiftworking practices will tend to be disliked by the workforce. The men and their wives will have organized their lives around the old system, so "change for change's sake" will necessarily be seen as an unwelcome disruption. There is, however, a steady trend towards a more rapid rotation—which has been due (in part at least) to pressure from labour. So it is probably true to say that, on the whole, rapidly rotating systems are preferred.

The three shifts of a rotating system may start at various hours, which have social pros and cons. Starting the early shift at 5 am or even 4 am instead of 6 am will give the workers on the late shift more time in the evening to spend with their families or in the pub. But the additional pressure on sleep time for the early shift probably outweighs the advantages gained. The American practice of starting shifts at 8 am, 4 pm and midnight has the double advantage of giving two shifts out of three a clear evening, and a more civilized starting time for workers on the early shift. Socially, a late shift which finishes at midnight is little worse than one which finishes at 10 pm—but it may be that a 12–8 nightshift is more arduous than one which starts and finishes earlier, since the nightworker will not get to bed until well into the day. In some places, the timing of shifts may also be constrained by the availability of public transport.

People may suffer fewer ill-effects from a shift system which rotates in a clockwise direction—that is, early, late, nights, rather than vice versa. This is probably because the internal body clock has a natural length of more than 1 day. (You probably also suffer less jet lag after flying from east to west across time zones than from flights from west to east.) In a study of policemen, Orth-Gomer (1983) found that a change from anticlockwise to clockwise rotation had a favourable effect (at least in the short term) on both subjective well-being and coronary risk factors.

Night shifts should always be worked at the end of a rotation rather than the beginning, so that they are followed by rest days for people to catch up. Short and simple rotas are preferable to long and complex ones, since they allow people to plan social activities in advance. It is also an advantage if rest days fall on weekends as often as possible.

The two most commonly encountered rapidly rotating *continuous* shift systems are called the "metropolitan rota" and the "continental rota". The metropolitan rota involves a continuous sequence of two early shifts, two late shifts, two night shifts and two rest days. It takes 8 weeks to complete its cycle. The continental rota is as follows:

| | *Mon* | *Tues* | *Wed* | *Thurs* | *Fri* | *Sat* | *Sun* |
|---|---|---|---|---|---|---|---|
| *Week 1* | E | E | L | L | N | N | N |
| *Week 2* | R | R | E | E | L | L | L |
| *Week 3* | N | N | R | R | E | E | E |
| *Week 4* | L | L | N | N | R | R | R |

(where E = early, L = late, N = night, R = rest)

The continental rota yields a 3 day free weekend every 4 weeks, whereas the metropolitan gives a 2 day free weekend every 8 weeks. The continental seems therefore the more civilized of the two—and the one spell of 3 nights in a row which arises every 4 weeks is unlikely to be a significant physiological disadvantage. In practice, the continental rota is probably more or less as good a system as can be achieved without taking on additional staff.

The metropolitan and continental rotas both require four working teams. By increasing the number of teams, it is possible to shorten working hours (e.g. have four 6 hour shifts, etc.), reduce the number of nights worked by any individual, and so on. These of course incur substantially increased labour costs. It is generally recognized, however, that large numbers of people, in manufacturing industry and elsewhere, work shift systems which could be substantially improved without extra cost. The only barrier to doing so appears to be inertia.

Good ergonomic practice in work design is doubly important in organizations which function through the night. Rational workstation layout and interface design, adequate pauses for rest and refreshment, and a varied pattern of work activities are all especially relevant.

## The "On-Call" System

Junior hospital doctors do not work shifts in the ordinary way. They are "on call" at night if, when and where they are required—generally after they have done a full day's work on the wards. The on-call system is also encountered in industry—amongst site engineers, for example. The difference with junior hospital doctors is that when they are "on call" they may well be working most or all of the time.

At the time of writing, the average working week of junior hospital doctors in the UK is 86 hours (*Daily Telegraph*, 29 December 1988). But stories of junior hospital doctors being on call for 160 hours, out of the total of 168 hours that there are in a week, are not uncommon. Poulton *et al.* (1978) have shown quite convincingly that the performance of junior hospital doctors, in standardized tests of mental function, deteriorates as a result. And there are well-attested reports of house officers falling asleep during operations (*Daily Telegraph*, 29 December 1988).

# Chapter Nine

# Accidents, Errors and Interfaces

"To err is human"

Alexander Pope (quoting Seneca)

"Accidents will happen in the best regulated families"

Micawber in *David Copperfield*

"If something can go wrong it will"

Murphy (Attrib.)

*An accident is an unplanned, unforeseen or uncontrolled event—generally one which has unhappy consequences.* Accidents characteristically have multiple causes. It is the way in which these causes come together (often in improbable combinations) that makes the accident unexpected.

If the consequences of an accident are particularly serious, we may refer to it as a *disaster* or a *catastrophe*. In everyday usage the distinction is usually one of scale. The Greek word from which "catastrophe" is derived literally means an overturning. In the terminology of the classical drama, it referred to the final twist of fate which led to the conclusion of a tragedy. (The last scene of *Hamlet* is a good example). "Catastrophe theory" is a recently developed branch of mathematics which deals with discontinuities in the behaviour of complex systems. Large-scale industrial accidents are often catastrophic in both of these senses.

At the present time, some 500 people or more are killed each year in accidents at work (in the UK), compared with 5,000 or more fatal accidents on the road and another 5,000 in the home. (The exact figures vary from year to year in each case.) At first sight, these figures seem to tell us that work is relatively safe—at least when you compare it with other human activities. To some extent this is true. Work is (or should be) more organized and better regulated than domestic activities or driving. But less than 40% of the population are exposed to the risks of work, and the working population does not include those people who are most at risk—the very young, the very old, the infirm, etc. Furthermore, most people work during the relatively safe daylight hours and when they are more or less fully alert; and people are more likely to be at risk due to the influence of drink and drugs, away from the working environment. Taking these factors into account, work seems surprisingly dangerous.

Work is, however, getting steadily safer. According to the Robens report of 1972, the annual number of accidental deaths in the UK per 100,000 workers was 17.5 in the first decade of this century, compared with 4.5 per 100,000 in the 1960s. In 1981 it was 2.1 per 100,000 and in 1986 it was 1.7 per 100,000 (HSE, 1989). The overall downward trend is thought to be partly due to working activities being better regulated and partly to changes in the nature of work, such that fewer people are involved in its riskier varieties. The fatality rate in specific branches of industry has remained fairly steady through the 1980s—so whatever happened earlier, the current decline must be due almost entirely to the latter cause.

There are two principal ways of looking at the accident problem:

- *Theory A* Accidents are caused by unsafe behaviour (and some people are more prone to behave unsafely than others). Accidents may therefore be prevented by changing the ways in which people behave.
- *Theory B* Accidents are caused by unsafe systems of work. Accidents can therefore be prevented by redesigning the working system.

Theory A is a case of *fitting the person to the job*; Theory B is a case of *fitting the job to the person* (p. 120). Until now, Theory A has had the upper hand, probably because industrial accidents tend to have been studied by psychologists—and psychologists tend to prefer psychological explanations for things they study. Accident prevention programmes based on Theory A tend also to be cheaper.

## Accident Proneness

The concept of accident proneness was formulated by Greenwood and Woods (1919) on the basis of statistical analyses of accidents occurring in munitions factories during World War I. It has been going in and out of fashion ever since.

Greenwood and Woods found that a small number of people had a disproportionately large share of the accidents which happened in these factories: that is, they were having significantly more accidents than you would expect on the basis of chance alone. In other words, the accidents were not completely random events. The finding has been confirmed in subsequent studies (for example, by Newbold, 1926); and although the statistical techniques used in these studies have been debated at length (see Porter, 1988), the principal point at issue is how we should interpret the non-random distribution of accidents, which clearly does seem to occur.

One possibility is that some people possess certain physical or psychological characteristics which make them more likely to have accidents than other people. These characteristics might be specific to a particular working situation or they might be of a more general nature. (That is, an individual who is at risk in one job might or might not be at risk in others.) They might be permanent features of the individual's personality or they might be transient (but recurrent) mental states. Alternatively, some people may be exposed to greater risks than others (due to the nature of their job or the circumstances in which they work); having one accident may increase (either temporarily or permanently) the likelihood that you will have another; or there may be biases in reporting. In practice, all of these hypotheses are probably true up to a point, but none is likely to be the sole explanation of the statistical data. In any given situation, one or other may predominate (see McKenna, 1983, for a discussion of these possibilities).

## Personal Characteristics

Accident involvement is highest amongst workers in their teens and early twenties. The frequency of accidents drops sharply and levels off in the mid-twenties. It remains low or continues to fall slightly through the middle part of working life before climbing again somewhere between the mid-forties and early sixties. Young people are likely to be impulsive and undisciplined, but the effects of age and experience are confounded. By its very nature, industrial work is often fraught with more or less foreseeable dangers; the ability to negotiate your way around these hazards is in part a learned skill. The available data indicate that part (but not all) of the excess of accident involvement in young workers is attributable to inexperience (Hale and Hale, 1972).

Extroverts are likely to have more accidents than introverts—perhaps because they are also more prone to impulsive behaviour. But there is no evidence for an association between low intelligence and accident involvement except at the very bottom end of the scale (Tiffin and McCormick, 1970; Hale and Hale, 1972).

Some people seem to be clumsy—"always bumping into things". An individual's safety is often contingent on his capacity to make a rapid appraisal of a complex set of perceptual data and to organize an appropriate motor response. A person with defective vision would presumably do this more slowly. People with defective vision have been found to be accident prone in some (but not all) working situations (Tiffin and McCormick, 1970; Hale and Hale, 1972). But the problem may be cognitive rather than sensory. Accident-prone drivers are likely to exhibit a collection of psychological traits called *field dependence*—that is, they may be relatively poor at extracting salient information from a complex perceptual field (Goodenough, 1976).

People differ in their capacity to cope with situations which make multiple demands on their attention. Porter and Corlett (1989) developed a questionnaire for measuring accident proneness by asking the subject to report how frequently he was involved in minor everyday mishaps—like bumping into things, breaking things, knocking things over, and so on. These mishaps tended to occur when the individual's attention was not on the task at hand. Accident-prone subjects (as identified by the questionnaire) performed significantly less well in a complex experimental task involving divided attention).

It has been suggested that accident repeaters might have deeper psychological problems. Hill and Trist (1953) found that accident repeaters in a steelworks also took more unauthorized time off for other reasons. They argued that accidents and absenteeism had similar psychological causes—both being examples of "withdrawal behaviour". This interpretation is based upon the psychoanalytic concept of the accident as a subconsciously motivated event (Freud, 1901). Like much of psychoanalytic theory, this proposition does not lend itself to empirical verification. Hill and Trist's findings could also be interpreted on the basis of differences in task demands: in a study of people doing much lighter work in a photographic processing plant, Castle (1956) found no relationship between accidents and uncertified absence.

The social aspects of the accident problem cannot be ignored, however. Speroff and Kerr (1952) found that accident-free steelworkers tended to be popular amongst their workmates, whereas accident repeaters tended to be unpopular. Verhaegen *et al.* (1979, 1985) found that people who had been *actively* involved in industrial accidents had high absenteeism rates compared with those who had been *passively* involved or not involved at all. And the former group considered the work to be less inherently dangerous and admitted to taking more risks. You could argue, therefore, that people with negative attitudes to working life are likely (a) to be less popular with their workmates; (b) to absent themselves more often; and (c) to take more risks. This is a more parsimonious explanation than one based on a subconscious urge to self-destruction, and to a non-Freudian it is psychologically more plausible.

## Transient States

The influence of drink and drugs both slow the individual's reactions and change his attitude to risk. Smith (1989) has noted, however, that a dose of influenza can result in a 50% impairment of performance in a reaction time test, compared with the impairment of around 5% which results from enough alcohol in the bloodstream to get you banned from driving in the UK. Hersey (1936) found that the typical industrial worker feels low and miserable for about 20% of the time—but more than 50% of accidents occur during these periods of negative mood. The extent to which these mood states may be regarded as manifestations of fatigue is open to question (p. 160). Both drivers and pilots are more likely to have accidents when going through major life events (Selzer and Vinokor, 1974; Alkov and Borowsky, 1980).

We all experience days when things go well and days when things go badly for no apparent reason. You would expect a certain amount of clustering on the basis of chance alone; and a "run of bad luck" might in some cases be explained in terms of cumulative stress effect. Thus, "when sorrows come, they come not single spies, but in battalions" (*Hamlet*, IV, v). But can we discern a more systematic pattern in our good and bad days?

Some women become more accident prone at the time of their menstrual periods. Dalton (1960) studied women admitted to hospital because of accidents. Of those with normal menstrual cycles, more than half had been involved in accidents during the 4 days preceding or following the onset of menstruation—as against the 29% you would expect in any consecutive 8 days if accidents occurred at random throughout the 28 day cycle. The odds against this distribution occurring by chance alone are more than 1000 to 1. The relationship held for both active and passive accident involvement, although the significance level was lower in the latter case.

The theory of the so-called "biorhythms" is based on the speculations of the psychologists Swoboda and Fliess at the turn of the century. It proposes that human beings are under the influence of three bodily cycles, commencing on the day they are born, and having periods of 23, 28 and 33 days, respectively. Our bad days are alleged to occur when these cycles interact in certain particular ways. A number of studies have shown that this theory has no empirical basis whatsoever (McKenna, 1982). Why should it have become so popular therefore? Like many other varieties of pseudoscience, it purports to explain something of direct personal importance that we have all experienced but do not understand—and for which we seek a simple explanation, which would make the complex patterns of our lives seem more orderly.

**Ergonomic Aspects**

Early investigations conducted by the Industrial Fatigue Research Board showed that the excessively long hours worked in munitions factories during wartime were associated with a disproportionate increase in the number of accidents—due presumably to the effects of fatigue (Vernon, 1936). An association between accident frequency and the amount of overtime worked has also been reported in light industrial jobs under peacetime conditions (Powell *et al.*, 1971).

Fatigue reduces our capacity to deal with the hazards of work. These may be inherent in the nature of the working tasks. Winsemius (1965) studied the operators of alphabet punching machines used in the manufacture of address books, etc. The task is repetitive, with a short cycle time. The operator works in a standing position and the machine runs continuously so long as she keeps her foot on a pedal. When punching the last few pages of a book, the operator's fingers occasionally get trapped in the punch. This typically occurs when there is a disturbance of the normal working rhythm: some pages of the book may stick together and the operator needs to make an additional movement to free them. To do this safely, she should stop the machine by taking her foot off the pedals. She does not want to do so because it takes a considerable effort to restart the machine—the pedal is big and heavy, and supports a considerable proportion of her body weight, so a major change of posture is required. The accidents which occur in this process do indeed result from unsafe behaviour (Theory A), but the unsafe behaviour is the direct consequence of bad ergonomics (Theory B).

Saarhi and Lautela (1979) compared firms in the light metal working industry which had high and low accident frequencies. The structure of the labour force was similar in both groups of firms. The firms which had more accidents were inherently more dangerous in terms of having more unguarded machinery, more things to fall on you, more oil and rubbish on the floor, and so on. But the jobs of the individual workers in the more hazardous plants also tended to be more varied, less repetitive, less pre-planned and more mobile, and required the assimilation of more complex information. The more uncontrolled and unpredictable the working environment, the greater the demands it places upon the individual's information-processing capacity and the more he is at risk. Thus those features of task design which make working life more satisfying (p. 136) may also make it less safe. This is a disturbing paradox.

## Human Error

Accidents are commonly blamed on human error. (Or, given that accidents have multiple causes, human error is recognized as being a significant contributory factor.) Notable examples include the Bhopal disaster, the nuclear accidents at Three Mile Island and Chernobyl, and the losses of the space shuttle *Challenger* and the *Herald of Free Enterprise*. Human error is said to be a major causative factor in up to 45% of critical incidents in nuclear power plants, 60% of aircraft accidents, 80% of marine accidents and more than 90% of road traffic accidents. For industrial accidents on the shop floor the figure is probably in the upper part of this range.

What does the expression "human error" tell us? The word "error" is derived from the Latin *errare*: to wander or to stray. It is used in this way in the *Book of Common Prayer*:

> "We have erred and strayed from Thy ways like lost sheep .... We have left undone those things which we ought to have done and we have done those things which we ought not to have done. ..."

*An error is an incorrect belief or an incorrect action* (the one commonly leads to the other). In reaching a diagnosis of "human error" in the case of any particular accident, we are essentially making two assertions: firstly, that it was caused by the actions of a human being rather than by mechanical malfunction or some other impersonal agency; secondly, that the consequences of these actions were unplanned. (Acts of wilful destruction are not "errors" in the usual meaning of the word.) But real life may not be this simple—the distinction of *agency* and the distinction of *volition* may both be blurred.

We tend to think of error as an intrinsic part of the human condition (as illustrated by the epigraphs of this chapter). But a diagnosis of human error may mask serious deficiencies in the design of the working system, in the absence of which the accident would not have happened. Of itself, the human error diagnosis tells us little about how an accident occurred and less about how it could have been prevented. The bogus connotations of inevitability which are implicit in the concept may have subtle political overtones—and may be used as a means of shifting responsibility.

The cognitive mechanisms of human error have been discussed at length elsewhere by Reason and Mycielska (1982) and Reason (1989). I shall not attempt to deal with this aspect of the subject here. Instead, we shall consider the problem of human error by way of example; and in particular we shall discuss the role of human error in the catastrophic failure of complex working systems.

### Chernobyl

At 1.24 am on 26 April 1986, two explosions and a subsequent fire at the Chernobyl No. 4 reactor caused a massive release of radioactive material. About

30 plant and rescue workers died as a direct result, and the surrounding area (including the nearby town of Pripyat) will remain uninhabitable for the foreseeable future. Estimates of the number of cancer deaths likely to result in the longer term run into tens of thousands.

Soviet sources were quick to blame the catastrophe on "human error". (And the men who were said to be responsible were subsequently sent to labour camps.) By emphasizing this particular aspect of what in reality was a very complex event, they implied that such an accident could happen anywhere. Western authorities countered this suggestion by blaming the design of the reactor. They pointed to certain instabilities in its behaviour, the lack of an adequate containment structure, and so on—thus implying that "it couldn't happen here". Well, they would, wouldn't they? In reality, both sets of factors were involved.

The catastrophe occurred during the course of an experiment, the intention of which was to test whether the reactor's turbines could be used to power cooling pumps and other safety-critical systems, during the short period following an emergency shut-down, before the back-up diesel generators could be brought into operation. The experiment was planned for the afternoon on which the reactor was due for its annual maintenance shut-down prior to the Mayday holiday. There was a limited "window of opportunity", which would not be repeated for another year. Thus everyone involved was under pressure to get it right first time.

The reactor was known to be unstable when running at low power. The tests were to have been carried out with the reactor running at 25% full power (at which it would have been adequately stable). At about 1 pm the run-down of power commenced and the emergency cooling control systems were shut off (this was necessary to perform the experiment). But an unexpected increase in the demand for power led to the experiment being postponed until about 11 pm. Some time after midnight, one of the operators failed to enter a critical command into the system and the power plunged to a very low level. This took some time to correct. By about 1 am they had managed to get the reactor running at a steady 7% full power and they decided to continue the experiment—despite the fact that they were within the range of conditions in which the reaction process was known to be unstable. During the next 20 minutes they made a series of bad decisions involving the violation of various safety procedures and the shutting off of automatic safety mechanisms. These ultimately led to the uncontrolled surge of power which blew the top off the reactor. (For further details of the Chernobyl catastrophe, see Reason, 1987.)

The individuals involved had made a major error of judgement. They took a considered risk in deciding to press on with the experiment when it should have been abandoned and, in doing so, they drove the system beyond the boundary conditions of its stability. This is something like taking a corner too fast on a wet day and getting into an uncontrollable skid.

## Risk and Error

In general, we take risks in the belief that we can "get away with it". Sometimes this belief turns out to be erroneous. But insomuch as accidents tend to involve the coming together of multiple adverse circumstances, they will be relatively rare. Thus the belief that we can "get away with it" will tend to be borne out by experience. Successful risk taking may thus lead to *illusions of invulnerability*. Biased or erroneous sources of information may reinforce these illusions. Or we may attend to those sources of information which support our erroneous preconceptions, and ignore those which do not. Thus the appraisal of risk becomes increasingly and systematically biased. The history of military incompetence provides numerous examples (Dixon, 1976).

At a little before midnight on 14 April 1912, the White Star liner *Titanic* was on her maiden voyage across the Atlantic. She was steaming full ahead at 22 knots when she struck an iceberg which tore a massive hole in her side. She sank with the loss of 1,513 lives. Icebergs were a known risk at that time of year in the North Atlantic. The *Titanic* was said to be unsinkable. The captain was sailing too fast, in dangerous waters at night, in the erroneous belief that this was indeed the case.

The speed/error trade-off (p. 154) may be regarded as a species of risk taking, in which we accept a higher probability of error in the interests of expediency (although this is not necessarily an entirely voluntary choice). We "take short cuts" or "bend the rules" in order to get the job done more quickly or with less effort—driving provides obvious examples. Reason (1989) distinguishes between errors *per se* and *violations* of safe working practice or the norms of safe behaviour (and he places the incorrect decisions which led to the Chernobyl catastrophe in the latter category). There are many shades of psychological grey between the truly unplanned accident and the wilful premeditated criminal act.

The law recognizes this grey area in the concept of negligence—the failure to exercise a reasonable degree of prudence in the face of foreseeable danger (p. 22). But how blatant does a given danger have to be before we regard it as foreseeable? And how much prudence is it reasonable to expect under the circumstances? As individuals we often betray an extraordinary lack of foresight into the possible consequences of our actions. Consider, for example, the case of the man who thought it would be a good idea to cut his hedge with the lawn mower. It has been shown that the perceptions which industrial workers have, of the ways in which they could be injured in their work, are determined principally by their knowledge of the ways in which injuries have occurred in the past (Powell *et al.*, 1971). It seems, therefore, that our appraisal of risk is likely to be based on hindsight rather than foresight.

A working group is commonly permeated by a set of collectively held attitudes which colour the judgement of its members. These make up the *social ethos* of the group; and the group's collectively held attitudes to risk make up its *safety culture*. In some working situations the norms of safe behaviour are set very low: poor housekeeping, sloppy workmanship and the violation of officially laid down safety procedures may become endemic, and may be condoned by shop floor management in the interests of meeting production goals or to avoid controversy. (Firms with unsatisfactory safety cultures often have unsatisfactory labour relations too—and both are a symptom of poor management.)

High-level management decisions may be equally biased by shared beliefs which act to reinforce illusions of invulnerability. Janis (1972) calls this *groupthink*. Organizational deficiencies which isolate high-level decision makers from individuals lower in the hierarchy and limit the free flow of information may contribute to this bias. (The tendency will be more pronounced in rigidly authoritarian hierarchies in which the individual at the top is surrounded by yes-men. Thus are political dictators and gurus made—but you can find examples in all walks of life.)

Organizational deficiencies and groupthink contributed to the disastrous decision to launch the space shuttle *Challenger* on 28 January 1986—which led to the death of its seven crew members and a major setback for the American space programme. NASA management were subsequently said to have lost touch with "the engineers who held the spanners" and to have ignored the latter's misgivings. Financial considerations added to the pressure to keep the flights on schedule. John Glenn (a former astronaut) is quoted as saying that the NASA ethos had changed from one of "can do" to one of "can't fail".

## Risk Homoeostasis

The theory of risk homoeostasis, as first set out by Wilde (1982), proposes that people react to changing circumstances by modifying their behaviour in such a way that the level of risk to which they are exposed remains constant. That is, people behave as if they had an internalized target level of risk, which they attempt to maintain—by risk seeking or risk avoidance as circumstances may demand. In other words, risk-taking behaviour is controlled by a negative feedback mechanism—as a thermostat controls temperature.

This theory has been most extensively explored with respect to road safety, where it represents a fundamental challenge to much of the conventional wisdom. The theory predicts, for example, that drivers who wear seat belts, or motorists who wear helmets, will drive more recklessly as a result. As secondary safety (the chances of surviving an accident) is increased, primary safety (the chances of avoiding an accident) is reduced. According to the theory, the two trends should cancel each other out—and most safety measures of the conventional kind will be doomed to failure.

It is not surprising, therefore, that the risk homoeostasis theory has been extremely controversial. Much of this controversy turns on relatively abstruse statistical arguments (often based on data which are incomplete or confounded with other factors). So the overall status of risk homoeostasis theory is difficult to assess. The case *for* the theory has been summarized by Adams (1985, 1988) and Wilde (1989); and the case *against* has been summarized by McKenna (1982, 1985, 1987). The subject of risk perception, as applied to the prevention of occupational accidents, has recently been reviewed by Simpson (1988b).

Behavioural compensation of the kind predicted by the theory does indeed occur under some circumstances. For example, Rummer *et al.* (1976) found that drivers who were provided with studded tyres to give extra grip in icy conditions responded by driving faster—thus negating the potential safety benefit. Similarly Perrow (1984) has noted that the provision of sophisticated navigational equipment on merchant ships has been used to increase their economic performance, rather than to increase their safety. (So ships are more likely to arrive on time under adverse weather conditions; but seagoing remains a relatively hazardous activity overall.) But the theory also predicts that drivers should reduce their speed when adverse weather conditions make driving more hazardous and they do not.

There is extensive evidence that people are poor "intuitive statisticians"—and that their estimates of risk are particularly inaccurate when it comes to low-probability events like accidents (Slovic *et al.*, 1981). This seems a good reason for regarding risk homoeostasis theory (at least in its classical forms) as psychologically implausible.

The concept of an internalized target level of risk may be invoked to "explain" various types of risk-seeking behaviour—hang-gliding, rock climbing, driving too fast, etc. But a variety of other psychological explanations would do equally well. It seems more straightforward to regard risk as the downside to certain activities which we are prepared to accept in the interest of expediency—or for thrills, social end-gaining, self-expression, and so on.

This trade-off is particularly important where technological innovations, leading to improvements in the design of man–machine systems, may be consumed as *either* performance benefits *or* safety benefits. Expediency in the pursuit of short-term goals, the minimization of human effort and the pressures of a competitive economic environment will all tend to favour the former option. Thus the systems will continue to be operated at the limits of their safety.

## Error Ergonomics

The examples we have considered so far have all involved errors of judgement in the appraisal of risk. We shall now return to those cases in which the erroneous beliefs or actions of the individual concerned can be shown to be the direct consequence of ergonomic deficiencies in the design of the working system, and in particular of deficiencies in the design of the man–machine interface. We call these *system-induced errors*.

### The *Herald of Free Enterprise*

At about 6 pm on 6 March 1987, the cross-channel car ferry *Herald of Free Enterprise* capsized and sank just after leaving Zeebrugge harbour. Of the 459 or more people on board, at least 189 died. The principal features of the catastrophe are summarized in Table 9.1.

Car ferries like the *Herald* are inherently riskier than most other ships (by virtue of their top-heaviness and the large unobstructed spaces near to the water line); and circumstances on that day were particularly unfavourable. There was a high tide which made it difficult to match the *Herald's* bows to the loading ramp at

*Table 9.1* *The* Herald of Free Enterprise

CONTRIBUTORY FACTORS:

*ENGINEERING*
- Inherent instability of roll-on/roll-off ferries
- Zeebrugge loading ramp
- Ballast pumping system

*ERGONOMICS*

*Primary safety*
- Absence of critical information displays
- Procedural vagueness
- Time stress
- Shift system

*Secondary safety*
- Accessibility and usability of life-jackets
- Lack of emergency lighting, etc.

*ADMINISTRATIVE/ORGANIZATIONAL*
- Pressure for fastest possible turn-round
- Management's failure to heed warnings given by serving officers

*MISCELLANEOUS CONTINGENCIES*
- High tide at Zeebrugge (negative)
- Sandbank (positive)

Zeebrugge. The ramp had originally been designed for single-deck ferries—and since the *Herald* had two car decks, her ballast had to be pumped forward so that her bows would lie low in the water. The pumps were not powerful enough to correct this before she put to sea. She set out with her bow doors open and the resulting inrush of water flooded her lower car deck, destabilizing her and causing her to capsize. Had she not come to rest on a sandbank, the resulting loss of life would probably have been greater.

Our principal concern here is with the antecedents of the fatal error of omission. The assistant bosun, who was directly responsible for closing the doors, was asleep in his cabin (where he remained until thrown out of his bunk as the ship turned over). There was no foolproof system for ensuring that the safety-critical task of closing the bow doors would be performed in his absence. It seems that the captain was to assume that the doors were safely closed unless told otherwise—but it was nobody's particular duty to tell him. The written procedures which dealt with these issues were woefully unclear. As the report of the official enquiry subsequently observed: "Clear instructions are the foundation of a safe system of operation".

The likelihood that such an error could occur was doubtless compounded by the pressure which the captain and crew were under (applied by company management) to achieve the fastest possible turn-round in port. The crew of the *Herald* were required to be on board continuously for 24 hours at a stretch—and then had 48 hours ashore. Their sleep patterns would necessarily have been erratic therefore. (Working hours in the company subsequently became the focus of an exceedingly acrimonious industrial dispute.)

There was no information display (not even a single warning light) to tell the captain whether the bow doors were closed or not. This elementary deficiency in the design of the man–machine interface stands out above all the other factors as the weak link in the chain of the working system. The captain, who is ultimately deemed to be "responsible" for the safety of his ship, had no direct means of knowing whether his ship was safe or not. Two years earlier, the captain of a similar vessel owned by the same company had requested that a warning light should be installed, following a similar incident when he had gone to sea with his bow doors open. Company management had treated the request with derision. Following the loss of the *Herald*, bow door warning lights were made mandatory on roll-on/roll-off car ferries. The catastrophic loss of the *Herald of Free Enterprise* was the direct result of a system-induced error of the classic kind. (For further details see the report of the official enquiry—DoT, 1987.)

### Three Mile Island

The accident which occurred at the Three Mile Island nuclear reactor at Harrisburg, Pennsylvania, on 28 March 1979 started during a routine maintenance operation. A small quantity of water found its way into the workings of a certain pump. This initiated a complex knock-on effect involving various other parts of the reactor system. Certain valves which should have been open were closed—probably as a result of an oversight on an earlier occasion. The indicator light for one of these valves happened to be obscured by a cardboard maintenance tag (rather like a luggage label). In dealing with the relatively minor crisis which ensued, the operators made a wrong move which made the situation very much worse than it needed to have been. They cut back the emergency water supplies to the reactor's cooling system in the erroneous belief that it was in danger of "going solid" (overfilling). An indicator light on the control panel told them that the relief valve of the cooling system was closed when in fact it was open. (The indicator responded to the electrical mechanism which operated this valve rather than to the valve itself—it was easier to build it that way.) Large quantities of radioactive water escaped through the open valve and flooded down a relief pipe into the basement of the reactor building.

No loss of life occurred in the accident, and the amount of radioactive material released into the environment was relatively small; but the reactor was a write-off and the surrounding buildings were massively contaminated. It is said to have cost the owners of the plant (and their insurers) something of the order of $1000 million.

The incident occurred at 4 am on a Wednesday morning, the time of day at which human beings are at their lowest ebb, on the third night of a weekly rotating shift system. So the operators' body rhythms would not have adapted to the inverted pattern of rest and activity, but they would have built up a significant sleep deficit during the two preceding periods of disturbed daytime sleep (p. 170).

The crisis developed with alarming rapidity. The control panel lit up like a Christmas tree, adding to the operators' confusion. In the first two minutes of the incident more than 100 separate alarm signals went off. (Most were just warning lights on the panel—about five or six would probably have been auditory alarms.) Many of these were interlinked, but the operators had no means of cancelling the ones which were unimportant. (For further details of the Three Mile Island incident, see Perrow, 1984.)

The control room of a typical nuclear power plant will have a display panel that is 8 feet high and as much as 100 feet long, covered with instruments of various types—many of which look identical except for their labels, which are only legible close up. Anomalies of layout are common—controls may be widely separated from their associated displays. Unimportant instruments may occupy central positions, and important ones may be inaccessible. There may be as many as 3,000 status indicators for valves, pumps and other components. In most plants a green light indicates that an electrical switch is open or that a mechanical valve is closed—a red light indicates the opposite. Sometimes this is reversed. Either way may indicate normal function, depending on the state of the system (Sheridan, 1980).

Since the Three Mile Island incident, there has been a major international effort to improve ergonomic standards in the nuclear industry—particularly with respect to the design and layout of control rooms.

## The Zeebrugge–Harrisburg Syndrome

The loss of the *Herald of Free Enterprise* and the Three Mile Island incident have important factors in common. Both are particularly clear examples of a particular combination of circumstances which we shall call the *Zeebrugge–Harrisburg Syndrome* (Pheasant, 1988a). Its principal features are summarized in Figure 9.1. A deficiency in the design of the man–machine interface makes the operator's task more difficult than it needs to be. This renders him more likely to make errors of omission and commission, incorrect decisions, and so on. The outcome may be catastrophic. The team of operators who were on duty at Three Mile Island on the night of the accident were experienced, highly trained and well motivated. Why were they unable to figure out what was happening? They did not know because they had no means of knowing. The display interface did not provide them with the information they required in a way they could comprehend. The messages they received were scrambled. In the words of Confucius: "If language is not correct, then what is said is not what is meant; if what is said is not what is meant, then what ought to be done remains undone."

The likelihood of error increases when the operator is under stress due to abnormal pressure of work—or when his working capacity is diminished because of fatigue, etc. The accidents at Chernobyl and Three Mile Island both occurred at around 1 am. The Bhopal disaster and the fire and chemical spillage which led to the massive contamination of the Rhine in November 1986 also occurred during the hours of the night shift. The time of day may be regarded as a contributory factor which reduces the individual's ability to cope with abnormal circumstances as they arise. (A surprisingly large number of these catastrophes also seem to occur

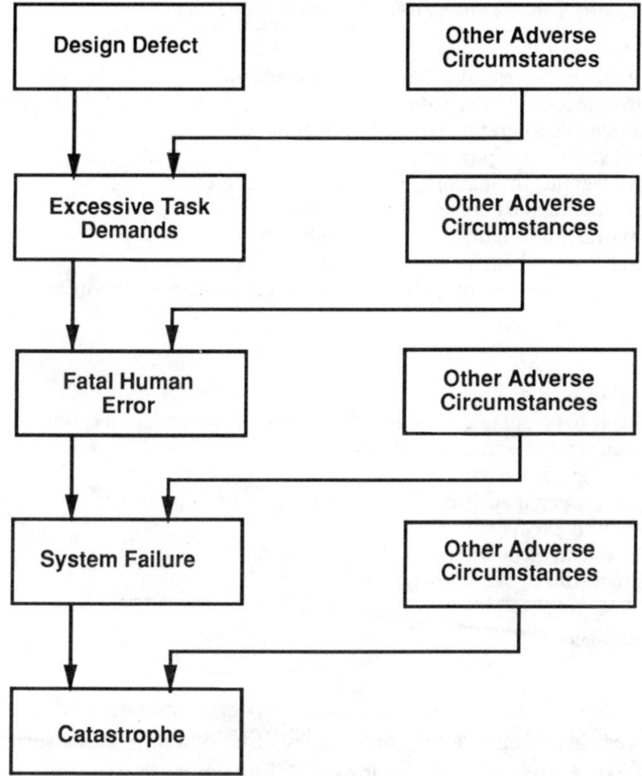


Figure 9.1 *The Zeebrugge–Harrisburg syndrome—after Pheasant (1988a)*

in the spring and early summer—but it would be difficult to determine whether this apparent clustering is significant.)

In the case of both the *Herald* and the space shuttle *Challenger*, high-level management failed to act on the misgivings of the people at the "sharp end" of their organizations. The operators at Three Mile Island had also complained about the design of their panel. Unheeded prior warnings seem to be a recurrent feature of these catastrophes.

## Interface Design

Experience suggests that certain characteristic deficiencies of interface design turn up in a wide variety of contexts. These are summarized in Table 9.2. The lack of a suitable warning display on the bridge of the *Herald of Free Enterprise* was an example of the first item on the list (A1) and the erroneous decision to cut off the emergency water supplies at Three Mile Island was an example of the second (A2). The incident at Three Mile Island also involved a number of other display problems (A4, A8, A9, etc.).

The problems at the chemical plant, described in the Prologue, where the

*Table 9.2 Checklist: Some Common Deficiencies of Interface Design*

| | |
|---|---|
| A—DISPLAYS | |
| A1 | Display not there (necessary information unavailable) |
| A2 | Display gives incorrect information |
| A3 | Display unreliable (therefore ignored by operators) |
| A4 | Display not visible/conspicuous |
| A5 | Display not legible (graduations, labelling, lighting, etc.) |
| A6 | Display not intelligible (ambiguous, unclear, etc.) |
| A7 | Display contrary to stereotype or convention |
| A8 | Confusion between inadequately differentiated displays |
| A9 | Display of irrelevant or poorly prioritized information leads to confusion |
| B—CONTROLS | |
| B1 | Control not accessible |
| B2 | Control too difficult to operate |
| B3 | Control likely to be operated accidentally (due to design and/or location) |
| B4 | Control contrary to stereotype or convention |
| B5 | Control not compatible with displays |
| B6 | Confusion between inadequately differentiated controls (too close, too similar, inadequately labelled, etc.) |
| B7 | Confusion due to inconsistencies of control layout |
| B8 | Inadequate feedback of control operation |

After Pheasant (1988b).

operators had to run up and down the stairs in order to operate a valve and monitor the results, were also due to a missing display. The outcome was a series of minor catastrophes rather than a single major one—but the basic pattern is the same. A similar problem was implicated in the chain of events which led to the great New York City blackout of July 1977, in which 8 million people lost their power supply, the subway came to a halt and looting broke out in the darkened streets. It started with an "act of God" which nobody could reasonably have foreseen: two power lines were taken out by separate lightning strikes. The operator who was trying to cope with this situation made an incorrect diagnostic decision, based on the erroneous belief that a certain part of the system was operational when it was not. The flashing light and warning buzzer which could have told him otherwise were in another room (items A1/A4 on the checklist). As the crisis mounted, he turned a master switch in the wrong direction—and the city was plunged into darkness. An ergonomist would want to know about the design of the switch, the way it was labelled, the direction in which it turned, and so on.

The massive leak of toxic gas from the Union Carbide pesticide plant at Bhopal in India in the early hours of 3 December 1984 is reckoned to have been the worst industrial accident in history. At least 2,500 people were killed and 200,000 injured. The chain of events which led to the catastrophe was complex, and circumstances are such that the full truth may never be known. Notable from an ergonomic standpoint was the fact that certain temperature and pressure gauges were known to be unreliable—so plant workers ignored the early warning signs of the abnormal situation which was developing (A3 on the checklist).

On 3 March 1974 a DC-210 crashed just after taking off from Orly, Paris. It was one of the worst crashes of aviation history—346 people were killed. A terrorist bomb was originally suspected. It later emerged that a cargo door had blown open, depressurizing the aircraft suddenly. A baggage handler had failed to fasten the door securely. He was working under pressure due to an air-traffic controllers' dispute in London—which created a narrow window of opportunity in which the aircraft could take off. Evidence from a number of previous incidents shows that the door had major design defects and was difficult to operate—excessive force was required (B2 on the checklist). The baggage handler was required to move the platform on which he stood and look through a small window using a flashlight, to see whether an inconspicuous marker was aligned (A4, A5). Instructions for locking the door were printed next to it, in two languages—neither of which the baggage handler understood (Perrow, 1984). This was the classic combination of interface design, pressure of work and prior warnings which had not been acted upon.

The potential consequences of deficiencies in the design of displays and controls have been recognized since the early days of ergonomics at the end of World War II, when Fitts and Jones (1947a, b) conducted their classic investigations of pilot error. They asked military pilots to describe critical incidents in which they had been involved. These reports were divided into categories. One particularly important category involved the misreading of multi-revolution display instruments such as the notorious "killer altimeter". This instrument has three pointers which rotate like the hands of a clock. The longest pointer indicates 100s of feet, the next indicates 1000s and the shortest indicates 10,000s. Thus it is easy to misread the altitude by 10,000 feet when coming in to land. As one of the subjects said in his interview: "pilots are pushing up plenty of daisies today because they read the altimeter wrong". The BEA Viscount crash at Prestwick in Scotland in 1958 was caused in just this way.

The solution to this particular problem is an instrument which displays altitude to the nearest 1000 feet on a digital counter and has a single rotating pointer for the 100s. The counter-pointer altimeter combines the accuracy of a digital read-out with the dynamic "feel" of an analogue dial. (Digital displays which are changing rapidly are difficult to follow, and changes of direction, velocity, etc., are not immediately apparent.) The counter-pointer altimeter has been adopted in most modern aircraft—but the old three-pointer instrument has not completely disappeared from use. Green (1989) notes that it is still found in many helicopters (although these rarely fly high enough for it to matter); and it is still used as the back-up altimeter in the Hawk jet trainer.

The largest single category of incidents reported by Fitts and Jones (1947a, b) was that of *substitution errors*—in which the pilot had confused one instrument for another. (These accounted for 50% of all the incidents involving controls and 13% of incidents involving displays.) The two particular controls which were most frequently confused were the two identical switches which operated the wing flaps and landing gear, respectively. In most military aircraft of the period these were immediately next to each other (problem B6 on the checklist). Norman (1988) notes that this unsatisfactory arrangement is still found in one popular type of light aircraft—and is still causing problems. For example, pilots may raise the undercarriage whilst driving down the runway prior to take-off.

A system-induced substitution error was involved in the incident at the Lowermoor water treatment works, near Camelford in Cornwall, on 6 July 1988, in which the drinking water supplies of some 20,000 people were massively contaminated with aluminium (and the fish populations of two local rivers were wiped out). A relief driver, delivering a lorry load of aluminium sulphate (on the wrong day), had been given somewhat vague instructions to pour it into a certain tank. He found two unlabelled tanks next to each other and dumped his load into the wrong one. He had been supplied with a key which fitted both.

Generically similar errors may be found in a wide range of everyday situations—as in the case of a young man who filled his ear with superglue instead of ear drops. (They were in similar, long-necked tubes.) Product designers often favour neat rows of identical pushbuttons, labelled as inconspicuously as possible so as not to detract from the visual symmetry of the objects (as in sophisticated video and audio products, car dashboards, etc.). In this situation, the ergonomics of the product may be in direct conflict with the aesthetic (or, more properly, stylistic) preferences of the designer.

Substitution errors become particularly probable where two actions, performed in a similar context, have elements in common: a young man walks into the bathroom carrying a pair of socks, lifts the lid of the WC and throws them in. His intention had been to throw them into the linen basket, which stands next to the WC and also has a lift-up lid.

Substitution errors are also common where inconsistencies in the layout or function of controls arise from a lack of standardization between the products of different manufacturers—or even between different products made by the same manufacturer. Thus in a new motor car you may revert to habits which you have acquired in another vehicle—perhaps the first you ever drove—and operate the windscreen wipers when you intend to indicate that you are turning left. (But you are more likely to do this if it is raining.)

There is a similar lack of standardization in the layout and operation of the special function keys on the keyboards of desk-top computers and word processors. Barber (1987), for example, notes that the first five keyboards he found within a short walk of his office had five different arrangements for the keys which move the cursor around the screen. In this particular case the consequences of an error are both minor and immediately apparent. But more severe problems may arise where the keys involved invoke conflicting functions of a more critical nature, or where the consequences of such an error are less apparent or less easy to rectify. On some keyboards, for example, the DELETE and RETURN keys are adjacent to each other—which makes it very easy to lose lines of text. On others the SHIFT key and CONTROL key are adjacent—thus when attempting to key, say a capital "O", you may inadvertently key "CONTROL O". If this is a non-existent command, the machine merely bleeps at you—but if it actually means something, then you may accidentally invoke some system function, the nature of which you do not in the least understand.

## Compatibility

Let us suppose that, as individuals, we each possess a *mental model* of the outside world—on the basis of which we plan our actions and organize our behaviour. If our mental model is at variance with reality, then our actions will be erroneous. Or if your mental model, as the designer of a product or a system, is different from my mental model, as its user (perhaps because our experiences of the world have been different), then you may unintentionally design something that I find impossible to use (Figure 9.2).

People share certain expectations concerning the ways in which everyday objects are likely to operate. For example, we expect that turning the volume control on a radio clockwise will make it louder rather than quieter; and we expect that to undo the top of a pickle jar we need to turn it anti-clockwise. Expectations of this kind are called *motion stereotypes*—and an arrangement of displays or controls which conforms to the expectations (or mental models) of its users is said to be *compatible*.

Most of our motion stereotypes are probably learned rather than innate; but children may acquire them at a remarkably early age (Ballantine, 1983), and some of them are deeply entrenched in our cultural traditions. The hands of a clock move in the same direction as the sun "moves" around the earth and the shadow moves around a sundial. The clockwise-for-increase stereotype is found in the right-handed screwthread—and our occasional encounters with a left-handed screwthread can be quite puzzling. There is a device for uncorking bottles which

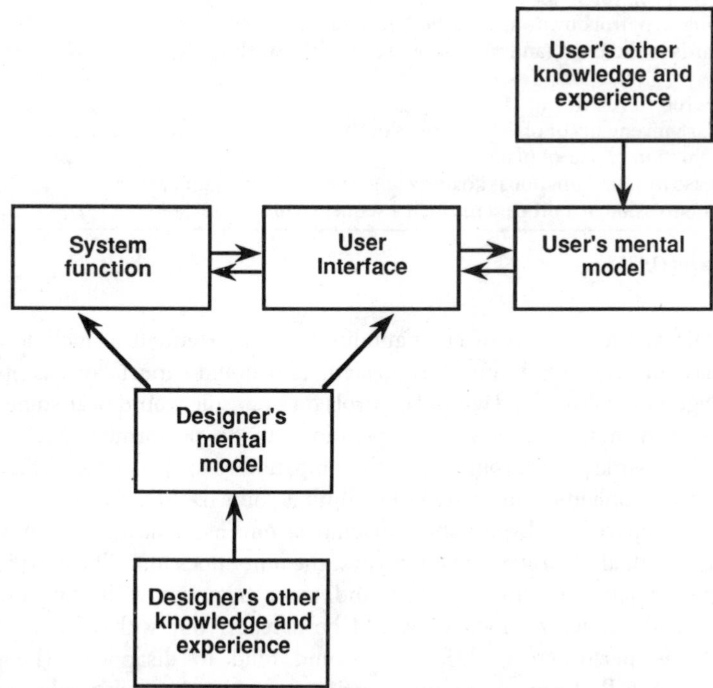


Figure 9.2 *Conflicting mental models of the system*

consists of a conventional right-handed corkscrew mounted inside a mechanism for drawing the cork which employs a left-handed screwthread. Mechanically it is very efficient—but its operation is by no means self-evident to the inexperienced user.

Motion stereotypes have been extensively investigated by ergonomists (Loveless, 1962; Murrell, 1969; Pheasant, 1986). The better-known ones are summarized in Table 9.3. But there are numerous exceptions—and some of these stereotypes are stronger or more universal than others. The difference between the British and American conventions for light switches and other vertically mounted levers (both of which are strongly established in the country concerned) is particularly troublesome. Bathroom taps (and other devices which operate valves which control the flow of fluids) violate the basic clockwise-for-on stereotype (because the valves are closed by ordinary right-handed screwthreads). This does not seem to give us any real problems—except in rare instances where the taps concerned are in unusual orientations, or make combined on–off/temperature controls, etc. (Norman, 1988, dwells on these cases at some length). Perhaps taps have their own device-specific stereotypes; the same might arguably be said of keys and door handles (Pheasant, 1986).

*Table 9.3 Direction of Motion Stereotypes*

**Actions of:**
(i) turning a control clockwise;
(ii) moving a control forward or to the right;
(iii) moving a control downwards in the UK (and some other parts of the world), but upwards in the USA (and some other parts of the world).

**Are expected to result in:**
(i) similar movements of physical objects or the pointers of displays (occurring in any combination or pair of planes);
(ii) increase in some functional quantity (e.g. speed, volume, time, etc.);
(iii) the progression of a process through a sequence of stages.

After Pheasant (1986).

By analogy, the concept of compatibility may be extended to include a much larger class of problems in interface design. We should expect, for example, that the arrangement of the displays and controls on a console would bear some orderly relationship with the sequence of operations to be performed. McCormick's principles of workspace layout deal with compatibility in this sense. The concept has also been applied to the design of computer software (p. 246).

Let us compare two hypothetical machines: one has controls which are fully compatible with all the relevant stereotypes; the other does not. The inexperienced user's performance would be faster and more accurate with the compatible machine, and his learning curve would be steeper; but with practice (and/or training) the performance difference would tend to disappear. (People are adaptable, etc.). Performance would deteriorate under stress with both machines;

but with the non-compatible machine the deterioration would be greater and the error rate higher. In emergency situations the user would tend to revert to stereotype—thus the chances of an accident would be greater with the non-compatible machine.

# Chapter Ten
# Visual Work

The design of the visual working environment and the visual tasks which an individual is required to perform are important because:

- the visual elements of the task effectively determine the posture of the head and neck;
- visually demanding work and/or adverse viewing conditions may result in visual fatigue and eyestrain;
- the visibility, conspicuousness, legibility and intelligibility of visual displays may be critical both to working efficiency and to safety;
- the aesthetics of the visual environment are important for the quality of working life.

## Light and Vision

### Units

The principal units now used in the measurement of light are the candela (cd) and the lux (lx). The energy output of a light source is known as its *luminous intensity*. The unit of luminous intensity is the candela—which, for the sake of simplicity, you could think of as the output of a standard candle.

The quantity of light energy which *falls on a surface*, like a wall or a table top, is called the illumination or *illuminance.* The unit of illuminance is the lux. One lux is the illumination you get from one standard candle at a distance of 1 metre. Some typical illuminance levels for different situations are given in Table 10.1.

The quantity of light energy which is *reflected back from a surface* is called its

*Table 10.1 Typical Illuminance Levels (lux)*

| | Illuminance (lux) |
|---|---|
| Outdoors, noonday summer sunlight | 160,000 |
| Outdoors, average clear day | 50,000 |
| Outdoors, average overcast day | 5,000 |
| Brightly lit office | 1,000 |
| Well lit office | 500 |
| Domestic living room | 50 |
| Candlelight/good street lighting | 10 |
| Moonlight | 0.5 |

*luminance.* This is the physical quality which corresponds to what we would ordinarily call the brightness of the surface. But the apparent brightness of an object depends upon physiological factors, such as the dark-adaptation of the eye, as well as on its luminance. The unit of luminance is the candela/square metre ($cd/m^2$).

The light falling on a surface, and the light reflected back from it, are related by a constant for that surface, called the reflectance. Hence:

$$\text{luminance } (cd/m^2) = \frac{\text{illuminance (lx)} \times \text{reflectance}}{\pi}$$

($\pi$ only appears in the equation because of the units we are using; if we measured luminance in a non-standard unit called the *apostilb* (asb), the $\pi$ would not be necessary: thus 1 $cd/m^2 = \pi$asb.) Typical values for the reflectance of some common materials are given in Table 10.2.

It obviously makes sense to measure the physical brightness of lamps or windows in the same units that we use for reflective surfaces. An unshielded

*Table 10.2 Approximate Reflectance Values for Some Common Materials*

| | Reflectance (%) |
|---|---|
| Fresh white plaster | 95 |
| White paint or good-quality white paper | 85 |
| Light grey or cream paint | 75 |
| Newsprint, concrete | 55 |
| Plain white wood | 45 |
| Dark grey paint | 30 |
| Good-quality printers' ink | 15 |
| Matt black paper | 5 |

After Pheasant (1987).

*Table 10.3 Illumination Levels and Luminances*

| Environment | Illuminance (lux) | Object | Reflectance (%) | Luminance ($cd/m^2$) |
|---|---|---|---|---|
| **Outdoors** | | | | |
| Clear day, noon summer | 150,000 | Newspaper | 55 | 26,000 |
| | | Grass | 6 | 2,900 |
| **Well-lit office** | 500 | Fluorescent lamp | – | 10,000 |
| | | Window (average day) | – | 2,500 |
| | | White paper | 75 | 120 |
| | | Desk top (dark colour) | 30 | 50 |
| **Outdoors at night** | | | | |
| Good street lighting | 10 | Parked car | 45 | 1.5 |
| | | Asphalt road | 6 | 0.2 |

200 W frosted light bulb has a luminance of about 50,000 cd/m$^2$. (Think of an enormous light bulb, 1 metre square, with 50,000 candles packed inside.)

Table 10.3 (which is based in part upon Boyce, 1981) might help you to build up a mental picture of what these figures represent.

## The Eye

The structure of the human eye is summarized in Figure 10.1. Light entering the eye through the *pupil* is focused by the *lens* onto the *retina*, which contains light-sensitive cells called the *rods* and *cones.* These send information up the optic nerve to the brain, which assembles a mental image of the outside world. The cones are densely concentrated in a small area surrounding the central axis of the eye, called the *fovea.* The rods are found mainly in the periphery, and have their greatest concentration in a circular band some little distance away from the fovea.

The rods operate at lower light levels than the cones; but they are not colour-sensitive and they have fewer neural connections to the brain, so they are not capable of making fine visual discriminations. In daytime we principally use the cones to build up a detailed colour picture of the world. As darkness falls, the colours of the day appear to fade—as the rods gradually take over from the cones. (The red end of the spectrum fades first and the blues may be enhanced by comparison.) In moonlight or less, we are dependent upon the rods—to build up a relatively crude picture of the world in shades of grey.

Dark-adaptation (the increase in sensitivity which occurs when we go from bright daylight into a darkened room) is partly due to dilation of the pupils but

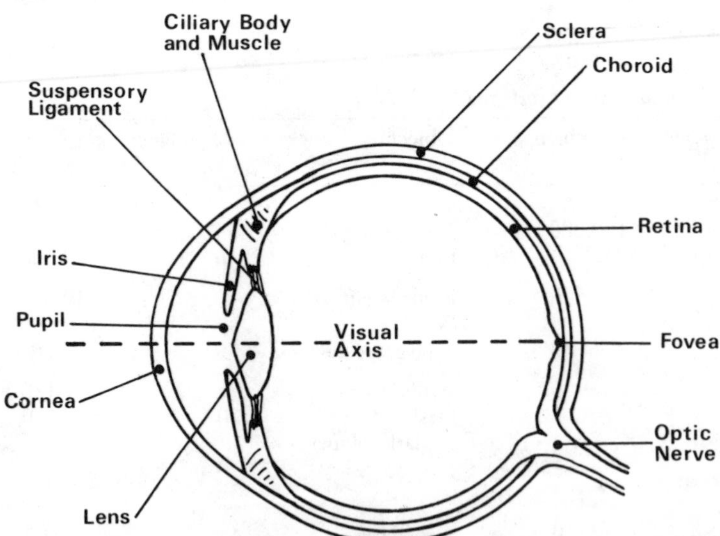


Figure 10.1 *The structure of the human eye.*

mainly to certain chemical changes in the rods and cones. The former occurs in 15–20 seconds; the latter may take anything up to 40 minutes or an hour (depending on the magnitude of the change in light levels). During this time we gradually find ourselves able to see our surroundings in more and more detail. The process is rapid at first and slower subsequently. The full range of dark-adaptation may increase the sensitivity of the eye by a factor of 100,000.

After a period of darkness, bright light may be painfully dazzling. Light-adaptation is more rapid than dark-adaptation—most of it occurs within the first second of exposure to a bright light and it is complete within about 2 minutes.

### The Visual Field

The degree of visual acuity required for fine discriminative tasks, such as reading or recognizing a face, is limited to the narrow cone of *foveal vision*, within an angle of 2° around the central axis of the visual field (the *line of sight*). You can prove this for yourself quite easily by carefully fixing your gaze on one word on a page or one face in a crowd. Surrounding words or faces will be little more than a blur. At a typical reading distance of 400 mm, the cone of foveal vision is equivalent to a circle of 14 mm in radius. Visual acuity falls off rapidly as we get further from the central line of sight. A faint signal, like a twinkling star which you can just about see, will appear somewhat brighter if you look away from it slightly—so that its image falls on the part of the retina (just adjacent to the fovea) where the more sensitive rods are concentrated.

In the periphery of the visual field, we are relatively good at detecting rapid movements "glimpsed out of the corner of the eye"—but little else except indistinct patterns of light and shade.

### Eye Movements

The "spotlight" of foveal vision may only fall on one small area at a time. The eyes of an alert person are therefore constantly in motion—scanning the visual environment, fixating on various sources of potentially interesting visual information—so that the brain may build up a composite picture of the person's surroundings. The eyes are rotated in their conical sockets by the action of the six tiny extra-ocular or orbital muscles.

In reading, the eye proceeds along the line of print in short jumps called *saccades* (from the French, meaning a sudden jerk), fixating for about a quarter of a second at a time. The average length of a saccade is about the radius of the foveal region—2° of visual angle or about 14 mm at a 400 mm reading distance. This is equivalent to about 8 printed characters (of ordinary size) or one and one-third words. Thus it takes about 75 forward moving saccades and fixations to read 100 words of ordinary text. The length of the saccades and the duration of the fixations will depend on the clarity of the print, the complexity of the text and the skill of the reader. The eye will sometimes backtrack to go over a difficult part again.

## Accommodation and Convergence

The process by which visual targets are brought into sharp focus on the retina is called *accommodation.* The lens is an elastic deformable structure which is held in place by the suspensory ligaments around its edge. Ciliary muscles are inserted on the suspensory ligaments. Normal-sighted people see distant objects clearly without accommodative effort—that is, without tension in the ciliary muscles. At viewing distances shorter than about 6 metres, the ciliary muscles are active. Their line of pull is such that they reduce the tension in the suspensory ligaments, thus allowing the lens to become more spherical and bringing the near object into focus. *The shorter the viewing distance, the greater the tension in the ciliary muscles.* This process reaches its limit at the *near point of accommodation.*

The lens gets stiffer with age. Greater muscular tensions are thus required for accommodation, and close work becomes more tiring; the near point recedes and it becomes increasingly difficult to make rapid changes in focus. In young adults the near point is typically about 120 mm from the eye (in children it is even closer). It reaches 180 mm by 40 years of age; 500 mm by 50; and 1000 mm by 60. This recession, which is called *presbyopia*, typically becomes a problem between the ages of 40 and 50, when it is no longer possible to bring fine print into focus at a distance close enough to read it. Reading glasses with converging (biconvex) lenses are thus required.

In near-sighted (myopic) people, the non-accommodated lens brings the image of distant objects into focus in front of the retina. Distant objects therefore appear blurred. Spectacles with diverging (biconcave) lenses are required for driving, etc. In far-sighted (hyperopic) people, the non-accommodated lens focuses distant objects behind the retina—so a degree of accommodation is required to see them clearly. The excess muscle tension can result in eyestrain (see below); and the range of accommodation is "used up" on distant objects, so the hyperope has a relatively distant near point.

It is true that focusing on infinity "rests the eyes" (i.e. relaxes the ciliary muscles); but it is not strictly true to say that "the resting eye focuses on infinity". In complete darkness the eye is focused at a distance which is a good deal short of infinity. This *dark focus distance* has a mean value of 660–760 mm in young adults (Liebowitz and Owens, 1975; Jaschinski-Kruza, 1988). Under normal viewing conditions the eye adopts a degree of accommodation somewhere between the dark focus distance and the true distance of the visual target. In other words, we are slightly myopic for distant objects and slightly hyperopic for near ones. The greater the difference between the dark focus distance and the true target distance, the greater the error in accommodation. Thus visual efficiency is maximal at the dark focus distance (Östberg, 1980).

The error increases in poor light (i.e. we shift towards the dark focus distance). This is sometimes called night myopia. It may be sufficiently pronounced to cause problems in night driving. Something similar happens when an individual looks into an unstructured visual field which is devoid of depth cues. Pilots flying in polar regions may suffer "white-out"—the earth below, the horizon and the clouds above are all a uniform white, and the light is diffused by snow crystals in the air. White-out was probably involved when the pilot of a New Zealand Airways sight-seeing aircraft flew directly into the side of Mount Erebus in Antarctica on a clear day, with 40 mile visibility, killing 257 people, including himself. (The accident occurred on 28 November 1979. There was also some confusion concerning the data which had been fed into the autopilot.)

As an object of fixation approaches the eye, it will be necessary for the two eyes to rotate (inwards) in their sockets, so as to bring the image of the object into focus on the fovea of each, and to superimpose the two visual fields on equivalent parts of the retina. Failure to do so results in double vision (as in the drunk). This is called *convergence. The closer the point of fixation, the greater the angle of convergence and the greater the tension in the orbital muscles.*

When viewing very close objects, the pupil also constricts. This is like "stopping down" the aperture of a camera to get better depth of focus.

## Visual Displays

The cardinal principles of display design may be summarized as follows:

- Display the information which the user requires for the task at hand—no more and no less.
- Displays must be visible, legible and intelligible—that is, it must be possible to see them, read them and understand the messages they convey.
- The displays which convey the most important messages should be the most conspicuous.

(Pheasant, 1986, 1987)

You may regard these principles as "just common sense"; but many of the catastrophic events described in the last chapter resulted from their violation in the design of high-risk systems.

### Location of Displays

*Visual Angle*

At rest, the eyes naturally assume a downward cast of about 15°. This is called the *normal line of sight* (Figure 10.2). The total range of movement of the eyes in their sockets extends 60° or more downwards from the horizontal. Weston (1953) observed people at work (in a variety of visually demanding activities) and found that they limited their downward eye movements to a little less than half of the possible range. Looking upwards, above the horizontal, is also rapidly fatiguing—the "front row of the cinema" effect.

There is, therefore, a general consensus that, for comfortable viewing, important or frequently used displays should be placed within 15° of the normal line of sight—that is, between the horizontal and an angle of 30° downwards and within

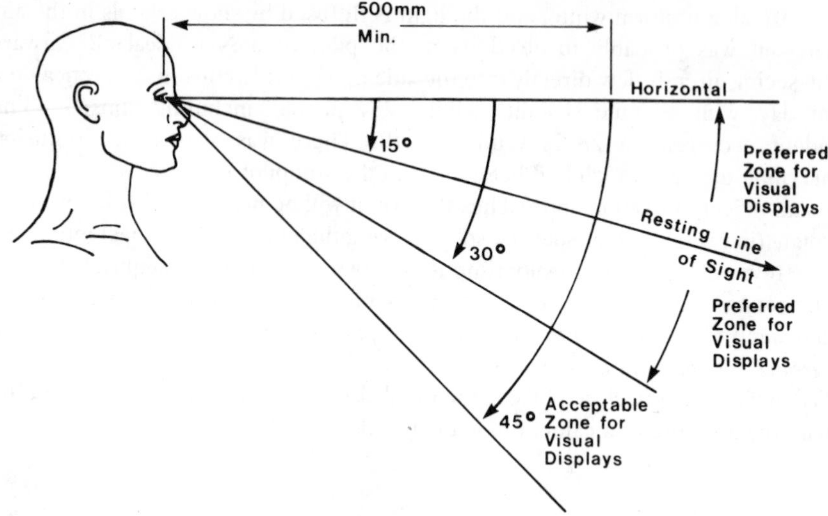


Figure 10.2 *Preferred location of visual displays—based on Pheasant (1986, 1987, 1991)*

15° either side of the straight ahead position. This defines the *preferred zone for visual displays.* On the assumption that tilting the head less than about 15° does not do much harm, locations down to 45° are considered acceptable.

These principles have been confirmed experimentally. Bennett (1977) found that subjects preferred angles between the horizontal and 15° downwards—and that dissatisfaction with the viewing angle increased more rapidly above this range than below it. In a particularly thorough field trial, in which VDU operators were given an adjustable workstation to use in their own offices for a week, Grandjean *et al.* (1983) found an average preferred viewing angle of 9° below the horizontal.

The evidence for the consensus recommendation would seem fairly conclusive therefore. But it has recently been challenged by Kroemer and Hill (1986), who reported a preferred viewing range of 29° below the horizontal for subjects sitting upright. This finding remains an unexplained anomaly. But it is worth noting that the discomfort caused by a display which is too high (i.e. above the horizontal) may be worse (or come on more rapidly) than one which is slightly too low.

*Visual Distance*

The shorter the viewing distance, the greater the muscular effort required for accommodation and convergence, and the greater will be the fixation of the posture of the head and neck. The greater the viewing distance, the more difficult it is to resolve the finer details of the display (or visual task). Displays should be bold enough (and sufficiently well lit) to be *legible* at a viewing distance which is *comfortable.*

People with reasonably normal vision tend to read average-size print at a distance of about 400 (±50) mm—bringing the page closer to the eye if the print is small or smudged, or if the light is poor, etc. This is partly a function of the ways in which you can comfortably hold a book—and the type sizes used in printing have probably evolved to suit the viewing distances required.

Recommended viewing distances for VDUs, given in standards and guidelines, etc., can fall anywhere between 350 mm and 700 mm (Helander and Rupp, 1984). The bottom end of this range is probably too close for comfort for many people. The average preferred viewing distance in the field trial of Grandjean *et al.* (1983) was 760 mm—with a range of 610–930 mm. *In the light of these findings we should regard 500 mm as the minimum acceptable viewing distance for VDU screens, and around 750 mm is preferable.* The information presented on screens should therefore be legible at the latter distance (see below). It may well be that in physiological terms 1,000 mm would be even better—but the character size required for good legibility at this distance might be unacceptably large (in terms of words per line, information per screen, etc.).

## Legibility

The legibility of a display depends upon the combination of a number of factors. The most critical of these are discussed below. In general, improvements in one factor may compensate for deficiencies in the others—as when you move the badly printed menu, in a dimly lit restaurant, right up to your face to increase the effective character size.

*Size* Under reasonable viewing conditions, a person with reasonable acuity can by definition resolve features which subtend an angle at the eye of 1 minute of arc ($1' = 1/60°$)—which is equivalent to 0.3 mm at a viewing distance ($D$) of 1 metre. Thus the minimum resolvable feature size is about 0.0003 $D$. (The exact figure depends on just how you decide to measure acuity—this point is discussed endlessly in the textbooks.) Within reasonable limits, both the legibility and the conspicuousness of a character or symbol will increase with its overall size—and more specifically with the size of whatever *discriminating features* are critical for its recognition. The capital E has five vertical discriminating features. In theory, therefore, text with a character height of 0.0015 $D$ should be just legible. In practice, the optimal character height for printed text is 0.005–0.0075 $D$. Larger characters may be required under adverse viewing conditions.

*Visual Form* Simple visual forms tend to be easier to recognize and interpret than complicated ones—partly because their discriminating features will tend to be larger, and partly because it is easier to organize simple visual patterns into unitary meaningful wholes (*Gestalten*). The Bauhaus slogan "less is more" is very relevant to display design.

*Contrast* Legibility and conspicuousness both increase with contrast—at least up to the point where the contrast becomes a source of *glare* (see below). Brightness contrast is generally defined as $(L_f - L_b)/L_b$—where $L_f$ is the luminance (or reflectance) of the figures (characters, symbols, etc.) in the display, and $L_b$ is the luminance of the background against which they are viewed.

Dark figures on a light ground are more legible than light figures on a dark ground—except in very bad light (less than 10 lux). Colour contrast increases conspicuousness but may not increase legibility, since the eye tends to create a shimmering white line along the boundary between areas of contrasting saturated colours (e.g. red and blue).

*Illumination* The legibility of a display will generally increase with illumination up to a certain point, beyond which further increases in illumination will have no effect (see below). The legibility of VDU displays is discussed in Chapter 12. For guidelines concerning the design of other visual displays see Pheasant (1987), which also includes a discussion of symbols and pictograms and of the layout of printed text.

## The Visual Environment

### Lighting Levels

The history of artificial lighting has been one of steadily increasing illumination levels. It is just possible to read and write by candlelight—which may give an illuminance on the page of about 10 lux. In the 1930s, when offices were lit with tungsten incandescent lamps, an illuminance of 100 lux would have been considered quite good. Today, rows of fluorescent tubes may illuminate offices to 1,000 lux or more. The availability of cheaper, brighter and more efficient sources of illumination has effectively solved one class of lighting problems—but it has created another. Fifty years ago, many people worked in places which were inadequately lit; today, people are much more likely to work in environments where the lighting levels are too high.

The finer the degree of visual discrimination, the more light you need. The ratio between the illumination level at which it is just possible to perform a task and the level at which performance is optimized may be something like 1:30 (Edholm, 1967). Within this range both performance and comfort will increase with lighting levels. The extent to which illumination affects performance will depend upon the visual demands of the task. For example, illumination will have a much more critical effect on the speed and accuracy of a copy typist working from handwritten originals than it has on those of an audiotypist. Similarly, lighting levels will affect safety in situations where the principal hazards can be avoided if you see them in time. But once there is enough light to *enable* a person to work with maximum efficiency, additional light cannot be expected to *improve* either his performance or his comfort.

A number of experimental studies have shown that satisfaction with the lighting of offices increases with the level of illumination, following a law of diminishing returns which flattens out somewhere between 500 lux and 1,000 lux (Boyce, 1981). Furthermore, Grandjean (1987) cites data indicating that the prevalence of dissatisfaction with lighting conditions (and of assorted visual complaints) climbs quite rapidly at illuminances of more than 1,000 lux. As Grandjean observed, it is not the excessive illumination itself that causes the problems but the fact that it tends to create reflections, glare, deep shadows, and so on. Thus an illumination range of 500–700 lux is appropriate for general office purposes—and an office illuminated to more than 1,000 lux can probably be regarded as over-lit (unless there are special task demands).

Desirable levels for VDU work are generally thought to be somewhat lower—say 300–500 lux (HSE, 1983). Left to their own devices, users of complex control rooms (where there is a mixture of VDU work and other activities) have been found to prefer even more subdued lighting levels, with illuminances on the working surfaces of only 250–300 lux (Wood, 1987). For VDU rooms, up-lighting (in which light fittings bounce light off the ceiling) may be an advantage, in that it minimizes reflections from the screens.

There are considerable national differences in recommended lighting levels. Table 10.4 is an attempt to summarize current recommendations. The ranges proposed by the *Commission International de l'Eclairage* (CIE, 1975) are essentially the same as the current International Standard (ISO 8995). The recommendations of BS 8206 for equivalent applications tend to fall close to the bottom end of these ranges—or occasionally lower. The precise descriptions of the tasks vary in the different sources—and the reader who is concerned with meeting the requirements of one of these standards should refer to the original. These figures apply to the illumination levels measured on the working surface or on an imaginary surface 850 mm above the ground.

## Daylight

There is no doubt whatsoever that most people like natural daylight in their living and working environments—and would prefer it to artificial light were they given the choice. Furthermore, most people enjoy direct sunlight in a room (at least in moderation), largely because it "makes them feel good". And most people place a high value on a good view—partly because it gives them aesthetic pleasure (perhaps by relieving the monotony, visual and otherwise, of their surroundings) and partly because a distant point of fixation relaxes the eyes. Conversely, windowless rooms are experienced as isolated, claustrophobic and depressing; and most people will only tolerate them under sufferance. (Boyce, 1981, reviews the empirical evidence for these common-sense beliefs.)

Large areas of glazing can give rise to problems of solar heat gain in summer (unless they face north), as well as problems of heat loss in winter. The windows are commonly the weak points in a building's insulation and they may set up unpleasant thermal gradients across the room. The energy cost of compensating for these gains and losses is offset (in part at least) by the cost of the artificial lighting. In practice, the heat gains seem to be the more difficult to correct—especially where the obvious expedient of opening the windows is not available, by reason of the noise outside or the wind speeds around the buildings. Windows may also be a source of glare—although this may be reduced by setting them in deep splayed surrounds which are painted white to soften the contrast between a bright window and a dark wall. On the whole, people seem prepared to tolerate a little discomfort in summer to gain the manifest advantages of good daylight throughout

*Table 10.4 Recommended Ranges of Illumination*

| | Application | CIE/ISO range (lux) | Examples | BS 8206 (lux) |
|---|---|---|---|---|
| General lighting for areas used infrequently or having simple visual demands | Public areas with dark surroundings | 20–50 | | |
| | Areas for occasional access | 50–100 | Walkways, cable tunnels | 50 |
| 20–200 lux | Rooms not used for continuous work | 100–200 | Storage areas, entrance halls | 100 |
| General lighting for working interiors | Tasks with limited visual requirements | 200–500 | Rough machining, lecture theatres | 300 |
| 200–2,000 lux | Tasks with normal visual requirements | 500–1,000 | Medium machining, offices | 500 |
| | Tasks with special visual requirements | 1,000–2,000 | Hand engraving, drawing offices | 750 |
| Additional lighting for visually exacting tasks | Very prolonged exacting visual tasks | 2,000–7,500 | Electronic or watch assembly | 1,000 |
| | Exceptionally exacting visual tasks | 5,000–10,000 | Micro-electronic assembly | 2,000 |
| 2,000–20,000 lux | Very special visual tasks | 10,000–20,000 | Surgical operations | |

After CIE (1975), ISO 8995 and BS 8206.

the year—and the problem can be reduced to some extent with blinds.

A building which is mainly artificially lit can be more or less any shape; a building which depends heavily on daylight cannot. The constraints imposed by providing for daylight penetration may have important implications for construction costs, efficient land use, and so on. In practice, these are likely to be the dominant determining factors—and people's preference for natural light commonly has to take second place.

The natural lighting inside a building is measured by the *daylight factor*, which is simply the ratio between illuminance due to daylight (excluding direct sunlight) at a certain point inside and the illuminance outside (under equivalent conditions). The daylight factor immediately adjacent to a window may be about 20%. This falls to 10% at a distance into the room which is equal to the window's height—and 2% at twice this distance (Boyce, 1981).

Given an outdoor illuminance of about 5,000 lux (a typical overcast day), it follows, therefore, that any room which is deep compared with its height must rely heavily on artificial lighting, at least in its inner recesses. Since nobody is going to make buildings taller just to get better daylight for their occupants, it also follows that this will be the case for most medium to large offices.

Many people living in northern climates become depressed in the winter; and for some people this depression may be severe. This seems to be a direct consequence of low light levels—and the symptoms may be alleviated by exposing the sufferer to 2,500 lux for about 1 hour per day (Wurtman and Wurtman, 1989). Outdoor light levels exceed this even on an overcast day—which is another good reason for a daily walk.

## Glare

Glare (or dazzle) is an unpleasant visual sensation which occurs when excessively bright objects (or, to be more precise, excessive brightness contrasts) intrude within the visual field. In some cases the glare may be severe enough to interfere with the performance of visual tasks (*disability glare*); in less severe cases glare may cause *discomfort* without functional impairment. It is rare for contrast levels in artificially lit environments to be high enough to cause disability glare, but discomfort glare is common in offices and factories.

The most common indoor glare sources are: direct view of light fittings; bright sky seen through windows; and reflections from polished surfaces (including VDU screens). The latter is called *specular glare* (Latin: *speculum*, mirror). Reflected light may also interfere with task performance when it is superimposed over the object of interest, e.g. reflections from VDU screens, glossy paper, spectacles; windscreens, etc. These are called *veiling reflections*. Bright objects in the visual field also "draw the eye" (the *phototropic effect*) and may thus be a distraction from the task at hand.

The physiology of *discomfort glare* is less well understood. The sensation of discomfort is thought to be associated with opposing action in the muscles which dilate and constrict the pupil—which occurs when different parts of the retina receive contradictory information concerning light levels (Hopkinson and Collins, 1970).

A number of formulae have been proposed as indices of discomfort glare. These have the following general form:

$$G = \frac{L_g{}^a S^b}{L_b{}^c O^d}$$

where $L_g$ is the luminance of the glare source; $S$ is its size (measured as the angle it subtends at the eye); $O$ is its angular position relative to the line of sight; and $L_b$ is the luminance of the background. $a$, $b$, $c$ and $d$ are constants. Details of these indices are reviewed in Boyce (1981).

The following recommendations for the reduction of glare in the working environment are based upon Pheasant (1987), Grandjean (1988) and the ergonomics literature in general.

- Uniformity of illumination within the working area is generally desirable, rather than highlights and deep shadows. Thus diffuse lighting, reflected from the walls and ceilings, is generally more comfortable than direct lighting—except where the latter is necessary for a particularly demanding task. Several low-power sources will generally be better than one high-power source, provided that they are located correctly (see below).
- The contrast ratio between the luminances of adjacent areas close to the centre of the visual field, such as the visual task and its immediate surrounds, should not exceed 3:1. For example: white paper ("task") has a reflectance of 75%; therefore the reflectance of the desk top ("surround") should not be less than 25%. But note that visual patterns significant for task performance (e.g. print on paper) should have a high contrast. The task should be brighter than its surround (because of the phototropic effect).
- The contrast ratio between the centre of the visual field and its periphery (or between different objects in the periphery) should not exceed 10:1.
- If the illumination on the visual task is set at 100%, then the illuminance on the walls should be 50–80% and on the ceiling 30–90%. The reflectance of the walls should be 30–70%, of the ceilings at least 60% and of the floor 20–30% (BS 8206).
- Light sources should not be placed within 60° of a person's principal working line of sight, or within 30° of a person's horizontal line of sight when his eyes are in the working position. All light sources should be shielded.
- Fluorescent tubes should be mounted at right angles to the line of sight (rather than parallel); but window surfaces should be parallel to the line of sight (rather than at right angles). Therefore, both desks and fluorescent tubes should be at right angles to the windows (this is particularly important in VDU work).
- Reflecting surfaces should be avoided as far as possible—particularly near to the centre of the visual field. Where this is not possible (as with VDU screens), care must be taken not to reflect bright images of windows, light fittings, etc., into the user's eyes.

## Stripes

Patterns of regular repeating stripes may cause curious visual effects which make them unpleasant to look at. The lines may seem to shimmer, flicker or blur, bend or wave; or illusory colours or vague shadowy shapes may appear amongst them. Many people find the experience unsettling—an effect which was extensively exploited in the Op Art school of painting in the early 1960s (as typified by Victor Vasarely and Bridget Riley).

The illusions are probably due to overstimulation of the visual cortex. People vary greatly in their susceptibility, as measured by the number of illusions they report when shown a particular pattern. Susceptibility increases with sleep deprivation—and probably with fatigue in general. Some people report visual discomfort ("eyestrain") or headaches as a result of looking at such patterns, and habitual headache sufferers report more illusions. Migraines may be provoked in susceptible individuals—and a very small number of people may suffer epileptic seizures. Stripes have a greater effect than chequers—and for chequers the effect increases with the height/width ratio. The effect is maximized when stripes are equally spaced and when the dark and light lines are equal in width. The effect increases with the contrast ratio of the stripes and the area of the visual field they occupy. The effect occurs when the stripes have a spatial frequency in the range of 1–10 cycles per degree, and is maximized at 3 cycles per degree (Wilkins *et al.*, 1984; Wilkins, 1986).

Expressing these figures in a more familiar way, we could say that the anomalous visual effects occur when striped patterns repeat themselves at an interval in the range of about 0.002–0.02 *D*, and are maximized at 0.006 *D* (where *D* is the viewing distance). The lines of ordinary printed text commonly fall in the upper part of this range, with an average of around 0.1 *D*, which may contribute to eyestrain in reading (Wilkins and Nimmo-Smith, 1984, 1987).

Striped patterns in the critical range of spatial frequencies are encountered surprisingly often in the interior decorations of public buildings—I have come across them in carpets and even in a shiny wall mosaic (presumably created by a non-susceptible artist). They can also occur accidentally. The striped metal tread of the escalators in some London underground stations is a notorious example.

## Fluorescent Lighting and Flicker

Flickering lights are also very unpleasant, probably for similar reasons, since the effects of flickering lights are exacerbated by having a striped pattern to look at; and the effects of a pattern are exacerbated by flicker (Arnold Wilkins, personal communication). The worst temporal frequency of flicker is about 16 Hz (just above the alpha rhythm of the brain). This is hazardous for photo-sensitive epileptics when used in disco strobe lights, etc.

Fluorescent tubes oscillate in luminous intensity at twice mains frequency (i.e. 100 Hz in the UK or 120 Hz in the USA). The amplitude is commonly about 40–50% of the peak value. An oscillation of 100 Hz is well above the critical fusion frequency of the eye (p. 157)—so the intermittency is not visible as such, although it may cause stroboscopic effects (particularly out of the corner of the eye). The 100 Hz oscillation is sometimes called "invisible flicker". A smaller oscillation at 50 Hz may be visible at the ends of the tubes (which should be covered); and it becomes pronounced in old tubes (which should be replaced).

Some people find fluorescent lighting disturbing—and believe that working under fluorescent lights gives them eyestrain and headaches. Their complaints are generally ignored by colleagues who are less susceptible to these problems.

Invisible flicker may be reduced by means of an electronic device called a "high-frequency solid state ballast", which cuts the amplitude of the 100 Hz modulation to about 7%. Wilkins *et al.* (1988) performed a double blind cross-over trial comparing fittings of this kind with conventional tubes in a working office environment. Subjects kept diaries of headaches and eyestrain over a 28 week period. *The incidence of headaches and eyestrain was reduced by more than half with the modified fittings, which almost eliminated the invisible flicker.* There was also less eyestrain and headache in rooms with good daylight. The special fittings cost about twice as much as conventional ones—but consume 40% less electricity and thus pay for themselves within 2 years (Arnold Wilkins, personal communication).

## Eyestrain

Eyestrain or visual fatigue (the terms mean the same) is caused by the performance of demanding visual tasks for long periods—commonly under unsatisfactory viewing conditions.

The symptoms of eyestrain are:

(i) aching or throbbing sensations in, around and behind the eyes; blurred vision, double vision, difficulty in focusing;
(ii) inflammation of the eyes and lids—giving hot, red, sore, watery eyes;
(iii) headaches (usually frontal), sometimes with dizziness or nausea; feelings of tiredness, irritability, etc.

The symptoms of eyestrain thus overlap with those of general fatigue (p. 156) and the so-called "sick building syndrome" (p. 161), as well as with referred symptoms from neck muscles (p. 85). All of these problems may be present in the same working situation.

The symptoms of eyestrain are mainly due to overuse of the muscles in and around the eye. The ciliary muscles employed in accommodation, the orbital muscles employed in convergence, the constrictors and dilators of the pupils and the muscles of facial expression may all be implicated. (We should expect the orbital muscles to be particularly susceptible to fatigue, since they are adapted for fast scanning eye movements and are therefore composed mainly of fast muscle fibres—p. 27.) Thus visual fatigue is in many aspects a special case of local muscular fatigue. People with uncorrected visual defects tend to be more susceptible. The inflammation around the eyes may be due to the fact that we blink less when concentrating.

Any working activity or visual stimulus which imposes an excessive demand upon the muscles of the eye is potentially a source of eyestrain. The most common is prolonged close work. Poor lighting and inadequate task contrasts typically result in a compensatory shortening of the viewing distance; and it is thought that, confronted with blurred images (e.g. on VDU screens), glare, veiling reflections, etc., we may tend to overaccommodate or make constant changes in accommodation in a vain attempt to bring them into focus.

Östberg (1980) showed that demanding VDU work causes an increase in the dark focus distance—coupled with an increase in the magnitude of accommodation errors for both near and distant objects. The latter effect may be regarded as a "*temporary myopization*"—which may result in a diminution in visual acuity (Haider *et al.*, 1980).

When looking into glare or when performing a demanding visual task (especially if we are stressed), we tend to frown and screw up our eyes—using the corrugator supercilii and orbicularis oculi muscles, possibly acting in opposition with the occipitofrontalis. This muscle tension is thought to be responsible for the characteristic frontal headaches of eyestrain. But some people also suffer genuine migraines when exposed to glare or other adverse stimuli. So other factors (presumably neurological) may also be involved—the work of Wilkins *et al.* (1984) supports this possibility.

The symptoms of eyestrain or visual fatigue, as described above, are typically short-lasting and rapidly reversible. But can the eyes be permanently damaged by overuse? It was once believed that people like lacemakers, who did very fine close work, became myopic as a result. But it is now thought that the apparent association was due to the fact that myopes are more likely to become skilled lacemakers (Hunter, 1978). The current consensus, therefore, is that visual work as such, however demanding, cannot permanently harm the eyes.

This consensus is not quite universal, however. People living in non-industrial societies tend to have lower prevalences of myopia. The differences are not solely due to ethnicity, since industrialized Far Eastern populations have prevalences similar to those of Europeans (Taylor, 1981). The prevalence of myopia in Western countries increases with educational level and thus with the amount of time in childhood spent reading (Angle and Wissman, 1980). Both sources of evidence point to the possibility that habitual use of the eye at a short focal length might lead to permanent myopization—but this interpretation is contentious (Bear and Richler, 1982; Taylor, 1983).

Could the prolonged use of VDUs lead to a long-term deterioration of visual function? A sizeable number of people have now been using VDUs for a decade or more—so the long-term effects (if there are any) should by now be apparent. At present, however, we do not seem to have sufficient hard data to answer the question conclusively either way.

## Chapter Eleven
# The Ergonomics of Seating

"All sedentary workers suffer from low back pain. We know when Plautus says: 'Sitting hurts your back; staring hurts your eyes'. I do not see what precautions we can prescribe for these workers, so long as the existing cause remains the same and they are driven by supplying themselves and their families with daily bread."

*De Morbis Artificum*, Bernardini Ramazzini (1713), quoting Plautus (*circa* 200 BC)

## The Physiology of Comfort

What is comfort? As far as we know, the human body has no peripheral nerve endings capable of registering a sensation of comfort as such. Comfort is a state of mind which arises in the *absence* of intrusive bodily sensations. Thus we do not notice our clothes unless they are too tight or they hamper our movements. Physiologically, comfort is the absence of discomfort, although the subjective experience of comfort may be modified by other influences.

The intrusive sensations which lead to perceived discomfort may be due to the distribution of pressure on the body's supporting surfaces or to the loading of musculoskeletal tissues. Consider the sensations which you experience sitting in a hard chair which is a little too high. If you have nothing much to occupy your mind, you may notice feelings of pressure beneath your buttocks and the upper part of your thighs. To ease the mild discomfort you experience, you fidget. You may cross and uncross your legs, thus changing the distribution of pressure beneath the buttocks (as shown in Figure 11.1). This helps restore the flow of blood to the

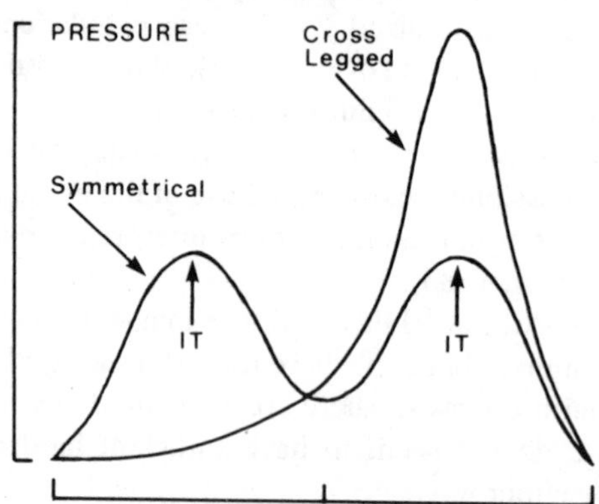


Figure 11.1 *Distribution of pressure across the buttocks of a sitting person, showing pressure "hot-spots" under the ischial tuberosities (IT). After Dempsey (1963)*

areas of compressed soft tissue. If your attention is engaged elsewhere, you will probably not be aware that you are doing this. But with time the sensations of discomfort become increasingly intrusive and fidgeting seems to bring less relief—until, however you sit, you still feel uncomfortable. The sensations you experience may begin to change in quality: your buttocks may start to feel numb, or you may get pains in your legs or notice that your feet have started to swell (all of which are due to restriction of circulation). Eventually you have to get up and walk around to relieve the symptoms.

We fidget to avoid the adverse effects which result from the prolonged mechanical loading of soft tissues and the associated disturbances of blood. The greater the tensions and compressions the chair causes, the more we fidget. Fidgeting is thus a protective mechanism.

Experiments have shown that people fidget more in chairs which they subsequently rate as uncomfortable (author's unpublished observations). But fidgeting is not usually a conscious process—at least, not until discomfort becomes extreme. Indeed, to some extent the purpose of fidgeting is to prevent the sensations of discomfort from reaching an intensity at which they would intrude into consciousness—although it is not completely successful in doing so.

We fidget more (and are subjectively less comfortable) when we are bored. Boredom apparently lowers the threshold of discomfort—perhaps because the gates of attention are not fully occupied with other stimuli. Conversely, when we are engaged in a test which demands our full attention, we may "forget to fidget". This might partly explain the association between back pain and mental workload (p. 67).

Whenever I give a lecture on furniture design, someone in the audience invariably complains about the seats. It is true that the seats in the lecture halls and seminar rooms of universities and colleges are commonly of a fairly basic nature compared with those in theatres and cinemas, for example, but they are no worse than those you find in cafés and restaurants. I suppose, therefore, that the subjective experience of discomfort is due as much to the lectures (and the need to sit relatively still for longish periods) as it is to the seats themselves.

The perceptual appraisal of comfort is subject to bias in a number of ways. Brown (1983) found that subjects who were asked to rank nine different upright chairs in order of preference for comfort sometimes gave different responses when they performed the experiment blindfolded. Thus the visual features of a chair were (in some cases) modifying the subjects' appraisal of its comfort. It seems unlikely that this effect would much outlast the stage of first impressions—and the ergonomic differences between the chairs in this particular experiment were probably not very great. But there are some chairs which are perceived as comfortable against all the biomechanical odds. In experimental studies I have sometimes observed that people fidget a lot in deckchairs—but subsequently report being comfortable. Perhaps deckchairs evoke comfortable associations (a summer afternoon on Brighton beach).

We have all noted the pleasurable sensation which occurs when we sink into a soft armchair—particularly if we are feeling tired. This is an example of what the psychologists sometimes call *hedonic contrast*: the relief of tension or discomfort is accompanied by a wave of pleasure, which is heightened by comparison. The pleasure is transient, and the pleasurable sensations we experience in these circumstances are a poor predictor of whether the chair will be comfortable in the long term. Indeed, it may well be that the very features of a chair (softness, depth of upholstery, etc.) which enhance these transient sensations of pleasure will be sources of discomfort in the long term, because the chair fails to provide postural support (Figure 11.2).

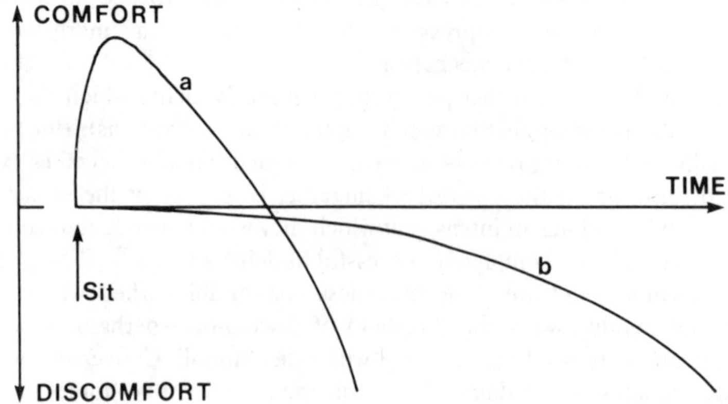


Figure 11.2 *Hypothetical curves showing sensations of comfort/discomfort as a function of time of sitting: (a) deep soft easy chair in which there is a transient sensation of comfort followed by rapidly increasing discomfort; (b) chair giving better support which does not evoke the transient effect but is less uncomfortable in the long term*

In the minds of many people there are "chairs which are comfortable" and "chairs which are good for you". This is a fallacy; it results from failing to distinguish between transient and long-term comfort. In general, a chair which is physiologically satisfactory will be comfortable in the long term, and vice versa. Indeed, the easiest way of finding out whether a chair is physiologically satisfactory is to see how long you can sit in it without feeling uncomfortable.

## Seat Design

An appropriate match between the dimensions of the seat and those of its user is a basic prerequisite both for comfort and for good posture—although it does not guarantee either. (In both cases it is necessary but not sufficient.) Some anthropometric data relevant to seat design are given in Table 11.1. The anthropometric aspects of seat design are discussed at length in my earlier books (Pheasant, 1986, 1991).

*Table 11.1 Anthropometric Data for Seat Design (all dimensions in mm)*

| | Men | | | Women | | |
|---|---|---|---|---|---|---|
| | 5th percentile | 50th percentile | 95th percentile | 5th percentile | 50th percentile | 95th percentile |
| Knee height | 490 | 545 | 595 | 455 | 500 | 540 |
| Popliteal height (floor to underside of knee) | 395 | 440 | 490 | 355 | 400 | 445 |
| Buttock–knee length (to back of knee) | 440 | 495 | 550 | 435 | 480 | 530 |
| Elbow height (from seat) | 195 | 245 | 295 | 185 | 235 | 280 |
| Mid-lumbar height (from seat) | 195 | 240 | 285 | 195 | 230 | 265 |
| Shoulder height (from seat) | 540 | 595 | 645 | 505 | 555 | 610 |
| Mid-cervical height (from seat) | 660 | 725 | 785 | 605 | 660 | 720 |
| Hip breadth (maximum) | 310 | 360 | 405 | 310 | 370 | 435 |

Estimated figures for British adults—after Pheasant (1986, 1991).

## Seat Height

The height of a seat should not exceed the popliteal height of its user (i.e. the height of the underside of the knees). For a resting chair, where the user may wish to stretch out his legs, a lower seat is preferable.

If the seat height is greater than popliteal height, the user will be unable to rest his feet firmly on the floor without undue pressure on the underside of his thighs. This rapidly becomes a serious source of discomfort. So the user will tend to perch on the front edge of the seat—thus losing the support of the backrest. A chair which is too high will be tolerated better if the front edge of the seat is rounded off.

In principle, an adjustable seat should avoid these problems—but in practice, people often set their working seats too high in order to get into a good position with respect to the table top. This is a very common problem in offices (see below). A seat which is too high is worse than one which is too low. So in a non-adjustable chair, the requirements of the short user take precedence. The popliteal height of a 5th percentile adult woman (wearing typical outdoor shoes) is 400 mm. Thus the height of a non-adjustable chair should not exceed 400 mm or 425 mm at the most.

An excessively low chair makes standing up and sitting down more difficult (see below) and it requires greater forward leg room for the comfort of tall people (which may be a problem in theatres, aeroplanes, etc.).

## Seat Depth

The depth of the seat (measured from the front edge of the backrest) should not exceed the buttock–popliteal length of a small user (5th percentile woman = 435 mm).

A seat which is too deep inevitably deprives you of the full benefit of the backrest. Either you must lean back in a flexed position with the lumbar region essentially unsupported, or you must sit forward and lose contact with the backrest altogether. Neither is satisfactory.

Seat depth is rarely a problem in upright working chairs. But serious problems arise with easy chairs. People sometimes complain that seat surfaces are too short; this is usually because the seat is too high or the backrest is too low.

## Upholstery

The ischial tuberosities are said to be specially adapted for weight bearing. The overlying skin has a rich arterial blood supply (Edwards and Duntley, 1939) and the tuberosities are sometimes coverd by a fibro-fatty pad. In both these respects they resemble the heel and the ball of the foot. The intervening gluteus maximus muscle moves out of the way as you sit down.

The pressure under the tuberosities of a large bony man sitting on a hard seat may exceed 60 psi (or about 3,000 mmHg or 400 kPa: Hertzberg, 1955). Padding the seat distributes this pressure over a wider area, leading to an increase in comfort (and in the length of time for which the seat is tolerable). Contouring the seat surface may have a similar effect—traditional Windsor chairs are surprisingly comfortable, despite the absence of upholstery. But excessively soft upholstery or inappropriate contouring may reduce comfort, because the pressure may be spread to sensitive areas such as the undersides of the thighs or the outer edges of the buttocks (beneath the greater trochanters). Excessively soft upholstery may also "bottom out" and may fail to provide proper postural support.

As a rough guide, the upholstery of a chair should be compressed by 1–2 inches when you press down hard on it with your hands. Upholstery materials should be porous (plastic chairs may become unpleasantly clammy in hot weather).

## Seat Angles

In the upright sitting position, the weight falls mainly on the narrow front portion of the tuberosities; as the pelvis tilts backwards, the weight is taken on the wider posterior part of the tuberosities; and finally the coccyx and/or sacrum come into contact with the seat surface. Contact between the seat and the sacrum is probably undesirable, since the upward force acting on the sacrum will tend to rotate it backwards at the sacro-iliac joints, and the ligaments of these joints are better adapted to withstand forward rotation. Åkerblom (1948) has shown that the sacrum is still clear of the seat surface with a pelvic inclination of 30–40°—but this may easily be exceeded, for example, when watching television in a backward slumped position.

The greater the rake of the backrest, the better it supports the weight of your trunk—and the less the compressive loading on your intervetebral discs (see Figure 11.4). *To be relaxed you must be laid back.* The greater the angle between the seat surface and the backrest, the greater will be the angle between the trunk and the thighs, and the easier it will be to maintain a lordosis. But an excessively laid back position may not be compatible with working activities. For many purposes a backrest with an angle of 100–110° with the horizontal is about right. This includes most seats for drivers and passengers in vehicles and for VDU operators. For easy chairs a steeper rake may be better.

As your back engages the backrest, the horizontal component of the force tangential to the backrest surface will tend to slide the buttocks forward in the seat. Branton and Grayson (1967) analysed time-lapse film of passengers on a train. There was a characteristic pattern of movements lasting 10–20 minutes, in which the individual slowly slid forward into an extreme slump and then regained the upright position in a single abrupt action. Dillon (1980) noted a similar pattern in people sitting in easy chairs. During the sliding phase, the individual may cross and uncross the legs, causing the buttocks to "walk" forward on the seat. The tendency to slide forward will be reduced if the seat surface has high-friction upholstery and a slight backward tilt (e.g. about 5°).

## The Backrest

Backrests may be roughly classified into low (supporting the lumbar region only);

medium (extending to mid-thoracic or shoulder level); and high (supporting the head and neck as well).

In general, the higher the backrest, the better the postural support it will give. For rest chairs, a backrest extending to shoulder height (95th percentile man = 645 mm) is desirable—and one which supports the head and neck is even better. A backrest which supports the lumbar region only is often recommended for working chairs, so as to leave the shoulder girdles free to move. The importance of this consideration may well have been overestimated (p. 228).

A backrest which is raked to more than about 100° should extend at least to mid-thoracic level (say about 500 mm above the seat surface), otherwise the upper part of the trunk becomes uncomfortably unstable. Conversely, a backrest extending to mid-thoracic level (or higher) which is close to the vertical tends to be unsatisfactory because it tends to tilt the upper part of the trunk slightly forward. (This happens in some dining chairs.)

In general, a backrest (of whatever height) should be contoured to the shape of the spine. Andersson *et al.* (1974) showed experimentally that a pad in the mid-lumbar region, to support the lordosis, reduces intra-discal pressure at any backrest angle (Figure 11.3). (Conversely, a pad in the thoracic region which pushes the spine into flexion increases the pressure.) Figure 11.4 shows a correctly designed backrest, extending to mid-thoracic level, as conceived by Åkerblom (1948). Note the space between the seat and backrest which accommodates the user's buttocks. The user may alter the posture of his spine by sliding his buttocks forward in the seat. Some degree of alternation between flexion and extension is probably desirable. A well-designed seat will offer the user lumbar support, but it cannot enforce lordosis, whereas a badly designed seat enforces flexion.

The lumbar pad of a chair should support you in the same place as you naturally

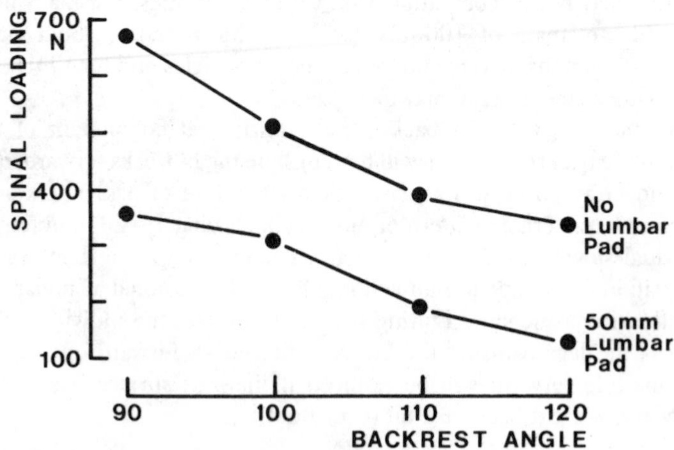


Figure 11.3 *Spinal compression (measured directly by needle-mounted transducer) at seat back angles from vertical (90°) to reclined (120°); with and without a pad in the lumbar region. Data from Andersson* et al. *(1974)*

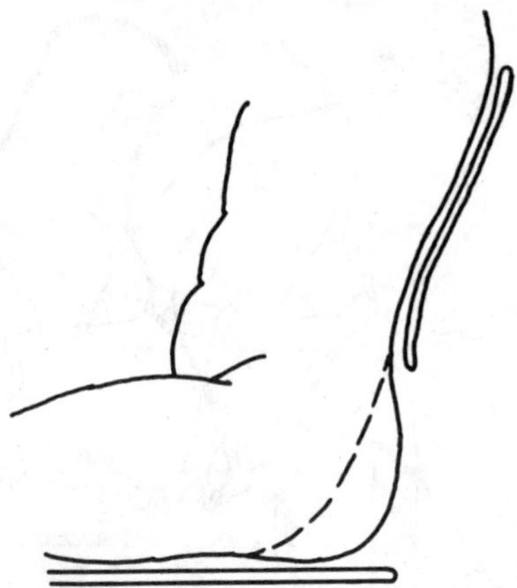

Figure 11.4 *The function of the backrest—according to Åkerblom (1948)*

support yourself with your hands to ease an aching back. The mid-point of the lumbar curve (L3) is relatively close to the seat surface, so the anthropometric variability of the distance between the two is not very great. Radiological studies have shown that the exact point at which the lumbar curve is supported is not very critical for the maintenance of lordosis (Andersson *et al.*, 1979). Furthermore, the lordosis has a relatively gentle curve. So it is possible to design a non-adjustable backrest with a lumbar support which suits most (if not all) users. The mid-point of its curvature should be about 230 mm above the seat surface. But an adjustable support would be better. The curvature of the lumbar pad should not be excessive. A curve which is too pronounced may sometimes be worse than no curve at all. Andersson *et al.* (1979) found that a pad which stands out 40 mm from the plane of the backrest gives a lumbar lordosis which is similar to that of the standing position.

Ideally, the upper part of the backrest of a high-backed chair should match the contours of your neck and the back of your head (occiput)—so as to support you in the same way as you do naturally when you clasp your hands behind your head. The height of the nape of the neck above the seat surface is more variable than that of the mid-lumbar point, and the curvature of the neck and occiput is much more pronounced than that of the lumbar region. So it is not possible to design a single contour which fits everybody. A movable cushion is a good solution.

## Standing Up

The action of standing up (and, to a lesser extent, sitting down) is commonly a problem for elderly people and those whose capacities are otherwise impaired (particularly if they are overweight). In a low deep seat, the centre of gravity of the body is well behind the feet. To stand up, the body must be propelled upward and forward, like a projectile, until a new position of balance is reached in which the centre of gravity is above the foot base (Figure 11.5). Difficulties arise when the

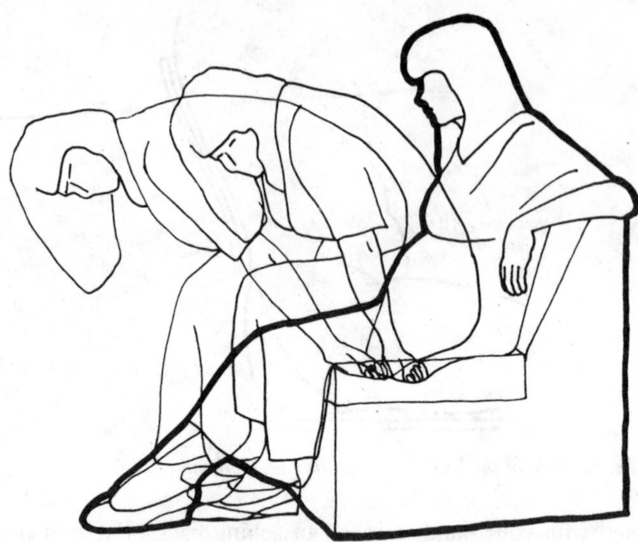

Figure 11.5 *Rising from an easy chair. Note the lowness and depth of the chair and the lack of clearance beneath the front edge. (Subject was in the 8th month of her pregnancy.) From an original kindly supplied by Jackie Nicholls*

power output of the lower limb extensors (particularly the quadriceps) is not sufficient to give the body weight the momentum which is required to carry it forward over the unstable position. In other words, the power/weight ratio is inadequate.

The problem increases with the lowness and backward tilt of the seat surface, the depth and softness of the seat, and the rake of the backrest. The problem may be reduced by providing adequate handgrips on the arms of the chair, and by ensuring that there is a clear space beneath the front edge of the seat (unobstructed by rails, etc.) so that the horizontal distance between the feet and the centre of gravity may be minimized.

## Rest Chairs

A rest chair should support your back, to shoulder level or more, in a reclined position, and help you to maintain your natural lordosis without muscular effort. Most armchairs and sofas fail to do this because they are too deep and too soft, and their backrests are inadequate. They are also often lower than they need to be—which makes standing up and sitting down needlessly difficult.

The seat depth of a full-sized armchair or sofa will typically be in the 500–600 mm range—although depths of 650 mm or more are by no means unknown. An average armchair therefore has a seat depth equal to the buttock-popliteal length of a 95th percentile adult man (550 mm). Most modern armchairs and sofas are therefore too deep for most people. It is quite impossible to sit "correctly" in such chairs—however conscious you may be of "good posture".

The tendency to slide forward in the seat becomes particularly pronounced in easy chairs which have a low backrest. The natural response to a backrest which is too low is to slide forward until your shoulders are fully supported and your head is resting on the top of the backrest. You are thus supported at two points: the upper back/shoulder region and the buttocks. The lower back is slung between the two in an unsupported state—so if its structure is not stiffened by muscle tension, it will necessarily bend into flexion until the ligaments and soft tissues take the strain. Tall people may find themselves tending to slide right off the front edge of the seat—and may thus complain that the sat is too short.

Why is the functional design of the modern easy chair and three piece suite so unsatisfactory? It is partly a matter of the features which sell an armchair in the showroom: transient comfort and the appearance of opulence which a deep, broad, bulky armchair tends to convey. Style is also involved. Designers from the early twentieth century onward have tended to regard a chair as well proportioned (visually, that is) if the depth of the seat is about equal to the height of the backrest. We see this in design classics by Marcel Breuer, Mies van de Rohe, and so on. These are really art objects: which is odd, when the modern movement in design is sometimes referred to as "functionalism". Paradoxically, the most functional easy chairs that you will find in modern furniture stores will often be ones which are copied from antique originals—like the William and Mary winged armchair (*circa* 1685) which is the basis of some of the best designs available today. In practice, people adapt to the furniture they have bought. They may use cushions to shorten the seats of the armchairs and supply the missing lumbar support. Or they may cease to use them as seats as such, but rather as padded surfaces for curling up on or for sprawling (Pheasant, 1986).

### Upright Chairs

Upright chairs are generally intended to be used at tables. The seat height of an upright chair will often be dictated by the height of the table at which it is to be used.

Dining room tables tend to be fairly high—commonly around 750–770 mm. Perhaps this is why dining room chairs are also rather high—commonly 440–475 mm. Taking popliteal height +25 mm as our criterion for maximum acceptable seat height (which is generous) and assuming that people wear 25 mm heels indoors (which is also generous) it is possible to calculate that:

- a 440 mm seat is too high for 4% of men and 36% of women;
- a 475 mm seat is too high for 31% of men and 82% of women (for the basis of these calculations, see Pheasant, 1991).

Is there any functional reason for a dining room table to be so high? Tables in cafés and restaurants tend to be lower—commonly around 720–740 mm (about the same height as an office desk). But the chairs are usually similar in height to those you find in the home. Ergonomically, a 740 mm table with a 400 mm chair would seem about right.

If the chairs concerned are really too high for a substantial proportion of the user population, why don't these people complain? Is it that people are so used to being uncomfortable that they find it unremarkable? Or perhaps you don't sit in a dining room chair for long enough for it to matter?

## Seating and Back Pain

Do people who suffer from back pain need special seats? Or are they simply less able to tolerate ergonomic failings in seat design which other people fail to notice? The available evidence supports the latter view.

People who suffer from back pain sometimes report that the combination of a steeply tilted seat surface with a relatively vertical backrest is particularly likely to trigger their symptoms. The acuteness of the angle between seat and backrest (less than 90° in the worst cases) forces the lumbar spine into flexion—and the tilt of the seat holds it there. This particularly unfortunate combination occurs in some stacking chairs. People who do not suffer from back pain also find these chairs uncomfortable; but the discomfort which they suffer is minor by comparison and they are unlikely to comment on it unless something draws it to their attention (like sitting on the chair during a lecture on ergonomics or taking part in an experimental trial). The same chair is thus a source of minor discomfort to some people and intense pain for others—and presumably for similar reasons. The people with back problems are less able to tolerate the degree of spinal flexion which the seat imposes.

In a lengthy series of fitting trials using an adjustable "seating machine", Grandjean (1973) found that the preferences of people with and without back problems were similar: both chose a reclined backrest with a contoured pad in the lumbar region.

Back pain is often a problem for pilots. Fitzgerald (1973) provided RAF aircrew who suffered from back pain with customized lumbar supports made out of fibreglass covered with polyether foam and a ventile fabric. These were moulded to the individual's own spinal contour and worn beneath his outer flying clothes. Of the 200 aircrew who were tested, 59% were subsequently symptom-free and 31% reported a marked improvement in their comfort.

Clinicians who believe that back pain is commonly caused by excessive lordosis (p. 103) recommend sitting positions which flatten the lumbar spine. This advice may be rejected on two grounds: (i) there is no epidemiological evidence of an association between hyperlordosis and back pain; (ii) there is good evidence that supporting the lordosis *both* reduces the compressive loading on the spine (Andersson *et al.*, 1974) *and* relieves back pain (Fitzgerald, 1973).

I possess quite an extensive collection of books and leaflets, prepared by well-meaning people, on how to avoid back problems. Some of these recommend a flat-backed sitting position; others recommend the maintenance of a lordosis with equal conviction. The patient who reads a number of these leaflets, with the aim of learning how to manage his back problems, may find himself in receipt of a range of conflicting advice.

The currently available evidence suggests that the pro-lordosis school of thought has the better case. To clarify the problem further, I decided to investigate the seating preferences of people with low back problems, using structured interviews and questionnaires. Fifty subjects, with back problems of varying degrees of severity, took part. (I am indebted to osteopaths Sheila Lee, Gwyneth Downey and Ronald Klein for their assistance in this study.) The subjects were shown two pictures: one showing a person sitting with a lordosis, the other showing him sitting with a flexed spine. The majority of subjects (80%) picked the lordotic position as the one which they would find more comfortable and more likely to relieve their symptoms (60% said it was much better and 20% said it was a little better). Some subjects (8%) said that the two positions were "about the same" (presumably implying that they would find them equally uncomfortable). But a significant minority (12%) expressed an active preference for the flexed position (8% said it was much better, 4% a little better). Other findings supported this overall picture: a sizeable majority (74%) said that it helped to put a cushion behind the lower part of the back, whereas minorities said that it helped to sit with the knees higher than the back (30%) or with the legs crossed (26%)—both of which tend to flex the spine.

People with back problems thus seem to fall into two different groups: the majority who show diminished tolerance of flexion and who therefore prefer to be supported in lordosis; and a significant minority for whom the opposite is the case. But I do not think that the existence of a minority of back pain patients whose symptoms are relieved by flexion has serious implications for chair design, since it is a great deal easier to flex your spine in a seat which provides lumbar support than it is to maintain lordosis in one which does not.

## Forward Tilt Seating

It has been argued that one way of reducing the spinal flexion which occurs in the sitting position might be to tilt the surface of the seat forward. The idea dates back to the work of the German anatomists of the last century, but its current popularity is due to Mandal (1976, 1981), who proposed that the seat surface of an office chair should be pivoted to tilt 15° forward or 5° backward as required.

In theory, a 15° forward tilt (which is the most that is reasonably practicable) should halve the backward rotation of the pelvis which occurs when changing from the standing to the relaxed sitting position (p. 104)—and thus reduce the muscle tension required to sit upright within a lordosis. It should also reduce the spinal flexion which is required for forward leaning tasks such as writing at a desk.

Bendix and Biering-Sørensen (1983) confirmed that tilting the seat forward to a maximum of 15° did indeed progressively increase the lordosis of subjects who were reading at a desk—although the effect was smaller than expected. Subjects preferred a 5° forward tilt to any other seat angle; and preferred a seat which was free to tilt between 5° forward and 5° backward compared with seats fixed in either position. (The subjects in these experiments all had "normal" backs.)

The trouble with forward tilt seating is that you may find yourself tending to slide off it. This is partly resisted by the frictional qualities of the upholstery. The advantage is less for women, who may have a low friction interface between their outerwear and their underwear—and may thus slide out of their skirts. To stop yourself from sliding, you press forward with your feet, which requires a steady tension in the calf and thigh muscles and may cause the chair to roll backwards on its castors (Figure 11.6).

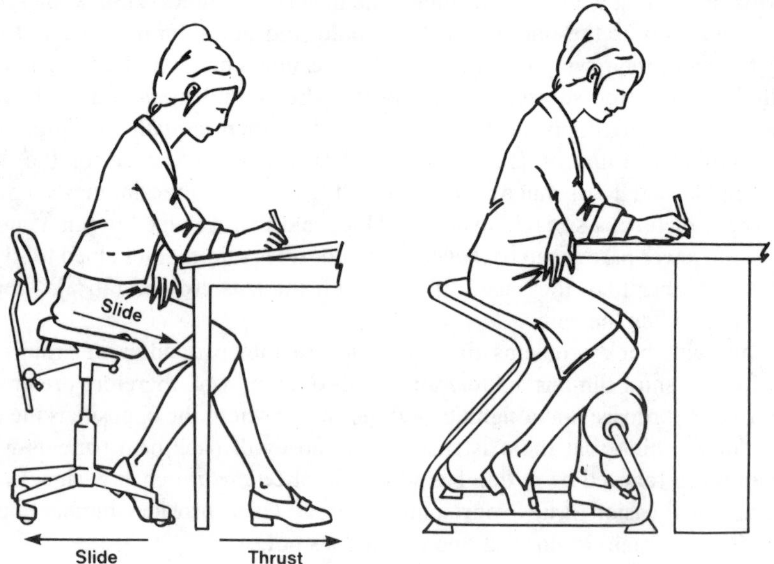


Figure 11.6 *The forward-tilting seat (*left*) and the knee-support chair (*right*). Reproduced from Pheasant (1986)*

The tendency to slide forward will probably be negligible at a 5° tilt (provided that the seat surface is not slippery)—but by 15° it may become quite marked. The difficulty experienced seems to get less with practice. Mandal (1981) states that it takes users 1–2 weeks to get used to one of his chairs. To overcome this learning effect, Bendix and Biering-Sørensen (1983) gave their subjects a forward tilt seat to use in their homes for at least 2 weeks before they took part in the experiments. It therefore seems that, from the point of view of subjective comfort, the disadvantages of a fixed 15° forward tilt outweigh the advantages to be gained from the slight reduction in spinal flexion.

These difficulties are avoided in the backless forward tilt seat with a knee support. Brunswic (1981) measured the lumbar lordosis in subjects performing typing and writing tasks in a knee-support chair and a conventional office chair (with a slight backward tilt). She found no significant difference. Bendix *et al.* (1988) compared a knee-support chair with a chair which tilted from 10° backward to 30° forward. Subjects performed office work and played solitaire to simulate industrial assembly work. The knee-support chair gave a more pronounced lordosis, although the spine was still considerably more flexed than in the standing position. But the subjects considered the knee-support chair more fatiguing to use for long periods.

Drury and Francher (1985) investigated the comfort of a kneeling chair for subjects who were typing or working at a computer terminal. They concluded that the knee-support chair was no more comfortable than conventional chairs in general—and could be less comfortable than well-designed office chairs. The discomfort was particularly pronounced in the knees and shins, and there was no compensatory improvement in the comfort of the back. They noted that the subjects often sat with flexed spines.

The latter observation is particularly interesting—and I have noticed the same thing myself. The spinal flexion which occurs in the sitting position is only partly due to the limited range of motion of the hip; people also flex their spines to rest their back muscles by "hanging on the ligaments" (p. 104). We should expect people to do this—at least for part of the time—in any chair which has no backrest (regardless of the seat angle).

It could be argued that the durations of these trials have been too short to allow the subjects to become accustomed to the novelty of the knee-support chair. This does not seem to be the case. Hart and Dillon (1987) tested a knee-support chair and an office chair with optional forward tilt, in an extremely sophisticated user trial in which 18 female subjects used each chair in their own offices for 3 weeks. Overall, the knee-support chair was rated significantly less comfortable than either the chair with optional tilt or the subjects' own office chairs.

All the trials cited so far involved subjects who did not suffer from back pain. Perhaps people with back pain would give a different result. (For people with a low tolerance of spinal flexion, the advantages of the knee-support chair, in terms of the improvement of lordosis, might outweigh the disadvantages.) Anecdotal reports suggest that a minority of people with low back problems find knee-support chairs extremely beneficial—but many do not. Atherton *et al.* (1982) tested a knee-support chair, together with 8 different conventional office chairs, on a group of 20 subjects with various musculoskeletal problems, half of whom had low back pain. Some of the conventional chairs were very sophisticated in terms of their adjustment and so on; others were quite primitive. Subjects used the chairs in their own offices for one day each. The knee-support chair tied, with two others, for fifth place out of nine in the overall rank order of preferences. The most popular chair was one which had a seat which could be tilted (using a lever control) between 5° backward and 6° forward.

The knee-support chair has three principal disadvantages:

(i) It has no backrest—thus, at best, the loading of the spine cannot be less than it is in the standing position.

(ii) It fixes the lower limbs, with the knees in flexion, well past the comfortable middle third of the range. (Depending on the design, the knee angle ranges from about 120° to 135°—out of a total range of 160° when kneeling with a fully compressed thigh.) This makes fidgeting difficult. You might expect it to restrict blood flow—although according to Lander *et al.* (1987) this is not the case.

(iii) Standing up and sitting down is difficult.

The results of the user trials cited above may be summarized as follows:

(i) Both ordinary seats with a fixed forward tilt and knee-support chairs have disadvantages, which for the majority of people (both with and without back pain) are likely to outweigh their benefits—at least insomuch as subjective preference is a valid criterion.

(ii) Chairs with an optional forward tilt (about 5°), under user control, have a high degree of user acceptability for office work.

## The Office Desk

The following anthropometric criteria are widely accepted:

- For *writing* (and general office work), the desk top should be about 50 mm above the user's elbow height.
- For *typing*, the home row of the keyboard (i.e. the middle row of letters) should be approximately level with the user's elbow.

Elbow height is measured in the upright sitting position with the shoulders relaxed and the arms held vertically at the sides. When writing, the forearms should rest comfortably on the desk top, with the shoulders relaxed and the arms abducted slightly and slightly flexed at the shoulder joints. When typing, the arms should hang vertically by the sides, with the shoulders relaxed and the forearms more or less horizontal. But note that, for VDT users, the latter recommendation has been challenged (p. 237).

The home row of a conventional electromechanical typewriter is generally around 70 mm above the table surface. Thus typewriter tables have traditionally been lower than writing desks, so that a secretary may turn from one to the other without needing to adjust her seat. But the distinction is fast becoming irrelevant as conventional typewriters are being replaced with the thin keyboards used with word processors.

*Office desks are made at standard heights; office workers are not.* An adjustable chair does not solve this problem completely. A person working at a desk is in contact with supporting surfaces at three points: the desk (or keyboard), the seat and the floor. To provide the best possible working posture for a range of users, two out of three of these points must be adjustable in height. Adjustable chairs are now the norm in the UK (although fixed upright chairs are still encountered—particularly in government offices). In Europe adjustable office tables are becoming increasingly common; but in the UK they remain rare. As a bare ergonomic minimum, all office workers should be provided with an adjustable chair—and taught how to adjust it (see below). Adjustable tables would doubtless be an advantage, although some postural problems can be solved by adjusting the floor level instead (by means of a foot rest).

At present, most office tables in the UK are manufactured to a standard height of 720 ± 10 mm (BS 5940). The same height is recommended in the draft version of a British Standard for VDU workstations which is in preparation at the time of writing. Both standards also include a specification for an adjustable table of 670–770 mm in height. Standards in other countries are generally similar. Older office desks are often higher than the current standards.

Suppose we say that the most comfortable height for an office chair is 25 mm below the popliteal height of the user (p. 216). Thus the optimal height of the table would be popliteal height + sitting elbow height + 25 mm. On average, this works out at exactly 720 mm for adult British men and women wearing outdoor shoes. Anthropometrically, therefore, the standard fixed-height desk is just about right for a hypothetical average person doing general office work (although the same person might have to adjust his or her chair a little too high, for his or her comfort, to use the word processor).

But experience suggests that the standard desk may be less than satisfactory for people who are markedly taller or shorter than average. A girl with short legs will have to adjust her chair too high for comfort, to get into a satisfactory working position with respect to her desk (especially when using a word processor). To compensate, she will tend to perch on the front edge of the seat—thus losing the support of the backrest. *Experience suggests that this is a common cause of back problems in office workers.* Alternatively, she may lower the seat so as to reduce the pressure under her thighs, and thus find heself working at a level which places an unacceptable strain on her shoulder muscles. *This is an equally common cause of neck and shoulder problems.*

Conversely, the tall man will find that his spine is unacceptably flexed however he adjusts his seat—either because he is stooping down over his work or because his knees are higher than his hips.

It is not possible to calculate precisely the percentage of people who will be affected by these problems; there are too many factors involved and the criteria which define an unacceptable match are not sufficiently distinct.

As a very rough guide, we might say that anyone taller than about 1,785 mm (5′10″) might well be more comfortable at a higher desk: that is, about 25% of men and a few very tall women. And anyone shorter than about 1,610 mm (5′3″) would probably benefit from a footrest: that is, about 50% of women and about 3% of men.

Footrests in offices are something of a rarity. Yet it is not unusual to see girls in offices with their chairs adjusted so high that their feet scarcely touch the ground; or with the keyboards of their typewriters and word processors at heights which cause them to elevate or abduct their shoulders to a marked extent. They seem to tolerate the discomfort involved and do not even complain (perhaps they feel that no one would listen). And when they start getting back, neck or shoulder problems, they rarely associate them with the design of their workstations.

### Leg Room

There should be clearance beneath the desk for the user's knees and for him to stretch his legs and change position; and there should be clearance between the seat surface and the underside of the desk for the thickness of the user's thighs: 200 mm allows a little leeway for a 95th percentile. Ideally it should be possible to cross the knees. It may be difficult to reconcile this requirement with the working height criteria given above, particularly for keyboard users, who must accommodate the thickness of the table top and keyboard in the relatively small space between the tops of the thighs and the undersides of the elbows. Thin keyboards reduce this problem. Kneehole drawers on desks should be avoided—these are uncommon on office desks (except antiques), but cash drawers are a problem at service tills, checkouts, etc. (See Pheasant, 1986, for a further discussion.)

### Sloping Desks

In Victorian times desks used to slope by up to 30°. Nowadays sloped writing surfaces are used only by draftsmen, etc. Many people believe that sloping desks are a good thing, because they encourage you to sit up straight. Hira (1980) observed students using a tutorial room which was equipped with desks which could be varied in angle between 25° and the horizontal. The students, who were unaware that they were being studied, showed a marked preference for angles of 10–15°. The principal disadvantage of a sloped writing desk is that things tend to roll or slide down it—so you need an adjacent horizontal surface where you can rest things conveniently. Reading stands are also a good idea—particularly for keyboards users (p. 237).

## Office Chairs

Typists' chairs have traditionally had low backrests—supposedly to leave the shoulders free to move. In the days of heavy-action mechanical typewriters this might have been valid; but with the advent of light-action electronic keyboards it is fast becoming irrelevant. (The traditional distinction between typists' chairs and executive chairs probably owed as much to status as it did to ergonomics.)

It is sometimes argued that the backrest of a chair is relatively unimportant in office workers, who lean forward much of the time. This is incorrect. In a study using time-lapse photography, Grandjean and Burandt (1962) found that people doing general office work were in contact with the backrest for 42% of the time. (They noted that most of the chairs which these office workers were using did not have particularly well-designed backrests and that adjustable backrests were used much more frequently than rigid ones.) The time spent in contact with the backrest would probably have been higher in a population of keyboard users.

Hünting and Grandjean (1976) found that a majority of office workers (75%) preferred a high-backed office chair to a conventional typists' chair; and a greater majority (89%) preferred a version of the high-backed chair which tilted back into a reclining rest position. (But although there was a general preference for the tilting version, many subjects complained that they did not have the option of fixing it if they wished to do so.)

The following recommendations may be made concerning the design of office chairs:

- The height of the seat should be adjustable (BS 5940 recommends a minimum acceptable range of 420–500 mm; Pheasant, 1986, suggests 380–535 mm). The front edge of the seat should be rounded to minimize the pressure under the thighs.
- The backrest should be about 500 mm in height.
- The backrest should tilt so as to support the user in both the working position and the resting position.
- The seat should tilt both forward and backward (5° in each case).
- It should be possible to lock both the seat and backrest in position. If the seat and backrest do not move separately, the angle between them should be about 105°.
- The seat base should be on castors, with a 5-point support (for stability).
- All controls should be easy to operate from the sitting position.

*Armrests* may be a great disadvantage in an office chair if they prevent the user from getting as close to the table as he would otherwise be able to do. On the other hand, some keyboard users like to rest their elbows as they work. If armrests are provided to satisfy this requirement, they should not extend more than about 200 or 250 mm in front of the seat back.

## Adjusting the Workstation

Office workers often fail to take full advantage of the adjustment mechanisms of their chairs. Sometimes this is because the mechanisms are excessively difficult to

*Table 11.2 Adjusting the Office Chair*

| | |
|---|---|
| **Stage 1.** | Adjust the chair so that you are in a convenient working position, with respect to the table, for the task concerned. (Ignore, for the moment, the height of the chair from the floor.) You may need to strike a compromise between the requirements of writing, keying, etc.<br>Check your working posture: Are your shoulders relaxed? Are you in contact with the backrest? Are you crouched over too much? Have you got adequate knee-room, etc.? |
| **Stage 2.** | Now see whether the seat is at the right height above the floor. Check that your feet are flat on the floor and your back is firmly against the backrest.<br>Can you feel undue pressure on the underside of your thighs?—if so, you probably need a footrest.<br>Are your knees noticeably higher than your hips—if so, you probably need a higher desk. |

operate (Kleberg and Ridd, 1987). Often the mechanisms are broken.

Adjusting a chair to get the most comfortable working position at a fixed-height table is best done in the two stages described in Table 11.2. If the chair and table are both adjustable, then the optimum combination may be found by trial and error—although it is probably better to adjust the chair first and then set the table to match.

Note that VDU users may prefer a reclined sitting position with the keyboard somewhat higher than the elbows and the wrists supported (p. 237). And users of chairs which tilt forward may prefer a relatively high seat.

Making a footrest is a simple job for a handy person. The height is best determined by trial and error (e.g. using books, etc.). Some people like them sloped. Its top surface should be big enough to allow for fidgeting, etc. (e.g. 400 mm square), and can be covered in carpet, etc. At a push, a desk which is too low may be raised on blocks, bricks, etc. (but check that it is safe and stable). When Ronald Reagan became President of the United States, he had the antique desk in the Oval Office of the White House raised, so that he could get his knees under it.

## Chapter Twelve

# Working with Computers

There was a time when people entered their data into computer systems using punched cards and paper tape. This was slow and inefficient—as those of us old enough to remember the consequences of dropping a card deck will testify. On-line interactive terminals were a rarity—outside a few specialized areas of application. The increasing availability of low-cost computing power in compact units of hardware has changed all this. From the early 1970s, visual display units (VDUs) have been an increasingly common feature of working life—in offices, shops and on the factory floor. As an increasing number of industrial processes are becoming computer controlled, the VDU is becoming the universal, general-purpose display medium and interface. The trend will certainly continue. It has been estimated that, by the end of the century, at least 50% of workers in the advanced industrial societies will be engaged in terminal-based tasks (de Matteo, 1985). This is probably an underestimate rather than otherwise—although by that time technological innovations (voice input, speech synthesis, new display media, etc.) may well have rendered the conventional hardware configuration of keyboard and screen obsolete.

The process of human–computer interaction may be analysed on three levels:

(i) the hardware interface;
(ii) the software interface;
(iii) the role of computers in the working organization.

## VDUs and Health

### Visual Fatigue

People who work with VDUs commonly suffer from eyestrain. Prevalences of 70–90% have been reported in people who do repetitive data entry and other screen-based clerical tasks—compared with 45% in general office workers who do not use VDUs. But part of this difference is due to the nature of the task rather than to the VDU itself, since both accounting machine operators and full-time typists also have prevalences which fall between these extremes (Table 12.1). The prevalence of eyestrain has been found to increase as a linear function of the number of hours spent at the terminal each day (Läubli and Grandjean, 1984; Grandjean, 1987).

*Table 12.1 Prevalence of Eyestrain in Sedentary Workers*

| | Prevalence (%) |
|---|---|
| **People Using VDUs** | |
| Data entry and other clerical tasks | 71–91 |
| General office work | 51 |
| **People Not Using VDUs** | |
| Accounting machine operators | 72 |
| Typing (full-time) | 60 |
| Clerical work | 50–60 |
| General office work | 45 |

Data from various sources cited by Laubli and Grandjean (1984).

## Musculoskeletal Disorders

Hünting *et al.* (1981) reported a similar pattern of prevalences for work-related aches and pains in the neck, shoulders and upper limbs. Data entry workers were again worst affected—but full-time typists had more problems than people performing non-repetitive VDU work. The prevalence of "almost daily pains" in the data entry workers were: neck 11%; shoulder 15%; right arm 15%; right hand 6%. These figures are almost certainly conservative: much higher prevalences have been reported elsewhere.

In a sample of data entry workers studied by Ryan and Bampton (1988), 44% experienced pain or discomfort in the neck or shoulders more than three times per week; and 39% in the forearms or hands. More than half (57%) had symptoms at one or other of these sites. Ong (1984) recorded similar figures for data entry workers in an airline terminal (neck 41%; right arm 35%; right hand 38%). In a study by Smith *et al.* (1980, 1981), 82% of VDU operators complained of neck and shoulder problems (compared with 55% of a control sample of office workers); and 75% complained of back pain (compared with 56% of controls). The differences between sources may be attributable to differences in reporting criteria or to differences in workload or task design.

The association between musculoskeletal problems and VDU use is "dose-related". de Matteo (1985) cites data from a large-scale Canadian survey, showing that the prevalence of low back pain climbs steadily with the number of hours per day of VDU use; and Knave *et al.* (1985b) found an association between musculoskeletal complaints and length of working hours.

The above data indicate that a substantial proportion of VDU users (in some samples, over half) suffer from work-related musculoskeletal aches and pains on a regular basis. There is epidemiological evidence that VDU users are also liable to musculoskeletal problems of a more severe nature. English *et al.* (1989) found that VDU operators and other keyboard users were statistically over-represented in a random sample of patients attending orthopaedic clinics with musculoskeletal disorders at various sites in the upper limb (compared with controls attending the same clinics for other reasons). These clinically more specific conditions presumably represent a progression of the ordinary work-related aches and pains, which occurs in a minority of cases.

The right forearm and hand of data entry workers are affected very much more frequently than the left—because the right hand is predominantly used for keying data (especially if the machine has a separate numeric keypad on the right-hand side of the keyboard), whereas the left is used to handle source material.

Why do people who work with VDUs suffer an excess prevalence of musculoskeletal problems, compared with other groups of sedentary workers? The working postures of VDU users (particularly those involved in repetitive data entry tasks) will in general be more constrained than the postures of people doing non-repetitive work at an office desk—with typists falling somewhere between the two, thus reflecting the pattern prevalences reported in the literature.

The constrained posture of the VDU user may sometimes result from the visual demands of the task—thus you may crane your neck at an awkward angle to avoid a troublesome reflection on the screen. Stammerjohn *et al.* (1981) found that those VDU users who reported musculoskeletal problems were also more likely to complain of glare and flicker from the screen; and Knave *et al.* (1985b) reported that office workers who suffer from musculoskeletal problems are more likely to have eyestrain and headaches than those who do not.

Keying rates in data entry work will also tend to be higher than in other office jobs. A proficient copy typist using an electromechanical machine is unlikely to exceed 7,000 keystrokes per hour averaged over the working period, and these will be fairly equally distributed between the two hands—although with the QWERTY keyboard layout (see below) there is a slight preponderance of both keystrokes and symptoms on the left (Läubli and Grandjean, 1984). But data entry operators are usually paid by results—in many industries, 12,000 keystrokes per hour is regarded as a basic minimum level of competence, and rates in excess of 20,000 keystrokes per hour are by no means uncommon—and these may be executed almost entirely with one hand. The data which the operator is required to enter are essentially an unbroken string of random numbers. The dehumanizing quality of such work rivals the worst excesses of the industrial assembly line. Note, however, that the association which is widely held to exist between high keying rates and musculoskeletal problems is by no means inevitable: Ong (1984) found that a package of ergonomic improvements simultaneously led to increased keying rates and a reduced prevalence of musculoskeletal aches and pains (see p. 18).

The dynamic contractions of the forearm muscle which are involved in the fast repetitive keying actions of a data entry task are likely to be superimposed over a background level of static muscle activity (in the extensors) which is required to stabilize the position of the wrist (both against gravity and against the force of the flexors). Co-contraction of the flexors and extensors may also be required for independent finger movement. The static loading will increase if the keyboard height is incorrect (see below) and/or if the wrist is held in a "cocked" (i.e. extended) position. It has also been suggested that users of light-action electronic keyboards might tend to "hold back" so as not to hit the keys too hard.

## Stress

The introduction of new technology into the workplace is commonly a stressful experience for all concerned. People have to adapt to new working methods. The machines may be seen as an implicit threat to job security. Computerization may result in the de-skilling of jobs. The performance of a newly introduced computer system may be unreliable, owing to bugs in the software or to errors made by inexperienced operators. The delays which occur when the system crashes may be an additional source of frustration, as work piles up in computer down-time.

The fragmented nature of some VDU-based tasks (repetitive data entry, etc.) makes them inherently more stressful than most of other types of office work. This will be exacerbated if operators are paid by results. In the worst cases, the software may be used to monitor the operator's performance (working hours, rest pauses, keying rates, data entry errors, etc.)—thus reinforcing her suspicion that she is an appendage of the machine.

We should expect to find a higher prevalence of stress-related disorders (both musculoskeletal and otherwise) in situations of this kind. Summarizing the results of a number of surveys, Grandjean (1987) concluded that VDU users as a whole were no more stressed than other people; but VDU operators engaged in certain fragmented repetitive tasks (data entry, etc.) showed clear evidence of raised stress levels, which were manifest in low reported job satisfaction, negative mood states and psychosomatic symptoms such as sleep disturbances and gastrointestinal dysfunction.

Conversely, Shotton (1986) has noted that reports of musculoskeletal and visual problems are rare in "computer junkies" (i.e. obsessional leisure time users)—despite the fact that they may program for extraordinarily long periods at a stretch at extraordinarily badly designed workstations.

Ryan and Bampton (1988) compared data processing operators with and without musculoskeletal symptoms in the neck, shoulder and forearms. People who reported these problems were significantly more likely to say that they did not have enough time for rest breaks, that they found work stressful, that they had to push themselves or that they were bored most or all of the time. They reported less cohesion in their peer groups, less autonomy and less role clarity in the organization (but they reported more support from their supervisors).

## Skin Rashes

There is evidence for an excess of skin rashes in VDU users—Knave *et al.* (1985a, b) found a prevalence of 35% in VDU users compared with 5% in controls (and the prevalence is higher in those subjects who also reported musculoskeletal and visual problems). The area over the cheekbones is most commonly affected; but also the forehead, neck, arms and hands. The symptoms are generally mild, and subside away from the workplace. They are thought to be caused by the combination of very low humidity and electrostatic effects (compare with sick building syndrome—p. 160).

**VDUs and Pregnancy**

The possibility that the use of VDUs during pregnancy might harm the unborn child is understandably a source of great concern to working women. At present the subject remains contentious.

The question arose when a number of clusters of miscarriages and birth defects were reported in groups of women working with VDUs during their pregnancies—mainly in the USA, but also in the UK, Canada and elsewhere. The existence of small clusters does not prove much either way. "Adverse pregnancy outcomes" are relatively common in the population at large: of the order of 10–20% of known pregnancies end in spontaneous abortion (miscarriage). Large numbers of women of childbearing age work with VDUs—and they presumably become pregnant as often as other working women. It follows, therefore, that a certain number of clusters of miscarriages are to be expected by chance alone. But the fact that an association *could have been* caused by chance does not of itself prove that it *was* caused by chance. The question of reporting bias also arises. Once the idea becomes established that VDUs might be a hazard during pregnancy, it is at least possible that a woman who has an early miscarriage might be more likely to report the fact if she works with a VDU than if she does not.

In the mid-1980s, three major epidemiological investigations failed to find any clear evidence for a connection between VDU use and adverse pregnancy outcomes. In a case-control study using the Finnish national register of congenital malformations, Kurppa *et al.* (1985) found no evidence for an association between VDU use and birth defects. In Canada, McDonald *et al.* (1986) found no significant difference in the reported frequency of spontaneous abortion between women who did and did not work with VDUs. (Although there was such a difference in a sub-sample of current pregnancies.) In Sweden, Ericson and Kallen (1986a, b) found no association between adverse pregnancy outcomes (perinatal deaths, birth defects, low birth weights, spontaneous abortions) and exposure to VDUs (high/medium/low usage or hours per week). But an association was found between spontaneous abortion and both self-reported stress and smoking—which in turn were associated with VDU use.

But the question is not closed. Goldhaber *et al.* (1988) report a case–control study of more than 1,500 women attending pre-natal clinics in Northern California. They found a significantly increased incidence of miscarriage in women who reported using VDUs for more than 20 hours per week during the first 3 months of pregnancy, compared with working women who did not use VDUs (relative risk = 1.8). There was a slightly elevated risk of birth defects but it did not reach statistical significance. The authors noted that it was possible that those women who had miscarried might have overestimated their exposure to VDUs. But they noted that self-reported exposure to pesticides was no higher in women with adverse pregnancy outcomes—and that if recall bias were operating, you might expect it to operate similarly in both cases.

This study reopens the possibility that there might be a link between miscarriage and heavy VDU use during early pregnancy. If there is such a connection, the underlying mechanism is obscure. Ionizing radiation can be ruled out—the emissions from VDUs are very much lower than currently accepted safety limits (see, for example, Bergqvist, 1984). The electrical mechanisms of VDUs do generate a certain amount of very-low-frequency (VLF: 30 kHZ–500 Hz) and extra-low-frequency (ELF: 500–5 Hz) electromagnetic radiation—but no more so than many other commonly used electrical devices. This is sometimes referred to as "electronic pollution" or "electronic fog". The biological effects of VLF and ELF radiation are in general poorly understood (Mackay, 1987).

At the time of writing the issue remains controversial in the extreme, and the details of this controversy are beyond the scope of the present book. There are, however, other possible explanations. For example, it might be that prolonged sitting in fixed working postures could have an effect (for example, by a restriction of blood flow to the pelvic organs); or the link (if there is one) might be due to stress. At present these possibilities remain speculative.

## Workstation and Task Design

The working posture of the VDU user is to a large extent determined by the height of the keyboard and screen relative to the seat, the most commonly encountered problem being keyboards which are too high and screens which are too low. Occasionally one also encounters screens which are too high (i.e. above eye level). But users are unlikely to adjust their seats such that the keyboard is too low (because their knees and thighs will meet the underside of the desk).

A screen which is too low requires postural compensations which lead to an unsatisfactory posture of the head and neck (p. 110). The prevalence of neck and shoulder problems in VDU users has been shown to increase with both the angle of the head inclination and the angle of rotation of the head in the horizontal plane (Hünting *et al.*, 1981).

A keyboard which is too high results in elevation of the shoulder girdles and/or abduction of the arms at the shoulders. The early electromyographic investigations of Lundervold (1951, 1958) showed that typing at a high keyboard results in increased tension in a number of muscle groups—particularly the trapezius and the muscles on the ulnar side of the forearm. Life and Pheasant (1984) showed that the torque about the shoulder increases with the height of the keyboard above elbow level, and that there is a decrease in reported comfort—particularly in the shoulder, upper arm and forearm.

If the arms are abducted at the shoulders, the keyboard user will be forced to make a compensatory adduction (or ulnar deviation) of the wrist in order to keep her fingers in correct alignment with the keys. Thus the ulnar deviation of the wrist will tend to increase as a function of keyboard height. (For an explanation of these anatomical terms see p. 54.) The prevalence of aches and pains in the forearms of VDU users (Hünting *et al.*, 1981) increases with the angle of deviation; and wrist deviation is known to be a risk factor for tenosynovitis (p. 92).

Users who do not touch-type pose a special set of problems. "Hunt and peck" typists tend to prefer a relatively high keyboard—so that they can see the keys more easily. They tend to have worse working postures (with respect to both the neck and the upper limbs) and to hit the keys harder than other, more skilled users. This may go some way to explaining the RSI clusters that have been reported in journalists.

The design of VDUs has improved considerably over the last decade or so. Early machines often had their keyboards and screens mounted close together in the same housing. Thus raising the seat to get the keyboard at the right level meant that the screen would be too low; and lowering the seat to get the screen at the right height meant that the keyboard would be too high. The majority of desktop computers now have a detachable keyboard—which is much better, since it enables the user to optimize the height of both devices and to position the screen so as to avoid glare, etc. It is also an advantage if the user can tilt the screen to his own requirements. Early machines also had relatively thick keyboards compared with those that are now the norm—which made it yet more difficult to adjust the seat correctly (p. 228). Hünting *et al.* (1981) noted that thicker keyboards were associated with a higher prevalence of upper limb problems.

Most keyboard users work from shorthand notebooks or some other kind of written or printed source documents, which are generally placed on the desk beside the keyboard. Reading this material thus requires both inclination and rotation of the head and neck. Life and Pheasant (1984) found that a reading stand for source material reduced the angle of head inclination in VDU users to an extent that was sufficient to reduce the loading on the neck muscles by about one-third. I have known the provision of a reading stand to result in dramatic improvements in the neck and shoulder problems of keyboard users.

Kukkonen *et al.* (1983) studied the musculoskeletal symptoms of a group of data entry operators in the computer department of a bank before and after a programme of ergonomic changes. Specially designed reading stands were provided for source documents, and the workers were given individual instructions on how to adjust their chairs correctly. Illumination levels were reduced to 300–500 lux (p. 205) and additional spotlighting was provided where necessary. The workers were also given a short course on the relevant principles of ergonomics, including instruction on the systematic use of micropauses (p. 164) and relaxation exercises (to help them recognize areas of muscle tension). The prevalence of "tension neck syndrome" (i.e. neck pain with palpable tenderness and hardenings) was reduced from 54% before the intervention to 16%, where it subsequently remained. There were also reductions in the prevalences of upper limb and back symptoms.

There are two basic schools of thought concerning the design and adjustment of VDU workstations. I call these the *upright approach* and the *laid-back approach* (Figure 12.1). Exponents of the upright approach (who until recently have been in the majority) advocate a working posture similar to the one which has traditionally been recommended for typing: trunk vertical, upper arms vertical, with elbows flexed to an angle of about 90° so that the forearms are more or less horizontal.

This view has been challenged by Grandjean *et al.* (1983), who conducted a

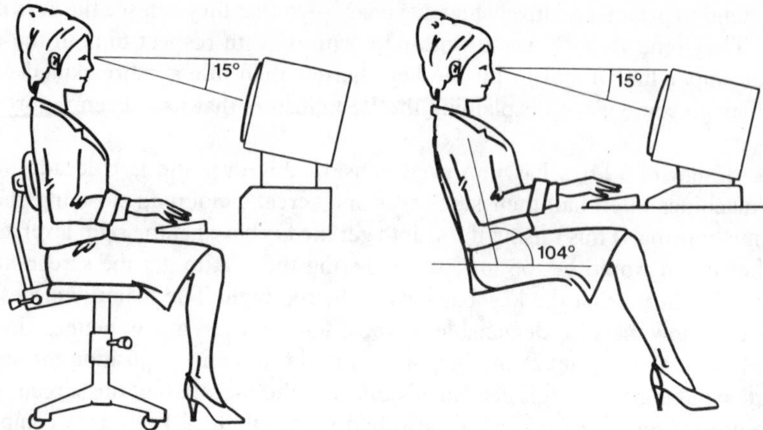


Figure 12.1 *The upright (*left*) and laid-back (*right*) positions for VDU users. Reproduced from Pheasant (1986)*

fitting trial in which VDU users were provided with a high-backed seat and a fully adjustable workstation in their own offices for a week. The majority preferred a laid-back position with the trunk at an angle of between 100° and 110° to the horizontal—only 10% of the subjects chose to sit upright. The arms were held forward of the trunk with the elbows at an obtuse angle (about 80° flexion) so that the forearms were inclined upwards slightly. The majority of subjects (80%) preferred to use a padded wrist support if it was available—and if it was not, 50% rested their wrists or forearms on the desk in front of the keyboard. The majority adjusted their screens to give a line of sight somewhere between the horizontal and 15° downward, with an average reading distance of 760 mm. (The screen must have been of good legibility.)

The principal advantage of the laid-back position is the extent to which the weight of the trunk is supported by the backrest. The data of Andersson *et al.* (1974) suggest that the postures adopted by Grandjean's subjects might decrease the loading on the spine by up to 45% compared with an upright sitting position. The associated upper limb posture is more problematic. The loading on the shoulder muscles increases with the horizontal distance of the hands in front of the shoulders. Supporting the wrist offsets this loading. A supported wrist is not likely to hamper the keying action significantly on a modern light-action keyboard—but Sauter *et al.* (1987) have noted a case of ulnar nerve compression in a data entry worker which appeared to have been due to prolonged contact with the hard edge of the keyboard and I have encountered one case myself which could (perhaps) be interpreted in this way.

One possible disadvantage of the laid-back position is that it requires a greater degree of flexion in the lower cervical and upper thoracic parts of the spine (to give a horizontal or slightly downward line of sight). But on balance it seems that the laid-back position is probably advantageous rather than otherwise—and there seems no special reason why it should *necessarily* be associated with a high keyboard. *Note that the laid-back approach to VDU use requires a seat with a relatively high backrest.*

Many modern workstations allow independent adjustment of the keyboard and screen—as well as varying degrees of seat adjustment. In principle, the greater the degree of adjustment that is available, the better it will be for the user (and Shute and Starr, 1984, have confirmed this in fitting trials). A certain amount of training will be required if the users are to get the maximum benefit from whatever degree of adjustment is available—but I do not believe that this training should be excessively prescriptive. On balance, it is probably best that the user should experiment with the workstation until he finds his most relaxed working position—and that he should learn to recognize and correct the subjective sensations of muscle tension, rather than following any specific formula for a "correct" working posture. The role of the trainer should be to facilitate this process—and to watch out for postures which are obviously incorrect. Having said this, a reclined trunk, combined with a keyboard at elbow height so that the shoulders are relaxed and the arms hang by the sides, would seem to be biomechanically most advantageous.

On 29 May 1990 the Council of the European Communities published a directive on "*minimum safety and health requirements for work with visual display equipment*". The directive includes provisions concerning keyboards, screens, workstations, the working environment, software design, and so on. It requires—for example—that the keyboard should be tiltable and physically separate from the screen and that there should be space in front of the keyboard so that the user may rest his hands; that the screen should swivel and tilt easily and that it should be possible to use a separate base for the screen or an adjustable table; that document holders should be stable, adjustable and positioned to minimize the need for uncomfortable head and eye movements; that the seat should be adjustable in height and that its backrest should be adjustable in both height and tilt; and that a footrest should be made available to anyone who wishes for one.

## The Keyboard

It is widely accepted that:

- *the keyboard* should be about 30 mm in thickness (measured to the middle row of keys) and should be tilted to an angle of 10–15° to the horizontal;
- *the keys* should be about 12–15 mm square, with a spacing of 18–20 mm between centres; should have matt, non-reflective and visually unobtrusive surfaces; and should be dished to aid location of the fingers;
- *the keying action* should have a resistance of 50–60 g, with a travel of 2–4 mm; and have a positive "feel" as the key engages (to reduce keying errors).

In practice most modern keyboards are likely to meet these requirements (with the exception of those provided for "briefcase" systems—which are in many respects a special case).

The standard layout of the alphanumeric keyboard, which reads QWERTYUIOP across the top row, is more than a century old. It was designed by

Christopher Sholes—inventor of the first practical typewriting machine, which was manufactured by Remington in 1874. The layout was chosen to allow a typist to work at speed without jamming the type bar mechanism—thus the pairs of characters which are most commonly used in combination were widely separated.

Ideally, the keyboard layout should be determined by the relative strengths of the fingers, the facility with which different patterns of movements can be executed, and the frequency with which different letters and combinations of letters occur in the language concerned. The most commonly used letters should be in the home row—and they should be allocated to the strongest fingers (index and middle). Common keying sequences should be executed by alternate hand-strokes, whereas consecutive use of the outer three fingers, repeated use of the same fingers and wide jumps (especially between the top and bottom rows) should be avoided.

The QWERTYUIOP keyboard does not meet these criteria particularly well. When typing in English, about 57% of the workload is allocated to the left hand. This imbalance is apparently large enough to make a difference—since Läubli and Grandjean (1984) have noted an excess of upper limb problems on the left in typists. The allocation to individual fingers does not match their relative strengths particularly well (see Figure 12.2), and only about 30% of keystrokes are made in the home row. But a reasonably high proportion of common character pairs are executed by alternate hand strokes—since this was implicit in Sholes's original intentions.

The best-known alternative to QWERTYUIOP is that of Alphonse Dvorak, which was based on an extensive programme of research (Dvorak *et al.*, 1936;

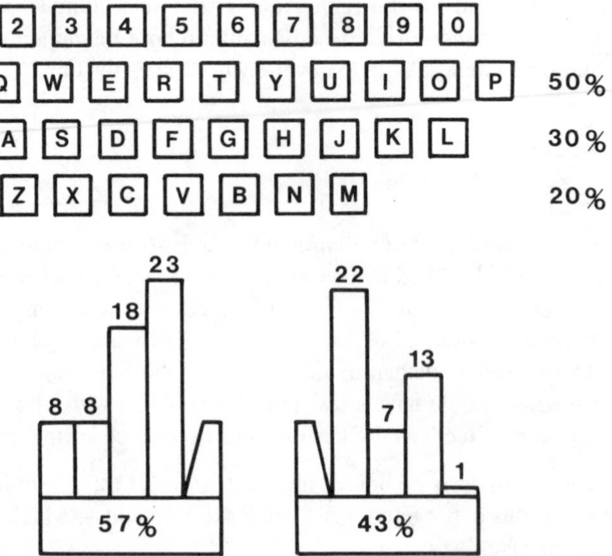


Figure 12.2 *Distribution of workload on the QWERTYUIOP keyboard—after Cakir* et al. *(1980)*

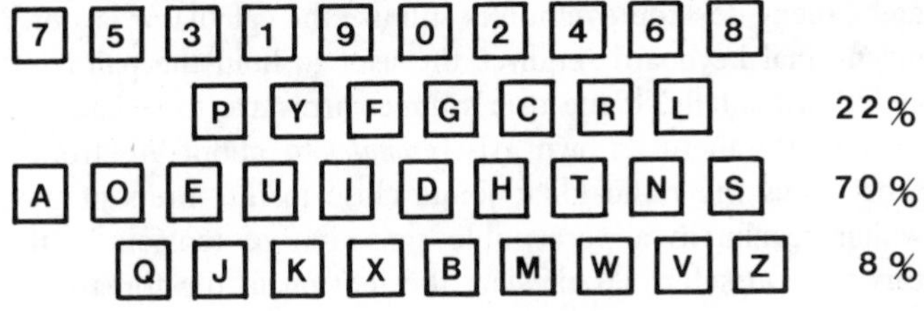


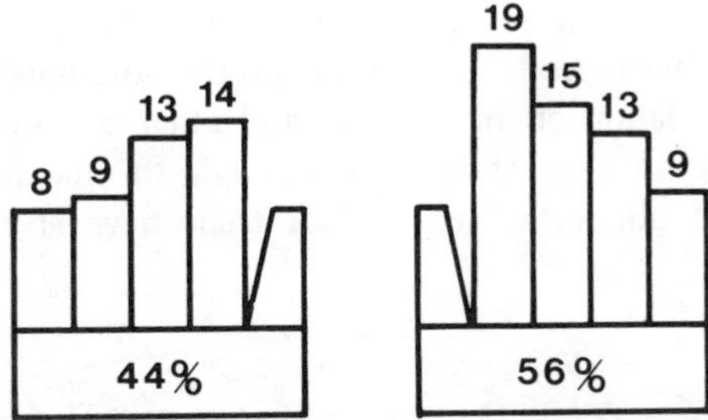


Figure 12.3 *Distribution of workload on the Dvorak keyboard—after Cakir* et al. *(1980)*

Figure 12.3). About 70% of keystrokes are made in the home row; and since vowels and consonants commonly alternate, a very high proportion of common pairs are executed with alternate hand strokes. The right hand makes 56% of keystrokes and the distribution of the workload reflects the relative strengths of the fingers extremely well.

Despite these manifest advantages, the Dvorak keyboard has never really caught on. The motor skills of experienced typists who are proficient at QWERTYUIOP cannot be easily unlearned; *negative transfer effects* make it difficult to adapt to a novel arrangement. It is likely, therefore, that attempts to improve upon the QWERTYUIOP layout, however sound in theory, will be doomed to failure in practice. (Although some people think that the advent of programmable keyboards may change this.)

Most VDU keyboards have a numerical keypad to the right of the alphabetical keys. This can be arranged in two ways, corresponding to the layouts which we find on a keyphone and a hand calculator:

| *Keyphone* | | | | *Calculator* | | |
|---|---|---|---|---|---|---|
| 1 | 2 | 3 | | 7 | 8 | 9 |
| 4 | 5 | 6 | | 4 | 5 | 6 |
| 7 | 8 | 9 | | 1 | 2 | 3 |
| | 0 | | | | 0 | |

The keyphone layout has been found to give higher speeds and fewer errors—and alternating between the two was slower than either (Conrad and Hull, 1968). The keyphone layout is so obviously the more sensible of the two that it is difficult to see why anyone should ever have thought of any other arrangement.

Unfortunately, many desktop computers still use the calculator layout.

The conventional keyboard requires the user to hold the palms of his hands approximately horizontally. If the user's upper arms are to remain vertical, each forearm must rotate about its own axis (*pronate*) by about 90° from the neutral position. This takes the radio-ulnar joints close to the limits of their range of motion—which results in a perceptible sensation of tension in the forearms (confirm this for yourself). To alleviate the tension in the forearm muscles, we tend to abduct our arms at the shoulders.

This problem was recognized by Klockenberg (1926), who suggested that the parts of the keyboard allocated to the right and left hands should be separated and set at an angle to each other in both the vertical and the horizontal planes (Figure 12.4), thus reducing pronation of the forearm and ulnar deviation of the wrist, respectively. The rows of keys should be curved, to match the sweeping movements of the hand about the wrist. User trials have shown that a split

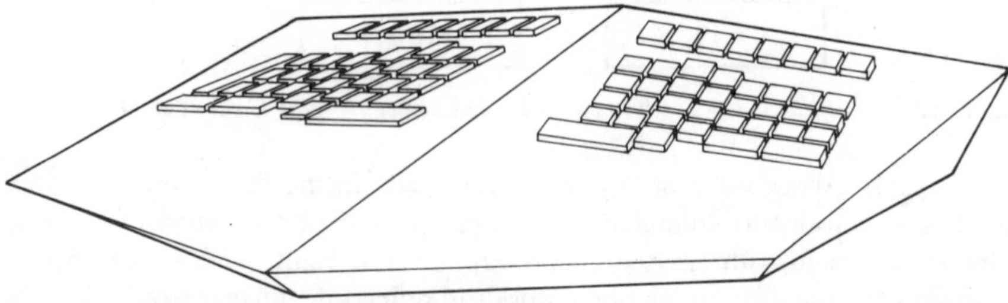

Figure 12.4 *The Klockenberg split-keyboard concept, as developed by Nakaseko* et al. *(1985)*

Figure 12.5 *The Maltron keyboard—photograph kindly supplied by Stephen Hobday*

keyboard results in a lesser degree of fatigue (Kroemer, 1972); reduces electromyographic activity in the neck and shoulder muscles (Zipp *et al.*, 1983); and is preferred by a majority of subjects (Nakaseko *et al.*, 1985). A commercially available split keyboard, produced by PCD Maltron Limited, is shown in Figure 12.5 (see Hobday, 1985, for further details).

## The Screen

The legibility of text displayed on a VDU screen is determined by the size, shape and layout of the characters and by the quality of the visual image. Taken together, these determine the distance at which it is possible to read the screen without undue effort. The shorter the reading distance, the more constrained the reader's posture and the more rapid the onset of visual fatigue—hence the correlation between visual and postural discomfort.

The image quality of VDU screens is improving all the time—particularly at the top end of the range. A good-quality screen now has a legibility similar to that of printed text of the same size. But a poor-quality screen is very much worse. The following comments will be limited to general principles: for further details see Cakir *et al.* (1980), Oborne (1985a), Pheasant (1986), Grandjean (1987), etc.

It is important that the image displayed on the screen should be sharply focused, stable and as free as possible from flicker and other disturbing visual effects. Instabilities such as "drift" (slow movements of characters) and "swim" (bending of lines) may result from technical deficiencies in the design of the VDU (e.g. unstable power supply) or from interference from other AC sources, etc.

The colour of the characters is relatively unimportant—although extremes of the spectrum (red and violet) are probably best avoided.

It is generally considered that a luminance contrast ratio, between characters and background, of between 4:1 and 15:1 is acceptable (e.g. HSE, 1983). The optimal ratio is probably about 10:1. (Figures quoted are for light characters on a dark screen.) The EC directive requires that the brightness and/or contrast should be easily adjustable by the operator.

The contrast between the screen and the source document (which is usually the brightest object in the visual field) should not exceed 1:10. (A maximum of 1:3 is sometimes quoted but this is scarcely realistic.) Other surfaces in the immediate surroundings of the VDU should have intermediate luminances.

The outer glass surface of a VDU screen produces a sharp *specular* (mirror-like) reflection of bright objects in the surroundings, and the phosphor produces a diffuse *veiling reflection* (p. 207). Degradation of the visual image may also result from grime, fingermarks, etc., on the screen, and scratches on the glass surfaces resulting from attempts at cleaning. The best way of dealing with reflections is to prevent them at source—by appropriate location of the workstation with respect to light fittings, windows, etc. (see p. 208). Tiltable screens help. Some anti-reflective filters reduce the sharpness of the visual image (see Grandjean, 1987).

Early types of VDU generally displayed light characters on a dark screen (known as *negative polarity*). The opposite arrangement (known as *positive polarity*) has definite ergonomic advantages. It resembles conventional print more closely and has been shown to give better readability and reduced visual discomfort (Bauer and Cavonius, 1980; Radl, 1980). The overall luminance of the positive screen is higher—thus the user's pupil will be more constricted, giving a greater depth of focus and reduced accommodative effort. Positive polarity displays are less susceptible to specular glare and more compatible with ordinary office illumination levels (i.e. there will be a lower contrast ratio between screen and source documents)—but flicker will be more conspicuous, so a positive polarity screen will need a higher frame frequency to overcome this problem.

The characters on a VDU screen are composed of patterns of dots. The more dots per character, the better the legibility. A $5 \times 7$ dot matrix is just acceptable for a character set of upper case and numerals only. For lower case, a matrix of at least $7 \times 9$ is required (excluding descenders). On a good-quality, high-resolution screen, the dots are small and closely spaced—thus the characters resemble solid letters.

The character size required—for comfortable legibility at an acceptable viewing distance—will depend upon the image quality, etc. Many sources quote a minimum acceptable character height of about 0.005 $D$ and a preferred range of 0.006–0.0065 $D$ (where $D$ is viewing distance). According to Schmitke (1980), a height range of 0.007–0.01 $D$ gives optimum readability.

For visual comfort and the avoidance of fatigue, a minimum viewing distance of 500 mm is required—and 750 mm is probably better. For adequate readability at a comfortable distance, therefore, a minimum character height of 3 mm is required—but 4 mm characters would certainly be better. In one survey cited by Grandjean (1987), four out of 18 types of VDU had characters smaller than 3 mm. People will adapt to adverse reading conditions by moving closer to the screen—thus characters having the minimum acceptable height at a given distance (say 500 mm) will in practice be read at a lesser distance (say 400 mm).

## Work Scheduling and Rest Pauses

Lengthy periods at repetitive data entry tasks are inherently undesirable—however well designed the terminal. The best solution to this problem is job enlargement. In some organizations, repetitive data entry tasks are, as a matter of policy, limited to half the working day. This is a sensible and enlightened approach—and the available epidemiological evidence suggests that (combined with basic workstation improvements) it should suffice to avoid most of the problems associated with excessive VDU use. But it presupposes that other tasks (preferably of a contrasting nature) are available to fill the remainder of the working day.

The EC directive requires that "the employer must plan the worker's activities in such a way that daily work on a display screen is periodically interrupted by breaks or changes of activity reducing the workload at the display screen".

In practice, rest pauses in VDU work are usually a matter for negotiation between unions and management. Negotiations often take place during the difficult period when new technology is a sensitive issue for all concerned—when stress levels are high and nobody may be in much of a mood to be rational—and when the ergonomic problems of the workstation, and the bugs in the operating system, have not yet been sorted out.

In principle we should not be extending rest pauses to make up for other deficiencies in the design of work. Some authorities take the view that (with the exception of certain data entry tasks) VDU work is not inherently more stressful than other types of work—and that rest requirements are therefore no greater (e.g. Grandjean, 1987). *If* VDU work were indeed just as varied as other office work, and *if* all the other ergonomic problems had indeed been sorted out, then this argument would seem reasonable enough. But frequently neither is the case.

The current consensus converges on the recommendation that for tasks involving extensive VDU use, between 5 and 15 minutes' rest in the hour is desirable. As a rough guide, 5 minutes' would be adequate if the working conditions were satisfactory and the task reasonably interesting, whereas up to 15 minutes' might be required if conditions were poor and the task dull or stressful, and 10 minutes' would be about average. It is probable that 10 minutes' rest in each hour is better than 15 or 20 minutes' at the end of the second hour. But in reality, all of these suggestions are provisional and should not be construed as having much in the way of scientific validity. Micropauses should also be actively encouraged (p. 164).

Haider *et al.* (1980) showed that an hour of continuous, visually demanding VDU work results in a measurable level of visual fatigue (p. 210). A substantial degree (about 75%) of recovery had occurred within 11 minutes, and recovery was complete after 16 minutes.

## Software Ergonomics

The extent to which working with computers is a rewarding experience, rather than a frustrating one, is dependent (at least in part) upon the design of the software—and in particular upon the rules and procedures which govern the exchange of information between the machine and its user. Simple transactions may be performed by selecting items from a "menu", or "pointing" with a "mouse": for more complex transactions, the user will generally be required to enter instructions into the machine in a formalized *command language.* This is where the problems generally begin—notwithstanding the fact that the products concerned are often advertised as "user-friendly". Given current technical limitations, you cannot reasonably expect to give instructions to your computer in the same way as you would to your secretary. But we could reasonably hope that the rules which govern the dialogue should be internally consistent, compatible with normal usage, and no more complex or jargonized than is absolutely necessary.

The technical capabilities of computer systems and software packages are improving all the time. But the complexities of their command languages remain such that many users fail to take full advantage of the facilities which the systems offer—whilst for a minority of enthusiasts the process of mastering (or mistressing) the system becomes an end in itself—and their involvement with the machine becomes disproportionate to the useful ends they achieve.

The following example (kindly supplied by John Long) is from an enquiry system for dealing with social survey data. The user wishes to "list couples, less than 25 years old, who have children". To do this, he must enter the following command:

*T<-<:S: (Age <25) & (CHILD >=1): >COUPLES

In addition to the complexity of its syntax, it requires 12 shift-return operations over and above its 36 characters (in lower case it would require 18 extra keystrokes). Thus the potential for error is considerable.

The most frustrating feature of command language is often the arbitrary nature of the rules which determine whether a statement is syntactically correct—and thus whether the machine does what you want it to do. For example, the use of spaces between characters or words in commands, or in the name that the user gives a file, may or may not be critical. It is even possible to encounter systems which are inconsistent in this respect—the spaces may be required in some operations and ignored in others.

The meaning of the special "function keys" on a keyboard is often idiosyncratic to the system or software package concerned. In one commonly used word processing package, the key "Ctrl-X" is used to move the cursor (the indicator which marks your place) down one line on the screen; in another package the same key deletes a line of text. Inconsistencies of this kind are an invitation to error when moving from one system to another (see p. 192 for further examples).

The messages which the machine displays to inform the user that he has gone astray are often unhelpful. Hence,

ERROR: SYSTEM FUNCTION 201

means nothing at all unless you are also told (and are able to understand) what system function 201 actually is. (People who do not work with computers may find it difficult to believe that things like this are not always explained in the instruction manuals which come with the machine.) And when detailed error messages are supplied, they may be oddly cryptic:

ERROR: NO KEYBOARD PRESENT PRESS F1 TO CONTINUE

or verging on the abusive:

YOU HAVE BEEN IDLE TOO LONG. PLEASE RESPOND

(Tom Stewart kindly supplied me with these two examples.)

Common sense suggests that, where possible, command language should follow the linguistic conventions of normal English. Linguistic compatibility has both syntactic and semantic aspects. Normal syntactic conventions would lead us to expect that the command INSERT X,Y is likely to mean "insert item X into location Y" rather than meaning "insert into location X, item Y". It is perhaps surprising therefore that Barnard *et al.* (1981) found relatively little evidence of performance differences, using dialogues which employed the "compatible" and "non-compatible" versions of statements of this kind. It seems that people can cope with either, provided it is used consistently.

Semantic compatibility may be more important. In normal English, for example, the term DUMP means to throw something away—whereas in computerspeak it means to display it on the screen. It depends on whether you think of the machine or the user as the starting point of the operation. The system designer construes the statement in the former way—but the user's mental model may be different. Computerspeak is often oddly violent in its overtones: KILL (for erasing a file), ABORT, TERMINATE, FATAL ERROR, ILLEGAL ENTRY, etc.

The use of contracted or abbreviated forms in command language saves keying; but this is a false economy if it creates a vocabulary of terms which bears so little resemblance to natural English that it has to be committed to memory as arbitrary strings of characters—rather than being recalled as encoded "chunks", as we do for natural words and phrases. Chapanis (1988) notes the following examples which come from a program which (can you believe this?) was intended for use in offices:

| | | | |
|---|---|---|---|
| .CM U5AUTH | aut$prof | nodeid | CRON |
| .im fn ft (fm) | GETUFIS | OFS OFSSMCNTL | |

Ledgard *et al.* (1981) compared a text editing program which used a highly abbreviated notation with one which used a command language much more like English. Thus LIST;* in the abbreviated notation became LIST ALL LINES in the more natural version; and FIND:/TOOTH/;-1 became BACKWARD TO "TOOTH". Experienced computer users performed better than inexperienced users, on both versions. But both groups of users performed better (in terms of both speed and errors) using the more natural version.

Chapanis (1988) compared a Dow Jones news retrieval system which used an abbreviated notation with a revised version using "English" commands. The latter gave very much better performance (twice as many correct commands per unit time).

Human–computer interaction remains fraught with apparently needless difficulties and frustrations. Technical limitations aside, one suspects that a variety of the first fundamental fallacy is often involved (p. 5). Software designers may find it difficult to put themselves in the shoes of a naive user—so dialogues which seem to them like models of logical order and clarity may seem to you and me like an impenetrable fog. Software is generally written by people who are totally absorbed in the world of computers—and most of them are under 25. For "computer junkies", the challenge of mastering the machines is a major source of satisfaction; for the rest of us it is an irritating chore which gets in the way of the tasks we wish to perform.

## Pausgymnastik

The programme of *pausgymnastik* described below is intended for office workers in general and keyboard users in particular. It was devised by Sheila Lee (osteopath) and Gunilla Kleberg (industrial physiotherapist). We call it the seven-way stretch. The exercises were chosen on the basis of the epidemiological and ergonomic evidence discussed earlier in this chapter, with the particular aim of stretching those muscle groups which are likely to be subject to static loading during keyboard work. The sequence takes about 5 minutes and all the exercises can be performed (if necessary) without leaving the office desk—although a change of scene would perhaps be better. An alternative and more extensive system of exercises, which is also based on sound ergonomic principles, is described by Gore and Tasker (1986).

Note that the following set of exercises are intended as a prophylactic measure—to help counteract the deleterious effects of prolonged VDU work. They may be less appropriate for people with existing musculoskeletal problems. Exercise is an individual matter—and it should always be matched to the patient's current clinical status. Different people have different views on the subject.

## The Seven-way Stretch

Begin by relaxing as completely as you are able: lean back in your chair, close your eyes, raise your shoulders and let them drop, unclench your fists, unclench your jaws. Mentally look for areas of muscle tension and try to relax them. Breathe slowly and deeply, using the diaphragm and abdomen rather than the upper part of your chest. (The abdomen should dilate slightly on the in-breath; this may require a little practice for some people.) Breathe away your tension.

With elbows on desk, rest the forehead on the hands. Using the fingers of both hands, gently pull the skin of the forehead sideways, inwards, upwards, downwards. (This stretches occipitofrontalis and corrugator supercilii muscles, which often become tense during visually demanding work.)

Hold each of the following positions for four complete deep breaths, stretching a little further with each breath.

*1. Sub-occipital Stretch* (Figure 12.6)
Clasp hands behind head. Bend head forward, gradually stretching out the upper part of the neck whilst breathing deeply—but keep the back and shoulder regions as straight as possible. Roll your head to one side, then the other, four times.

This stretches the posterior neck muscles, particularly those in the suboccipital region (semispinalis capitis, etc.). The rotation at the end gives an additional stretch to the splenius.

*2. Upper Trapezius Stretch* (Figure 12.7)
Grasp leg or seat of chair with right hand. Place palm of left hand above right ear (hand passes behind head). Pull head to left, laterally flexing neck. Repeat to the other side.

This stretches the upper part of the trapezius muscle.

Figure 12.6 *The sub-occipital stretch*

Figure 12.7 *The upper trapezius stretch*

*3. Head and Neck Rotation*
Balance head carefully with the shoulders dropped and arms folded behind your back. Turn head to left as far as it will go. Nod four times. Repeat to other side.

This stretches the sternomastoid.

*4. Backward Shoulder Stretch* (Figure 12.8)
Clasp hands behind back. Reach out backwards over chairback, whilst expanding chest, arching back (hyperextension) and lifting hands.

This stretches the muscles which protract the shoulders (pectorals, etc.).

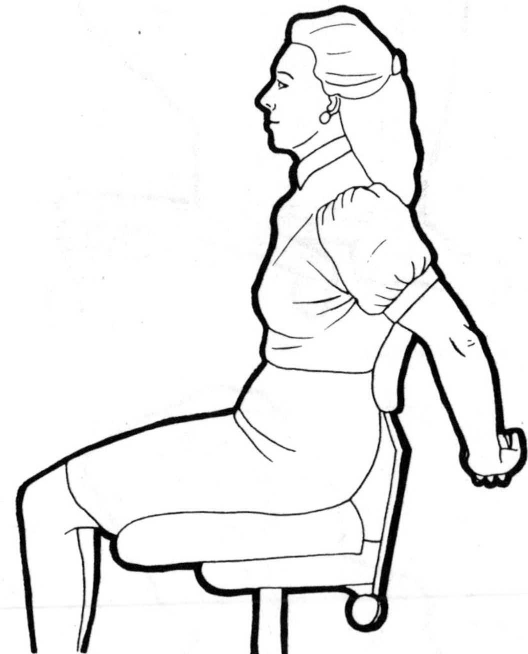

Figure 12.8 *The backward shoulder stretch*

*5. Lumbar Twist* (Figure 12.9)
Cross right leg over left. Place outside of left elbow against outside of right knee. Place right hand behind back on left hip. Turn head, neck and shoulders to the right as far as they will go, pressing with the left elbow. Hold for four breaths. Repeat other side.

This stretches the back muscles, particularly multifidus lumborum.

*Before each of the following wrist exercises, shake the hands vigorously.*

Figure 12.9 *The lumbar twist*

*6. Wrist Extensor Stretch* (Figure 12.10)
Raise your right hand in front of your face and turn it into pronation (so that the palm faces away from you). Place the palm of your left hand on the back of your

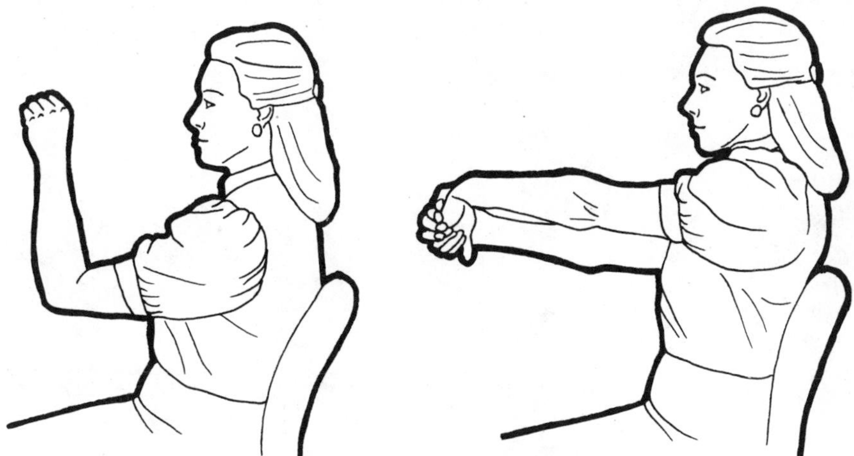

Figure 12.10 *The wrist extensor stretch*

right hand and gently press your right wrist into full flexion, straightening your right elbow as you do so.

This stretches the muscles of the extensor compartment of the forearm.

*7. Wrist Flexor Stretch*

Place palms together as if praying. Raise elbows (thus extending wrists). Separate palms so that only fingertips are pressing together.

This stretches the muscles of the flexor compartment of the forearm.

# Chapter Thirteen
# Driving

There are four good reasons why we should expect the high prevalence of back pain which we encounter in drivers:

(i) Drivers may spend considerable periods of time in a fixed posture with relatively little opportunity for fidgeting.
(ii) The driving position may be less satisfactory than other sedentary working positions—particularly with respect to the degree of spinal flexion required.
(iii) The driver's back is exposed to vibration and impact, and to dynamic stresses during acceleration, deceleration and cornering.
(iv) Driving may also involve long periods of intense concentration and may be psychologically stressful (at least for some people).

## Driving Posture and Workstation Layout

Overall, the design of seats in ordinary motor cars and in heavy goods vehicles is considerably better than it was in the past. But the seats in some industrial vehicles (earth-moving equipment, etc.) are still far from satisfactory. Most seats now provide a fair degree of lumbar support. But given that the position of the seat's lumbar pad is not adjustable, it may not be suitable for all users; nor may it be optimally placed to suit the greatest possible number. In general, however, the postural problems of the driver are more likely to result from the demands of the task and the layout of the controls than they are from the seat itself.

The backrest of the driving seat commonly has an angle adjustment. But steeply reclined sitting positions may not be compatible with the visual demands of driving or with a comfortable position with respect to the steering wheel. So in practice the backrest is unlikely to be set to an angle of more than about 10° from the vertical; and when driving becomes difficult or stressful (owing to poor visibility, heavy traffic, etc.), we tend to crane our heads forward and hunch ourselves over the steering wheel, thus losing the support of the backrest.

### Pedal Location

The fully depressed pedal should be well within comfortable reach; but the closer the seat is to the pedals, the greater the flexion that will be required at both the hip and the knee. Flexing the hip tightens the hamstring muscles much more rapidly than flexing the knee slackens them off. Thus, as the thigh passes the horizontal, the pelvis rotates backward and the lumbar spine is forced into flexion. This is particularly undesirable if the flexed spine is also subjected to dynamic compressive loading, when travelling over rough terrain, etc.

The degree of hip flexion required will decrease with the height of the seat above the pedals—but there are obvious practical limitations on this in the confined space of a vehicle interior. The strength of the pedal operating action is greatest when the pedal is approximately at seat height and the knee is only slightly flexed (Kroemer, 1971; Pheasant and Harris, 1982); and the available force decreases with seat height. Taking these various factors into account, the optimal position of the seat with respect to the pedal is one in which the thigh is not more than about 15° above the horizontal and the knee is flexed by about 70°, as shown in Figure 13.1 (Pheasant, 1986).

Some cars have the pedals offset from the middle of the seat (generally to the left in right-hand-drive vehicles). This imposes an appreciable torsion on the spine.

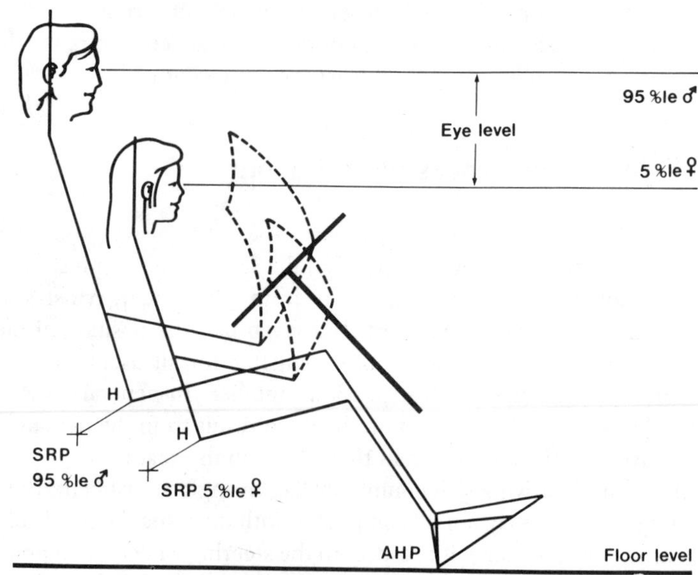


Figure 13.1 *The driver's workstation. Reproduced from Pheasant (1986)*

## Steering Wheel

It should be possible to hold the steering wheel in the recommended "10 to 2" position without having to reach out to move more than about three-quarters of arm's length (and without losing contact with the backrest). In this position, the hands should be somewhat below shoulder height.

A steering wheel that is too high increases the demands which are made on the shoulder muscles. At present there is a trend towards smaller steering wheels (perhaps because they give the car a "sporty image"). The smaller wheel reduces clearance problems and makes it easier to get into and out of the driving seat; but it increases the effort required to achieve a given operating torque. This is probably why women have greater difficulties parking in tight spaces. The effort of turning the wheel, with the arms in a raised position, can also cause a marked increase in blood pressure.

## Anthropometrics

The anthropometric problems involved in designing the driver's workstation are complex. The motor industry has certain more or less standardized techniques for solving these problems (see Pheasant, 1986, for a brief discussion). Anecdotal evidence suggests that these techniques are not altogether successful in catering for the requirements of extreme members of the user population.

The following comments illustrate the sort of mismatches which may arise for a short driver. They were provided by Ruth Haigh—an ergonomist who is 1520 mm tall, which is about 5th percentile for young women in the UK.

> "Whether I am able to drive a car or not depends on the adjustability of the seat. Obviously I need to adjust the seat as far forward as it will go in order to reach the pedals. However, in many cars the furthest forward position still does not achieve this end. Either I still cannot reach the pedals, or I can with the ball of my foot but my heel must be kept raised at all times, resulting in discomfort in the ankle and a potentially dangerous situation. Having moved the seat forward, I find that in some makes of car the gear lever is so far behind me that it is difficult to reach and operate. Alternatively my thighs become jammed under the steering wheel and I am just about hugging the steering wheel.
>
> I find that many seats are too deep; thus I am forced to slump to use the backrest or the edge of the seat cuts into my popliteal region.
>
> In some makes of car I find it difficult to see over the steering wheel, or the front of the body or the boot. Sun visors very rarely provide any protection from the sun for me."

Tall drivers report their own range of problems—commonly ones associated with headroom, knee room and an excessively cramped position (leading to extreme spinal flexion).

The five fundamental fallacies (p. 15) seem particularly relevant to the design of motor cars.

## Seat Belts

For maximum protection during impact, a seat belt should wrap around those parts of the body which are best able to take the strain. The diagonal strap should cross the shoulder—it should not slip down over the arm or ride up over the neck. In women it should pass between the breasts. The horizontal strap should pass across the bony pelvis—it should not ride up over the vulnerable parts of the abdomen. The fixing points of seat belts are standardized; thus extreme users tend to have problems. Short buxom women complain that the belt compresses the left breast (in a right-hand seat).

## Dynamic Loading

Both the motion of the vehicle and the task of driving subject the driver's spine to dynamic loading. The action of depressing the pedal will tend to rotate the pelvis backward—thus forcing the already flexed spine into further flexion. (The line of thrust, acting at the socket of the hip joint, turns the pelvis about the fulcrum of the ischial tuberosities.) Conversely, the action of lifting the foot off the pedal (which employs the iliopsoas muscle) will tend to extend the lumbar spine. Thus, when driving in heavy traffic, etc., the spine undergoes repeated cycles of flexion and extension, and is subject to a transient compressive loading in each phase. The spine will also be subjected to side bending and/or torsional moments when turning the steering wheel, applying the hand brake, and so on.

Sudden braking actions tend to throw the driver forward. Since the driver's buttocks will generally be firmly located on the seat, this will again tend to cause spinal flexion. Thus during braking the spine may be flexed from both above and below. Centrifugal force during cornering will tend to laterally flex the spine. Muscular efforts are required to resist these accelerations. Some people therefore like seats which provide lateral support.

Kelsey *et al.* (1984a) showed that the risk of low back problems in drivers increases systematically with the age of the vehicle. Since this trend was present for almost all makes of car, Kelsey concluded that it was unlikely to be due to design changes and was much more likely to be associated with the steady deterioration of the vehicle's shock absorbers. The driver's spine is subjected to a series of jolts and shocks as the vehicle runs over bumps in the road, pot-holes, "sleeping policemen", etc.—particularly at speed. The effect of a vertical impact will probably be magnified if the driver is leaning forward, since the impact will tend to flex his spine. A reclined sitting position with good lumbar support reduces this effect and distributes the loading over a wider area.

### Whiplash Injury

Whiplash injury occurs in a rear end collision—typically when a stationary vehicle is "shunted" from behind, taking the driver unawares. The impact is transmitted (through the back of the seat) to the driver's trunk and shoulders, which accelerate forward. The inertia of the head causes an initial forceful hyperextension of the neck—which is followed by a flexion phase as the head is catapulted forward. (Other incorrect sequences of events have been described in the past.)

It is in the initial hyperextension phase that the principal injuries characteristically occur: these may include disc lesions, damage to the anterior longitudinal ligament of the spine, various vertebral fractures and muscular injuries (particularly of the sternomastoid). There are sometimes signs of concussion, due to movements of the brain in the cranial cavity.

In practice, headrests on the seat are relatively ineffective in preventing these injuries. The head does not so much move backward as rotate on the neck as the trunk moves forward. This hyperextension will only be prevented if the head is in direct contact with the headrest at the moment of impact. This will rarely be the case, since most people drive with a reclined seat and an upright or protruded head. (For a further discussion of whiplash see Bogduk, 1986.)

## Vibration

Whole-body vibration is rarely encountered except in moving vehicles—so it is convenient to deal with the topic at this point. Vibration levels in ordinary motor cars are generally relatively low; levels in trucks and buses tend to be higher. The highest levels are found in tractors and other off-road vehicles and on earth-moving machinery such as bulldozers, etc. In a few industrial situations, the standing worker may be exposed to vibration transmitted through the floor from heavy machinery like presses and hammers. Vibration from hand-held tools is discussed in Chapter 15.

The vibration to which an individual is exposed may be described in terms of its intensity, frequency, direction and point of application. The intensity may be expressed either as the amplitude of the motion (i.e. the displacement of the vibrating object from its resting position) or in units of acceleration, most commonly in multiples of $g$ (where $g$ is the acceleration due to gravity).

All solid objects have a natural or *resonant frequency* at which they will vibrate freely when struck or otherwise mechanically excited. (The frequency is proportional to the square root of the stiffness of the object divided by its mass.) When an object is vibrated at or near its resonant frequency, the amplitude of its response is greater than that of the vibration to which it is exposed. Thus sopranos are said to be able to shatter wineglasses by singing top notes.

A moving vehicle may continue to oscillate vertically (at its natural frequency) for several cycles each time it passes over an irregularity in the ground. (The purpose of the suspension system is to damp out these oscillations.) There may be other sources of vibration within the vehicle itself, and different parts of the vehicle may resonate at different frequencies. These vibrations are ultimately transmitted to the driver via the seat.

The vibration in a moving vehicle (measured on its frame) is generally random in its wave form: that is, a spectrum of frequencies is present. Most of the energy in this frequency spectrum is generally concentrated in the 1–20 Hz range. Peak intensities for the vertical vibration of motor cars on roads rarely exceed 0.1–0.3 $g$; but for tractors and construction vehicles, etc., peaks of 1 $g$ or more are not uncommon. When the vertical acceleration exceeds 1 $g$, the driver is thrown into the air.

At frequencies of less than about 2 Hz the human body moves like a dead weight. At higher frequencies the body behaves like a complex system of masses and springs. The parts and organs of the body may move independently of each other—each part and organ having its own resonant frequency.

When exposed to vertical vibration in the sitting position, the human body has its principal resonance at about 5 ($\pm 1$) Hz. At this frequency, the amplitude of the motion of the head is amplified by 30–100% (compared with the motion of the seat surface). There may be additional smaller resonances at harmonics (i.e. multiples) of the fundamental frequency (Coermann, 1962, 1970; Wilder *et al.*, 1982).

The trunk and its parts resonate at frequencies in the 4–10 Hz range. The lumbar spine has its peak resonance at 4 Hz—at this frequency the amplitude of the vibration may be amplified by more than 200%. The resulting deformation of the intervertebral discs may be clearly seen in a cineradiograph. Electromyographic studies have shown pulsed bursts of activity in the back muscles, synchronous with low-frequency sinusoidal whole-body vibration (Dupuis and Zerlett, 1986). The abdominal organs (stomach, colon, etc.) resonate at 4–5 Hz and the thoracic cavity at about 7 Hz (causing chest pain and difficulty in breathing). Smaller organs resonate at higher frequencies. At 13–15 Hz (or thereabouts) the pharynx and larynx and the pelvic organs may resonate. Subjects may feel a "lump in the throat" and the voice may acquire a vibrato quality. There may be an urge to urinate, and some people are said to report sexual sensations. The eyes resonate at 20–25 Hz.

The damage which vibration causes to any particular part of the body will generally be greatest at (or near) the resonant frequency—where the amplitude of motion of the part concerned is maximized. The vibration which is transmitted to the occupant of a tractor or a construction vehicle typically has a dominant frequency in the 3–6 Hz range (Matthews and Knight, 1971; Wilder *et al.*, 1982). This is close to the resonant frequencies of the spine and abdominal organs.

Rosegger and Rosegger (1960) found strikingly high prevalences of radiological signs of degenerative changes in the spines of younger tractor drivers. More than 70% of tractor drivers in their twenties showed such signs, compared, for example, with 29% for a sample of craftsmen (average age 41) and 37% for building labourers (average age 51). Dupuis and Zerlett (1986) review a number of other epidemiological studies which have shown high prevalences of radiological signs of degenerative changes in the spines of people exposed to whole-body vibration at work, including truck and bus drivers, drivers of earth-moving equipment, aircrew (especially helicopter pilots) and concrete workers. The mechanisms of vibration-induced damage to the spine has been discussed by Sandover (1983).

Vibrations at frequencies and intensities similar to those to which tractor drivers are exposed have been shown to set up pressure waves in the colon and to cause oscillatory displacements of the stomach of the order of 25 mm or more (Coermann, 1970). Tractor drivers are therefore prone to suffer from stomach pain and gastrointestinal complaints (Rosegger and Rosegger, 1960), including haemorrhoids (Grandjean, 1988). Vibration may affect comfort and performance at relatively low intensities (lower at least than those which are believed to pose a risk to health)—for a discussion see Oborne (1987).

### Seat Design

Depending on its resonant characteristics, the vehicle seat may either attenuate or amplify the vibration of the vehicle. Griffin (1978) found that the seats of cars, trucks and buses could in the best cases attenuate the vibrations by 20% or more; but in the worst cases the vibration could be amplified by around 15%. An attenuation of 50% or more may be achieved by mounting the seat on its own suspension mechanism, as is sometimes done in tractors (Matthews and Knight, 1971).

### Vibration Exposure Limits: ISO 2631

Threshold limiting values for exposure to whole-body vibration are the subject of an international Standard (ISO 2631). Three threshold values of vibration intensity are specified as a function of vibration frequency and duration of exposure. In ascending order, these thresholds are for: (i) comfort, (ii) performance effects and (iii) health effects. The highest threshold value is regarded as the exposure limit for working activities. In each case, the lowest threshold values (for vertical vibration) are in the 4–8 Hz range.

ISO 2631 has been challenged on the grounds that the threshold values are based on inadequate empirical evidence—particularly with respect to the effects of the duration of exposure (Griffin, 1978; Oborne, 1985b, 1987). A more recent British Standard (BS 6841) therefore confines itself to methods of evaluating vibration exposure and does not quote limiting values. (Compare this with the problem of specifying a safe load for lifting—p. 306.)

Griffin (1978) compared vibration levels measured in motor vehicles with the limiting values specified in ISO 2631. The majority of cars would have passed the comfort limit after 8 hours on trunk roads or 10 minutes on country lanes; and the limit for health effects after 6 hours on country lanes. Vibration levels being higher in trucks and buses, the limiting levels would be reached in correspondingly shorter times. In the light of the epidemiological evidence (see Chapter 3), this does not seem unreasonable.

Village *et al.* (1989) found that vibration levels in load–haul–dump vehicles used in mining work, exceeded ISO exposure limits—sometimes by a factor of 2 or 3. The drivers of these vehicles had a high incidence of both back and neck problems—the latter being relatively uncommon in other groups of underground workers. The driver of this particular kind of vehicle sits sideways to the direction of travel; his neck is therefore subject to vibration and jolting whilst twisted at a right angle.

### Motion Sickness

Oscillatory motions of low frequency and high amplitude (particularly in the 0.2–0.5 Hz range) cause motion sickness. Vibrations in this range of frequencies are most commonly encountered during sea and air travel. Children and a few adults may also be affected in motor vehicles (probably owing to horizontal swaying motions, etc.—since vertical vibration is likely to be in a much higher frequency range). Symptoms progress from pallor and dizziness, through nausea and vomiting, to a condition of almost complete prostration in the worst cases. The symptoms generally pass off rapidly when the motions cease. Children are worse affected than adults; and young adults are worse affected than the elderly. Infants under 18 months are generally unaffected.

The physiology of motion sickness is complex. The condition is basically due to overstimulation of the vestibular apparatus—the organs of balance located in the inner ear. The vestibular apparatus consists of the semicircular canals (which respond to angular accelerations) and the utricle and saccule (which respond to linear accelerations). The latter are thought to be more important in motion sickness. Visual factors are also known to be involved. A conflict between these various sources of information concerning bodily motion (and the individual's expectations concerning the ways in which they ought to be related) may be the critical factors (Reason and Brand, 1975). Most seasoned travellers adapt to the stimuli which cause motion sickness, but a minority (about 5%) remain affected.

## Chapter Fourteen

# Hand Function and Tool Design

## Repetitive Manipulative Tasks

The principal ergonomic risk factors associated with musculoskeletal problems in production line workers (and other people who perform repetitive manipulative tasks) are: working posture, repetition rate, force application and wrist position. *The worst problems are often associated with repeated forceful gripping and turning actions executed with the wrist in a deviated position.* The turning action may involve various wrist and forearm movements: flexion/extension, ulnar/radial deviation, pronation/supination, or some combination of these. Gripping actions alone are less likely to cause problems—especially if the wrist is in a neutral position (see Chapter 4).

The workers on a modern production line are there to perform whatever tasks are left, that are too difficult (or too expensive) to automate. In practice, these often tend to be jobs which are awkward or fiddly, involving non-standard product, where parts have to be aligned at odd angles, and so on (Figure 14.1). This means that as assembly line work is automated, a higher proportion of those jobs that remain fall into the high-risk category, and the prevalence of musculoskeletal problems in the workforce increases.

High repetition rates do not solely occur in machine-paced tasks. Rates may also be high where the work is self-paced and payment is by results. Electronic monitoring is increasingly likely to be used to maintain high work rates in self-paced tasks: for example, some supermarkets now program the bar-coding machines at the checkouts to monitor the work rates of the operators.

Measures which entail the loss of output are unlikely to be acceptable to management—or to workers either, if payment is by results. A reduction in the repetitiveness of production line jobs may sometimes be achieved (without loss of output) by job rotation or job enlargement; but the extent to which this is a viable option is contingent upon the existence of contrasting work activities in the production process (p. 137). Some repetitive manipulative actions may be eliminated by the use of power tools, etc.

The deleterious effects of a given repetition rate are likely to depend upon the extent to which the job is ergonomically satisfactory in other respects. Thus higher repetition rates may be tolerated where forces are low and the wrist is in a neutral position. In practice it is often easier to change these other factors than to reduce the repetitiveness of the task. The force (or, more specifically, the muscle effort) which is required to perform a given action may often be reduced by designing tools that are easier to grip or have better mechanical advantage (see below). It is

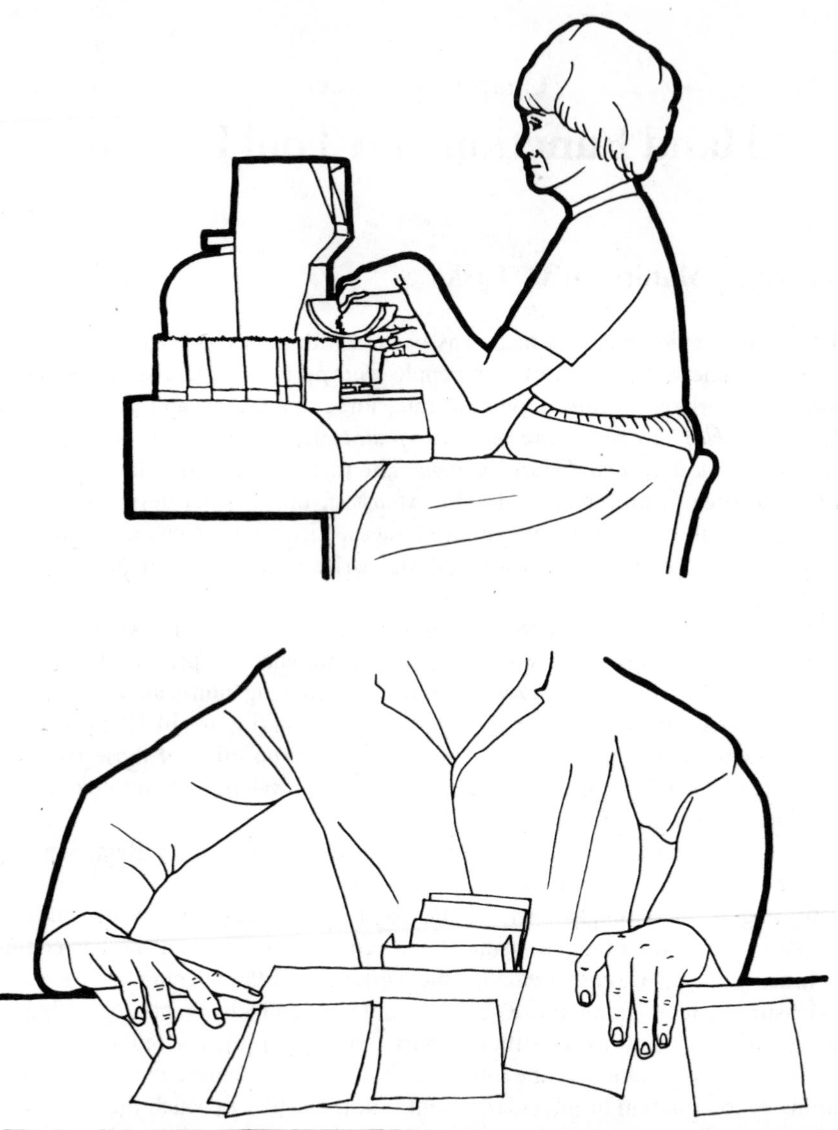

Figure 14.1 *Deviated wrist positions in repetitive industrial tasks:* top, *pinch grip with a flexed wrist, in a weighing task (subject suffered from carpal tunnel syndrome);* bottom, *movements of radial and ulnar deviation with an extended wrist in a packing task (note also abduction at shoulders). In both cases the working level is too high. From originals kindly supplied by Peter Buckle*

also important that tools should be properly maintained, have good cutting edges, etc.

The need for working with a deviated wrist may often be eliminated by:

- redesigning hand tools (see below);

- changing the height of the working surface (generally lowering it) or making it adjustable to individual requirements;
- tilting the working surface or workpiece towards the worker, tilting containers for ease of access, etc.;
- setting the workpiece at an angle to the worker's body. (Note that when the shoulders are in a relaxed neutral position, and the elbows are flexed to a right angle, the forearm makes an angle of 60° to the frontal plane of the body.)

## Hand Function

The hand is a complex and versatile mechanism. Each finger has four degrees of freedom and the thumb has five. The mechanism is operated by a total of 42 muscles (±1).

At rest, the fingers adopt a slightly flexed position and are partly spread. The little finger is usually the most flexed and the index finger the least flexed (Figure 14.2). The *position of rest* is determined by the passive tensions in the opposing muscle groups (particularly the long flexors and extensors in the forearm). Thus if the wrist is passively extended, the fingers will flex further and vice versa.

The functions of the hand may be broadly grouped into those which involve a gripping action (prehensile actions) and those which do not (see Figure 14.3). In *gripping* actions the hand forms a *closed kinetic chain*—that is, the parts of the hand are used in mechanical opposition to each other, exerting comprehensive forces on the object and holding it in place. In *non-gripping* actions, the hand forms an *open*

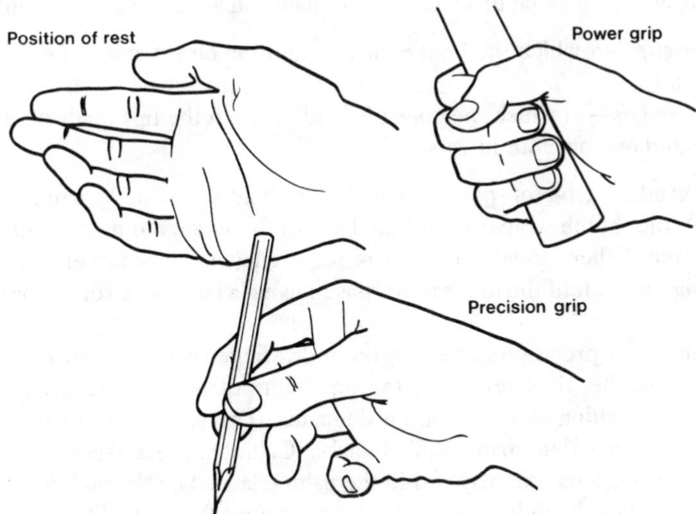


Figure 14.2 *The position of rest, the power grip and the precision grip. Reproduced from Pheasant (1986)*

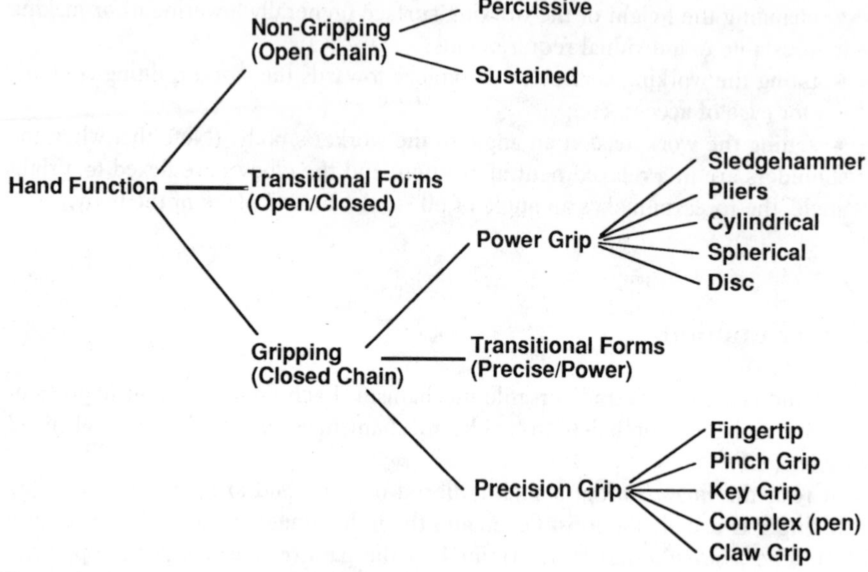


Figure 14.3 *A taxonomy of hand functions*

*chain*: these may be percussive (as in striking actions) or sustained (as in stroking actions). There are transitional forms in which the hand acts as an open chain which is on the point of closing: such as the *hook grip* (as used when carrying a suitcase etc.) and the scooping and cupping actions used to pick up a handful of small objects.

Gripping actions may be divided into two main categories (Napier, 1956):

*the power grip*—in which the fingers flex around the object and clamp it against the palm; and

*the precision grip*—in which the object is held between the tips (pads or sides) of the thumb and one or more fingers.

In the crudest type of power grip (as in holding a sledgehammer or a truncheon), the thumb wraps around the back of the fingers to provide additional gripping force. Where greater control is required (with less force), the thumb moves along the shaft of the tool handle—as in using a large screwdriver or a small hammer.

As the need for precision increases, the index finger may also move along the shaft of the handle, thus providing the possibility of both power gripping and precision manipulation as the occasion demands. Transitional forms of precise/power grip are used in many skilled tasks. Cabinetmakers' screwdrivers are designed so that the handle may rest in the palm (clasped by the middle, ring and index fingers) while the index finger and thumb control the shaft. The tenon saw is correctly held with "three fingers, one finger and a thumb". To hold a knife at table in a full power grip (i.e. like a dagger) would be considered crude; to hold it in a precision grip would be considered affected; but to hold it in a precise/power grip is both ergonomically and socially correct.

Precision grips may employ the tips of the thumb and the fingers (as in pulling on a thread), the pads (in a pinch grip) or the pad of the thumb with the side of the index finger (as in turning a key). Occasionally the fingernails are used. The more complex form of precision grip, which is used for writing, generally involves the pads of the index finger and thumb, and the side of the middle finger. The strongest (and least precise) variety of precision grip is the claw grip, which uses the tips or pads of the thumb and all four fingers.

Figure 14.4 shows a simple two-way classification of hand function based upon the degree of hand/object contact and the degree to which the hand is used in an open or closed chain configuration. Power grips are on the lower right-hand portion of the chart, and precision grips on the lower left. Reading the right-hand side of the chart from top to bottom, we see the chain closing as the fingers of the hand are increasingly flexed around the object. Finally, the thumb flexes over the fingers to form the fully flexed sledgehammer grip.

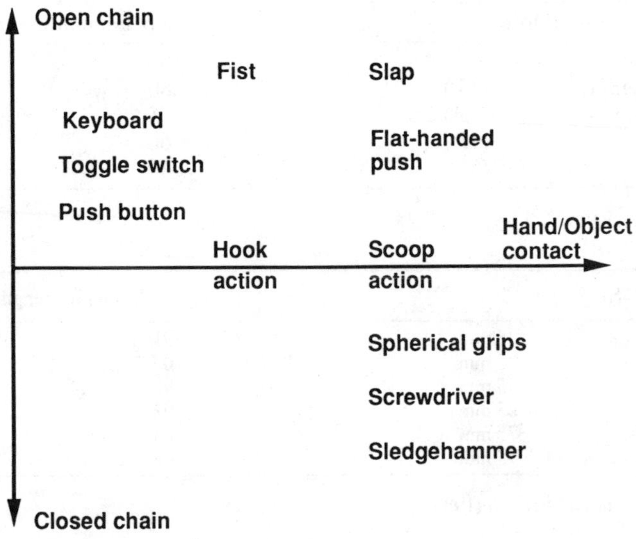


Figure 14.4 *Another taxonomy of hand functions*

The grip which a person adopts in a particular situation may be determined by the nature of the object and/or by the demands of the task. Some objects can only be held in one way; others may be held in various ways according to the circumstances. For example, in undoing a screwtop jar you use first a power grip, to overcome resistance; then you switch to precision grip to spin the top off the thread of the jar. When inserting a light bulb, you hold it first in a precision grip to find the correct alignment; then change to a power grip to press and turn it into place.

## Grip Strength

Grip strength is generally measured using a device (called a dynamometer) which fits into the palm of the hand. The dynamometer has two parallel beams. The fingers wrap around one beam and the other beam is pressed against the heel of the palm (the thumb is not involved). Thus the action resembles the variety of power grip which is used when holding a large pair of pliers or cutters.

The strength of a gripping action varies with the position of the wrist, as shown in Table 14.1(A). Strength is least when the wrist is fully flexed. Most of the force of finger flexion comes from muscles located in the forearm. The tendons of these muscles cross the wrist; thus, when the wrist is flexed, these muscles are shortened and therefore weakened (p. 38). To prevent the wrist from flexing when making a power grip, the wrist extensors must be active. (Prove this by palpating the back of your forearm.) Grip strength also varies with handle size—that is, with the separation between the two parallel beams of the dynamometer. The figures quoted in Table 14.1(B) are averages for a sample of young adult men and women.

*Table 14.1 Factors Affecting Grip Strength*

| A. WRIST POSITION | | % Maximum strength |
|---|---|---|
| Neutral | | 100 |
| Radial Deviation | (25°) | 80 |
| Ulnar Deviation | (45°) | 75 |
| Extension | (45°) | 60 |
| Flexion | (65°) | 45 |

Source: Rogers (1987).

| B. HANDLE SIZE | % Maximum strength |
|---|---|
| 35 mm | 91 |
| 45 mm | 100 |
| 55 mm | 98 |
| 65 mm | 92 |
| 75 mm | 83 |
| 95 mm | 68 |

Source: Pheasant and Scriven (1983).

## Gripping and Turning

The action of gripping and turning a handle of circular cross-section is analysed in depth in Pheasant (1986). The mechanics of the action are summarized in Figure 14.5. Experimental studies have shown that the optimum diameter for the exertion of thrust ($F$) along the axis of a cylindrical handle is about 40 mm. At this diameter the thumb and fingers can slightly overlap each other as they wrap around the handle, thus maximizing the compressive force; but the optimal diameter for exerting torque about the axis of the handle is greater—around 50–65 mm. This is because the decline in grip strength which occurs as the handle diameter passes

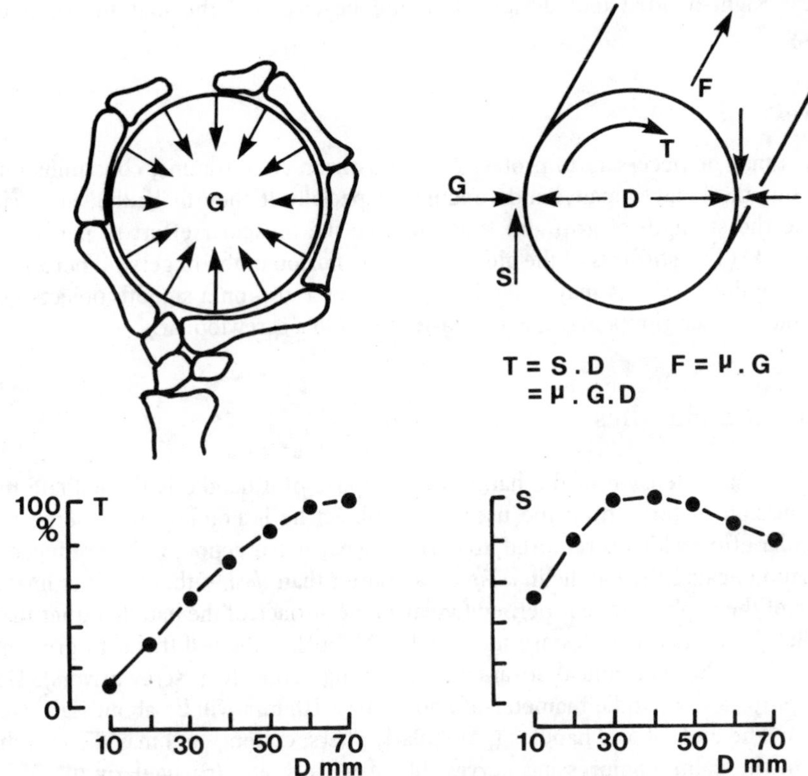


Figure 14.5 *The mechanics of the gripping and turning action, using a cylindrical handle. Note that torque (T) is greatest on the 70 mm handle, whereas both shear (S = T/D = μG) and thrust (F, not plotted) are greatest on handles in the 30–50 mm size range. Data from Pheasant and O'Neill (1975)*

the optimum may for a time be offset by the increase in mechanical advantage (that is, $D$ will be increasing more quickly than $\mu G$ is declining).

The optimal diameter for the exertion of torque on a *spherical* handle is about 65–75 mm; and for a *disc*-shaped handle it is as high as 90–130 mm. The increase in optimum diameter—from cylinders, to spheres, to discs—is probably a consequence of the opening of the kinetic chain of the hand.

## Handedness

About 9% of people are left-handed. The percentage seems to be increasing but it is not clear why (Fisher *et al.*, 1978). Some tools are difficult or impossible to operate left-handedly. Scissors often cause problems because of both the design of the handles and the set of the blades. The triggers of power tools may be badly placed for left-handed operation—as may the handles of tools which require both hands. Right-handed tool designers should be aware of the first fundamental fallacy.

### Gloves

Gloves may be necessary to protect the hands from cold, trauma, contamination, etc.; but they may impair hand function (especially if they fit badly); and may reduce the strength of grip and thus increase the muscular effort to perform a given task. (The stiffness of the glove resists the flexion of the fingers.) There are a few exceptions. Gloves may make it easier to get a grip on a smooth object—for example, rubber gloves may make it easier to undo a screwtop jar.

## Tools and Handles

Tools are an extension of the hand. The purpose of a handle is to facilitate the transmission of force from the user to the object he is holding. In general, the muscular effort which is required, to perform a particular action, will be reduced if the hand engages the handle in *compression* rather than *shear*—that is, if the line of action of the applied force is perpendicular to the surface of the handle rather than parallel. Many tool handles are too small. This makes them difficult to grip and may reduce the mechanical advantage in turning action (e.g. screwdrivers). For most purposes, a handle diameter of about 40 (±10) mm will be about right (see above). The ends of tool handles (particularly pliers, cutters, etc.) may dig into the base of the palm, compressing nerves, blood vessels, etc. (particularly the ulnar nerve) and causing numbness and tingling. To avoid this, the handle should be at least 100 mm in length.

Any unrounded edge, on the part of the tool which is in contact with the hand, can cause a pressure hot spot resulting in compression of tendons, nerves, blood vessels, etc., and abrasion, blistering or the formation of calluses or other "occupational stigmata". Finger shaping is generally undesirable. If the indentations match the fingers of the average user, they will be too small for a large user, and the ridges between the indentations may compress the nerves and vessels which run down the sides of his fingers. Conversely, a small user may find two fingers overlapping a single indentation.

*Surface Finish* If the handle surface is too smooth, it will slip in the hand; if it is too rough, it will be abrasive. Varnished wood gives a better purchase than either polished metal or smooth plastic (probably because it is resilient). *Blistering* is most likely to be a problem where the hand exerts a shearing force on the handle and/or when the handle slips in the hand.

*Triggers* The incorporation of finger- or thumb-operated on/off controls onto handles which need to be gripped firmly is a common source of ergonomic problems. Triggers should be guarded so that they do not trap the fingers.

*Wrist angle* A rod held in the hand makes an angle of 100–110° with the axis of the forearm when the wrist is in a neutral position—as shown in Figure 14.6. The

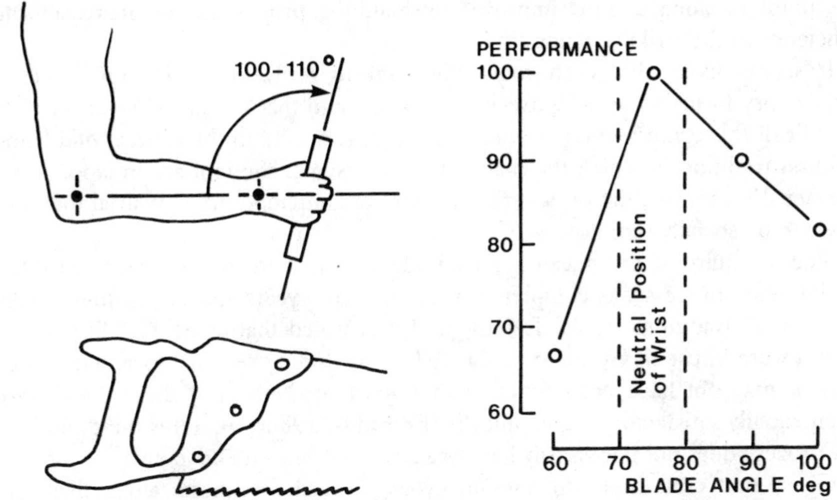


Figure 14.6 Left, *the neutral position of the wrist and the handle of a carpenter's saw.* Right, *results of an experiment in which cutting performance was measured at different blade angles. Performance is optimal when the neutral position is preserved*

angle between the "life-line" of the palm and the middle finger is about the same. This is because of the relative lengths of the metacarpal bones of the palm. The maintenance of this relationship is important in the design of saws (see below); and in tools such as power drills which have pistol grips.

*Anthropometrics* A rectangular aperture of 110 mm × 45 mm allows access for the fingers and palm of a large hand as far as the web of the thumb (as in a suitcase handle, etc.). A circular aperture of 30 mm in diameter allows the insertion, rotation and removal of a large finger or thumb (Pheasant, 1986).

For a further discussion of tools in general, see Fraser (1980, 1989), Pheasant (1986), Freivalds (1987).

## The Design of Craft Tools: Vernacular Ergonomics

Craft tools often have a long history. The collection at the Science Museum in South Kensington, London, includes an ancient Egyptian mason's mallet, which is believed to have been used for dressing stone in the building of the Pyramids. The design of this tool is not appreciably different from that of a modern carving mallet. It seems likely that a tool which has remained unchanged over so lengthy a period must be well adapted to its purpose. (Something approaching 200 successive generations of users have not found the need to modify the form of this tool to any appreciable extent.) The circular cross-section of the mason's mallet allows its user to strike the chisel at any angle without turning the tool in the hand or adopting awkward wrist positions. The inverted conical taper of the head increases the moment of inertia of the tool and brings its centre of gravity close to the point of impact. This improves its handling properties and increases the efficiency of the striking action.

It seems likely that craft tools will tend to evolve towards ergonomically satisfactory forms by an adaptive process similar to that of natural selection. We could call this *vernacular ergonomics*. The process occurs in the context of a living cultural tradition in which the maker and the user of the tool are in close social contact. By contrast, the designers and users of contemporary industrial tools are likely to be socially remote.

The antiquity of the mason's mallet suggests that in this particular case the evolutionary process was complete several thousand years ago. But is this equally true for all traditional tools? Lehmann (1962) noted that some 12,000 types of shovel were in use in Germany in the 1930s—including regional variations which may or may not have been functionally important. Not all of these could have been equally satisfactory ergonomically (Freivalds, 1986a, b). Thus the evolutionary process does not necessarily lead to a single optimized end-point.

Given that craft tools do tend to evolve towards ergonomically satisfactory designs (in some cases, at least), there are still a number of ways in which mis-matches may arise when the tools are used in the industrial context. A tool which was well matched to one particular set of functional requirements may be adapted for another purpose for which it is less satisfactory; or the nature of the circumstances in which it is used may change.

A tool which is well matched to the physical characteristics of a particular group of users may be ergonomically unsatisfactory if the user population changes. Manual trades involving the working of hard, unyielding materials (wood, stone, metal, etc.) have traditionally been practised by men—presumably because of the degree of physical strength which is required. Women are taking up these trades increasingly frequently, both professionally and in the home. Ducharme (1975, 1977) investigated the difficulties experienced by female engineers in the US airforce. The principal problems which these women reported are summarized in Table 14.2.

*Table 14.2 Percentages of Women in Engineering Trades Reporting Difficulties in Using Tools*

| | | |
|---|---|---|
| Crimping tool | up to 25% | – handles too far apart, too hard to squeeze |
| Metal shears | 22% | – too large, need two hands |
| Wire stripper | up to 19% | – handles too far apart |
| Rivet cutter | 17% | – too hard to squeeze |
| Soldering iron | up to 17% | – too heavy, handle too large |
| Tool chest | 16% | – too heavy |
| Jack plane | 16% | – too big |
| Soldering gun | 15% | – too heavy, can't reach trigger, difficult to hold |
| Flexible mechanical finger | 12% | – hard to manipulate |

Source: Ducharme (1975).

## The Axe

The variety of axe which is used most extensively in forestry work today evolved in England some time during the eighteenth century and rapidly became popular in the North American colonies (Salman, 1975). Its principal distinguishing features are its short, wedge-shaped blade and heavy, flat "poll" (the part of the head which lies behind the blade). It is generally known as the "wedge axe" or "American axe".

The shaft of the handle is oval in cross-section and has a characteristic curve which terminates in a reverse taper (known as a fawn's foot) for ease of grip. One hand grasps the handle near its base throughout the action; the other grasps it near the head, in order to lift the weight of the axe, and slides down the shaft during the swing, following a natural curve. The shaft is 36 inches in length—the traditional length of a man's arm from the breast bone. This dimension is what Drillis (1963) calls a "folk norm".

The objective in swinging an axe is to give it the maximum possible kinetic energy at the time of impact, and to convert as much of this energy as possible into the useful work of cutting wood. The kinetic energy increases with the mass of the head and the square of its velocity; but if the axe is too heavy for the user, he will not be able to give it the maximum possible degree of acceleration. The weight of the poll of the American axe also brings the centre of gravity of the tool closer to the point of impact, thus increasing the efficiency of energy conversion and minimizing the energy which is dissipated into the tissues of the user's hand and arm.

The American axe is significantly heavier than its predecessors, due principally to the weight of the poll. The price of iron was falling in New England at the time when the axe came into use—and the demand for axes was high as the forests of the interior were felled. Widule *et al.* (1978) and Corrigan *et al.* (1981) showed experimentally that the added weight of the American axe did indeed give greater kinetic energy in the swing, in the hands of an experienced (and presumably strong) male user. But the energy cost of tree felling (as measured by oxygen uptake) increased more rapidly with the weight of the axe than did the power output of the stroke. So although the user of the heavier axe would have been able to fell a tree more rapidly, he would not have been able to continue working for so long at a stretch. (Energy expenditure with the heavy axe was equivalent to 75% maximum aerobic power.) Given that lumbering involves relatively short periods of very intense activity, alternating with lighter work, the greater weight of the axe probably conferred an advantage overall.

## Saws

The handles of most types of carpenters' hand saws are set at an angle of 70–80° to the cutting edge of the blade. This particular design feature seems to have evolved in the seventeenth century—possibly in Holland (Goodman, 1964). It allows the neutral position of the wrist to be maintained during the cutting action. The angle is more acute for tenon saws than for the larger panel saws and rip saws, presumably because the former are used on a trestle or "sawing horse". Figure 14.6 shows the results of a simple experiment in which a group of design students were timed cutting uniform pieces of wood using an adjustable saw on which it was possible to set the handle at different angles to the blade. The experiment confirmed that the traditional angle does indeed optimize the speed of cutting.

Different cultures may evolve different solutions to the same design problem. (This is rather like the divergent evolution of different species of animals and plants.) European hand saws are designed to cut on the downstroke, by a pushing action. Chinese and Japanese hand saws cut on the upstroke, by a pulling action. Which is better? The pushing action may make it easier to hold the workpiece steady against a sawing block or trestle. At a constant level of work output, however, the energy expenditure using a European saw is about 30% higher (Bleed *et al.*, 1982). The blade of a saw which cuts on the pull stroke does not have to withstand compressive forces, so it is not likely to bend or buckle. Both the blade and the kerf (saw cut) can therefore be narrower, and less energy is expended in turning wood into sawdust.

## Pliers

Pliers, pincers and tongs are members of a large and important family of gripping, cutting and crushing tools which operate by a crossed-lever action. In mechanical terms, these are "levers of the first class". Other examples include wire cutters, wire strippers, metal cutting shears ("tin snips"), garden pruners and secateurs, nut crackers, and so on.

Ducharme (1975, 1977) noted that many of the tools which women find difficult to use are in this category, and suggested that this was principally because the spread of the handles was too great for the female hand. The data of Pheasant and Scriven (1983) suggest that this is not the case—since the optimum handle separation is similar for both men and women (Figure 14.7). Pliers and cutters generally have a mechanical advantage of around 6:1 or less. This is apparently enough to allow the typical male user to perform the tasks for which the tools are designed. Since women on average have a lesser grip strength, they will be more likely to experience difficulties.

Modern pliers have presumably evolved from blacksmiths' tongs. These were used for gripping the workpiece and holding it steady—but not usually for making twisting and turning actions. To turn an object about the axis of the forearm, using a conventionally designed pair of pliers, requires an ulnar deviation of the wrist close to the limit of the range of motion. Tichauer (1978) found a high incidence of tenosynovitis, carpal tunnel syndrome and tennis elbow in assembly workers in an electrical goods factory who were using pliers for twisting wires. The incidence of these conditions, over a 12-week training period, was cut from over 60% to only 20% by replacing the conventional long-nosed pliers which the workers had been using with a redesigned tool which had a bend in the handle so as to allow the wrist to remain in the neutral position.

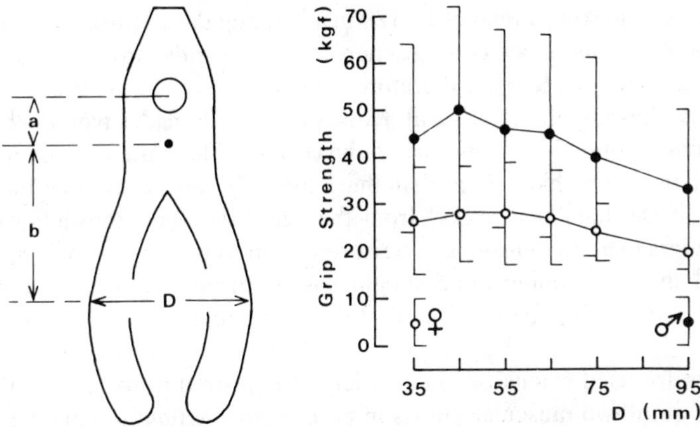


Figure 14.7 *Grip strength (G) as a function of handle separation (D). The effective crushing or cutting force of the tool = G × B/A. Reproduced from Pheasant (1986)*

## Screwdrivers

The first hand-cut screws date back to the fifteenth century. The first screw-turning tool was a T-handled box spanner. Screwdriver bits for carpenters' braces appeared later, and the screwdriver itself did not appear until the nineteenth century. Screwdrivers did not become common until after 1850, following the mass production of taper-pointed woodscrews (Fraser, 1980).

The torque available from a screwdriver depends principally upon the *effective diameter* of its handle (assuming that the blade is correctly matched to the slot of the screw). The *shape* of the screwdriver handle is of minimal importance. Pheasant and O'Neill (1975) tested a range of commercially available screwdrivers of different shapes and sizes, together with polished steel cylinders and cylinders which had been knurled to give better purchase, the diameters of which ranged from 10 mm to 70 mm. When the effects of handle size were taken into account, none of the styles of screwdrivers was any better (or worse) than any other; and none of the screwdrivers was better than a knurled cylinder of equivalent diameter, or worse than a polished cylinder.

The largest screwdrivers readily available at the present time have an effective diameter of not more than about 35 mm. But the optimum diameter for torque exertion would be 50–60 mm. The evolution of the screwdriver would seem to be incomplete. Current styles are presumably adequate for their purpose in the hands of a strong (male) user. Larger handles would reduce the muscular effort required and enable the weaker user to perform working tasks which are currently beyond his (or her) capacity.

## Knives and Scissors

Knives are used extensively in the food processing industry. A production line worker in a poultry factory can fillet 3,780 turkey legs per shift; each leg requires four knife cuts, making a total of 15,120 in a working day (Armstrong *et al.*, 1982). This works out at about one cutting action every 2 seconds throughout the working period. Accidents and RSI are therefore common.

Knife handles are often too small. Knives with small blades typically have small handles, but ergonomically this may not be correct. The more difficult it is to grip the knife, the more likely it is that the hand will slip down over the blade—especially if the handle is greasy from the carcasses. The conventional boning knife requires extreme positions of wrist deviation in use. Armstrong *et al.* (1982) suggested that the boning knife should have a pistol grip and a strap passing around the outside of the hand so that the grip could be relaxed between work cycles.

Scissors are used extensively in a variety of industrial processes. Conventional scissors require two muscular efforts in each cutting action: to open and to close them. One of these is avoidable by using spring-loaded self-opening scissors—at the cost of a small increase in the force required to close the scissors. But self-opening scissors can be designed to sit in the palm of the hand for a stronger, power-grip cutting action.

## Vibration Syndrome

Vibration transmitted to the hand and arm from power-operated tools may be responsible for a variety of vascular, neurological and musculoskeletal symptoms, collectively referred to as *hand–arm vibration syndrome.* Occupational groups affected include the users of chainsaws, pneumatic road breakers and chipping hammers, jackhammers, rock drills, grinders, rotary cultivators, hedge cutters, motor mowers, and the like.

The vascular symptoms of hand–arm vibration syndrome resemble the spontaneous blanching of the fingers, occurring in the cold, which is known as Raynaud's phenomenon (after the French physician who described it). This condition occurs in 5–10% of the general population, affecting women 8–9 times more often than men. It is usually idiopathic and runs in families, but it may be secondary to trauma or various systemic conditions. It is exacerbated by smoking. Similar vascular disturbances occur in thoracic outlet syndrome (p. 87) and have been reported in RSI patients (p. 78). Raynaud's phenomenon may also be caused by cold injury—for example, in people who work with frozen food products.

Raynaud's phenomenon due to vibration exposure is known as *vibration white finger.* It has been a prescribed industrial disease in the UK since 1985. The earliest manifestation is generally a relatively trivial tingling or numbness in the fingertips. The next stage (which may occur several years later) is a blanching of the fingertips in the cold. The episodes last from several minutes to an hour, during which manual dexterity may be severely impaired. The episodes are followed by a red flush (reactive hyperaemia) and there may be pain as the circulation returns. If exposure to vibration continues, the blanching episodes may affect progressively larger areas of the hand and may occur at higher temperatures.

The fingers may acquire a dusky, blueish appearance due to the permanent occlusion of blood vessels, and in the worst cases may become gangrenous. (For further clinical details see Taylor, 1974, 1985, 1988; Pykko, 1986; etc.)

In some working populations the prevalence of vibration white finger may be as high as 90%. Data from Taylor *et al.* (1975) are summarized in Table 14.3. In some of these groups, the observed prevalence was limited by the high rate of labour turnover; and it was estimated that virtually all workers would have been affected after two or three years' exposure.

*Table 14.3 Prevalences of Vibration White Finger in Occupational groups*

| Occupational groups | Prevalence (%) |
|---|---|
| Chain sawing | 48–89 |
| Grinding (hand and pedestal grinders) | 0–60 |
| Chipping (pneumatic tools) | 46–57 |
| Swaging (bending copper pipes by hand) | 50 |
| Controls (not exposed to vibration) | 4–19 |

After Taylor *et al.* (1975).

Workers who use vibrating hand tools commonly suffer from pains, numbness and paraesthesia in the arms and hands—as well as muscular weakness and fatigue—which lead to a decrease in manipulative dexterity. Prevalences of up to 80% for paraesthesia and 35% for muscular fatigue have been reported. The symptoms are thought to result from peripheral nerve damage. Areas of decalcification may develop in the bones of the hand, wrist and forearm—but the prevalence of these is probably low, compared with the vascular and neurological symptoms. Joint pains are also reported. Under some circumstances, the combination of very-low-frequency vibration and shocks with static muscle loading in an awkward working position—as in a miner using a rock drill—may lead to the premature development of osteoarthritis in the elbow joint (Pykko, 1986).

Vibration from hand tools is also known to be a risk factor for carpal tunnel syndrome (p. 94) and low back pain (p. 70)—perhaps because it requires a greater muscular effort to grip a vibrating tool and hold it steady.

The elimination of the vibration at source is easier to achieve in some tools than in others. Improvements in chainsaw design have resulted in a steady decrease in the prevalence of vibration syndrome in forestry workers (at least in the UK, Scandinavia and Japan); from 50–90% in the late 1960s to 20–35% in the 1970s and 10–15% in the 1980s. The problem is technically more difficult to solve in a tool like a roadbreaker, where the repetitive hammer action is an essential feature of its operation. Nonetheless, a considerable degree of vibration damping can be achieved (for example, by spring loading the handles). A pneumatic road-breaking drill which exposes the user to acceptably low levels of vibration has been available for around a decade, but such tools are not as yet widely used (Taylor, 1989).

Vibration from hand-held tools typically has a broad frequency spectrum (e.g. 2–2,000 Hz)—although there may be peaks at the impact frequency and its multiples. The tissues of the hand and upper limb act as a low-pass filter. Low-frequency vibrations are transmitted to the trunk and head. Subjectively, therefore, low frequencies are the most disturbing (Miwa, 1963). Current guidelines concerning acceptable levels of vibration exposure incorporate a system of weighting factors which places greatest emphasis on the 8–16 Hz range of frequencies. Frequencies above 100–150 Hz are almost completely absorbed by the tissues of the hand itself; lower frequencies are transmitted to increasingly proximal parts of the limb (Reynolds and Angevine, 1977). Resonance phenomena occur in the 30–80 Hz range (Teisinger, 1972). It may be that the musculoskeletal effects of vibration are principally associated with the lower parts of the frequency range, whereas the symptoms of vibration white finger are associated with higher frequencies. Measurement methods and guidelines concerning limiting levels of exposure are the subject of a British Standard (BS 6842).

The transmission of vibration from tool to hand increases with the gripping force which is applied (Pykko *et al.*, 1976). Unskilled workers who grip the tool tightly have been found to be worst affected (Teisinger, 1972). Ergonomic features which make a tool easier to grip and handle are therefore likey to be beneficial. Gloves may also attenuate the vibration—particularly at higher frequencies. But thick gloves may require the user to grip the tool more tightly, thus negating such beneficial effects as they may have. The best compromise is probably to wear relatively soft gloves which will protect the user from the cold without impairing his grip on the tool. Cold hands are thought to be more susceptible to the deleterious effects of vibration than warm hands.

## Chapter Fifteen
# Lifting and Handling

## Epidemiology

The present century has seen a steady increase in the automation of industrial work; but the number of injuries each year which are attributable to lifting and handling tasks remains high. Accident statistics from the annual reports of HM Chief Inspector of Factories and (subsequently) the Health and Safety Executive, going back to 1924, when such data were first collected systematically, show that despite considerable fluctuations in the overall figure, the percentage attributed to "handling goods" has remained relatively constant*. In each year between 1945 and 1977, handling accidents accounted for between 25% and 30% of the total (HMCIF, 1924–1974; HSE, 1975–1977). Handling accidents were by far the largest category in this system of classification. In 1977, when handling accidents accounted for 30% of the total, second and third places were held by falls (16%) and accidents associated with machinery (14%). Since 1978 a different system of classification has been in use. The nearest relevant category is entitled "overexertion, strenuous or awkward movements and free bodily motion". For the years from 1978 to 1982 the "overexertion, etc." category accounted for between 20% and 23% of the total (HSE 1978–1982).

It is widely believed that the back is the part of the body which is most likely to be injured during lifting and handling activities; and conversely, that industrial back injuries are usually attributable to lifting and handling. Neither belief is completely correct. Analysis of the "handling" and "overexertion, etc." categories reveal different distributions with respect to the parts of the body injured. Figure 15.1 is based on HMCIF (1964) and Troup and Edwards (1985). Only 28% of handling accidents resulted in back injuries, compared with 61% of overexertions. Figure 15.2 shows a breakdown of back injuries reported to HSE in 1979 (after Troup and Edwards, 1985). Sixty-nine per cent of these were overexertions—which included 51% attributable to "lifting, carrying, wielding and throwing" activities. In this year, back injuries accounted for about 18% of all reported accidents.

The distinctions between *handling injuries* and *overexertions* are worth emphasizing. The latter do not include contact injuries (being crushed or trapped by the load) or other "true accidents" such as those due to a loss of footing. In the

* Note that not all of these injuries would have been due to "true accidents" in the sense we have used the term elsewhere: many would have occurred during normal working activities without the intervention of unforeseen events.

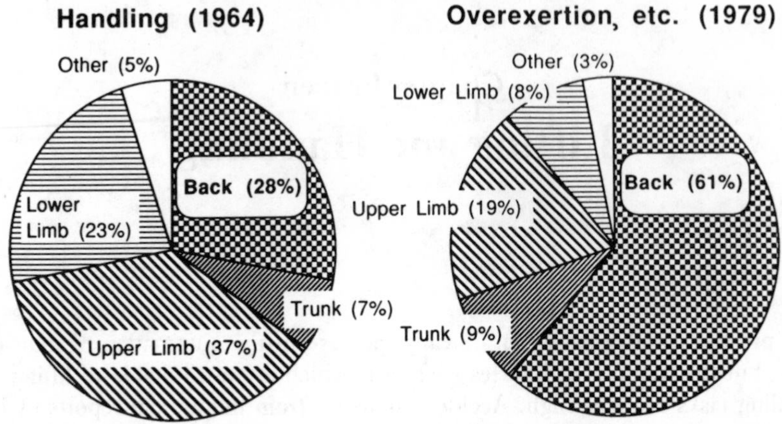


Figure 15.1 *Breakdown of "handling accidents"* (left) *and "overexertion injuries"* (right) *by part of body affected. Data from HMCIF (1964) and Troup and Edwards (1985)*

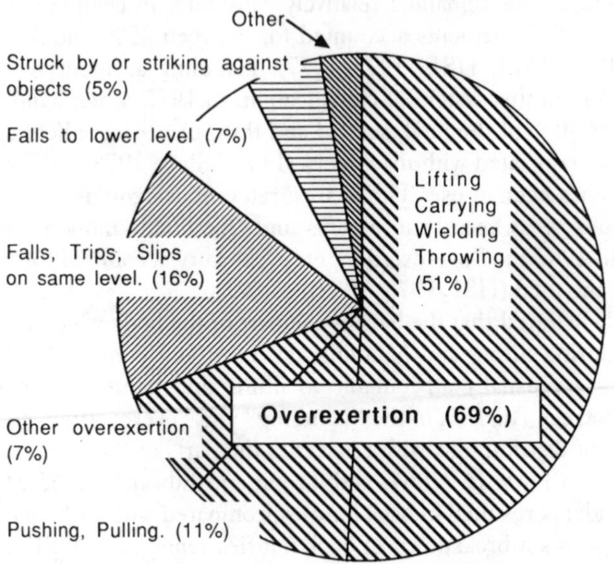


Figure 15.2 *Breakdown of reported back injuries by cause. Data from Troup and Edwards (1985)*

industrial context, the former category may define a more clearly circumscribed range of working activities, but the latter may be more relevant in ergonomic or clinical terms. Neither category is co-extensive with "back injuries".

## The Mechanics of Lifting

Lifting and handling tasks involve three kinds of risk:

- *the risk of accidental injury*—due to loss of footing, being trapped by the load etc.;
- *the risk of overexertion*;
- *the risk of cumulative damage.*

It may not be possible to separate these risks in any particular case: for example, if a person loses his footing, or he loses control of an unstable load, he may overstrain himself in trying to regain his balance; and a person whose back is vulnerable owing to cumulative overuse may overstrain himself lifting a load which is well within the capacity of a normal person.

Elementary biomechanical considerations predict that the risk involved in handling a load will increase as a direct function of:

(i) the weight of the load;
(ii) the distance of the load from the person's body (either forward or to the sides).

The product of the two is sometimes referred to as the *load moment.*

The risk of injury increases with the *distance* of the load from the body because:

(i) for a given weight of object, the loading on the spine and its muscles is greater (thus overexertion is more likely);
(ii) the strength of the lifting action is less and the load may therefore get out of control;
(iii) the body is out of balance (which may lead to loss-of-footing accidents);
(iv) the back may be flexed (particularly if the load is low off the ground).

In free lifting actions, the relevant distance is measured to the centre of gravity of the load. It follows, therefore, that a bulky load will be more difficult to lift (and more hazardous) than a compact load of the same weight. In exertions performed against immovable objects, and lifting actions in which the far end of the load is supported (including team lifts), the distance is measured to the person's hands.

Figure 15.3 was traced from a unique sequence of film. It shows a subject who

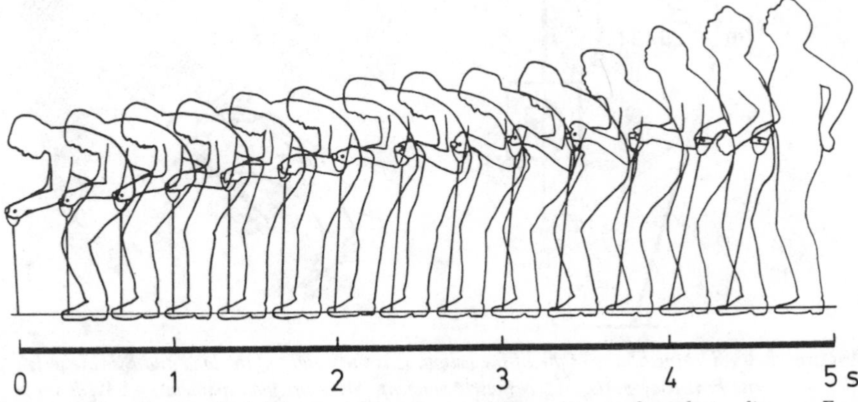


Figure 15.3 *A minor back injury occurring due to a lifting action performed at a distance. From an original kindly supplied by Professor D. W. Grieve and Ann-Marie Potts*

accidentally sustained a minor injury to his back by overexerting himself during the course of an experimental study of dynamic lifting strength. The subject was exerting himself against a hydraulic device which measured the force and the velocity of the lift. The line of action of the lift was 200 mm in front of his feet. Fortunately for all concerned, he made a speedy and uneventful recovery.

The strength of a vertical lifting action falls off rapidly with the distance the axis of the lift is in front of the subject's feet. This is partly because of the distance of the line of action of the force from the fulcra of the articulations, but it is also a matter of balance. If the line of action of a lift *falls within the person's footbase*, then the strength of the lifting action is determined solely by the capacity of the person's muscles to exert an upward thrust. But as the axis of the lift passes in front of the person's leading foot, he will be required to use the weight of his body to counterbalance the weight of the load, so as to prevent himself from toppling forward. When lifting a load outside the area of his footbase therefore, the person is at risk of overbalancing—particularly if the centre of gravity of the load should shift for any reason.

In general, a lifting action which starts at a distance and moves towards the person's body is likely to be easier to control than one which starts close to the body and moves away, since in the former case the individual will be lifting the load *into* the region of greater stability provided by his footbase and *into* his region of greater strength.

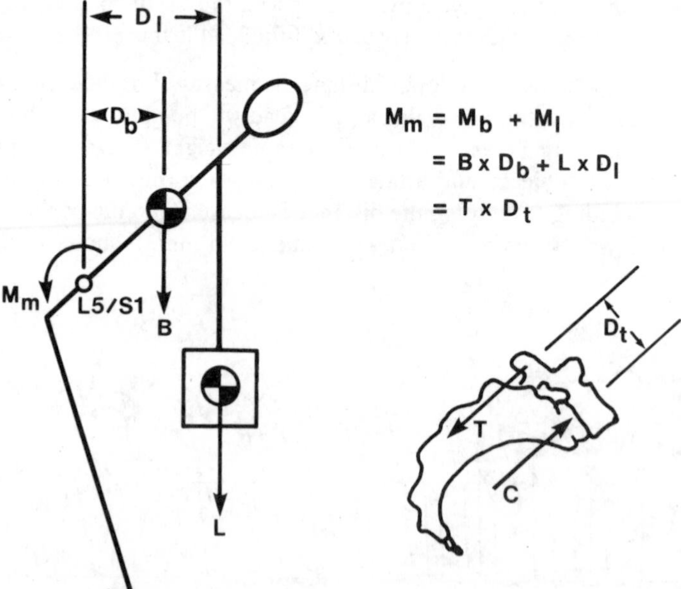


Figure 15.4 *The mechanics of the lifting action. L is the weight of the load and B is the weight of the superincumbent body parts.* $M_m$ *is muscle moment,* $M_l$ *is the load moment and* $M_b$ *is the body moment. Note that the tension in the muscles (T) is equal and opposite to the compressive loading on the spine (C). Therefore* $T = C = (B \times D_b + L \times D_l)/D_t$

When lifting at a distance, the extensor muscles of the spine must act to oppose: (i) the turning moment exerted by the weight of the superincumbent body parts; (ii) the turning moment exerted by the load (both of which are tending to flex the spine). Thus:

$$\text{muscle tension} \times \text{muscle leverage} = \text{body moment} + \text{load moment}$$

The turning moment due to the superincumbent body parts increases with the inclination of the trunk and the distance the hands are in front of the shoulders. The long superficial parts of the erector spinae group act more or less parallel to the spine at a leverage which is variously estimated at 50 or 75 mm (see below). So the tension in the muscles results in an equal (and opposite) compressive loading on the spine; and since the muscles are working at a mechanical disadvantage, the forces involved may reach ten times the weight of the load or more (Figure 15.4).

To this loading we must add the direct compressive effect of the weight of the superincumbent body parts and the load they carry. (Actually it is not quite as simple as this, since, as the trunk tilts forward, these will increasingly load the base of the spine in shear rather than compression.)

Figure 15.5 gives a rough idea of the magnitude of the forces which are involved

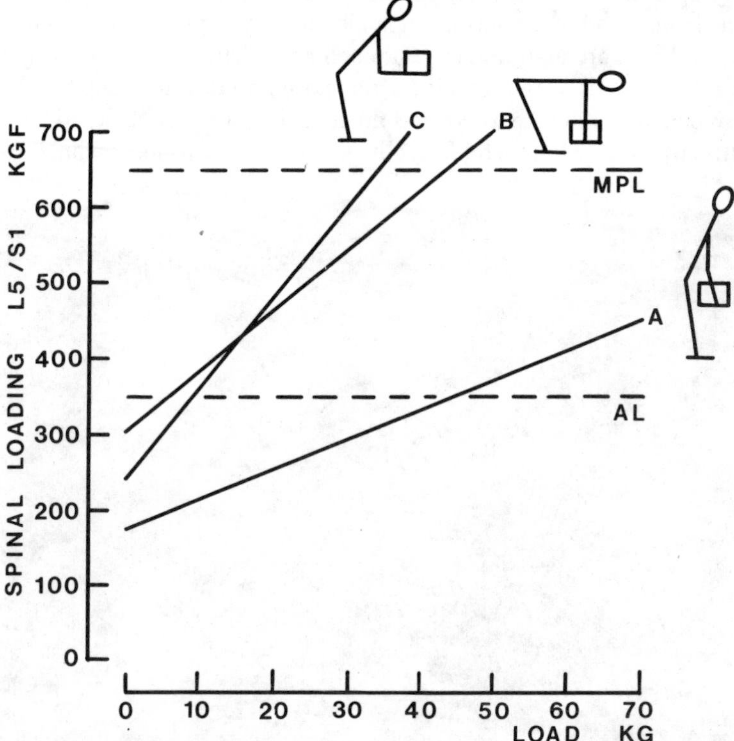


Figure 15.5 *Predicted levels of compressive loading on the spine, in three lifting postures, plotted as a function of the weight of the load in the hands. MPL is "maximum permissible limit"; AL is "action limit" as defined in NIOSH (1981)*

in three different lifting postures. In the first of these (A), the trunk is inclined by 30° and the load is 250 mm in front of the spine (L5/S1 disc). This is likely to be more or less the most advantageous position that can be achieved in practice. In the second position (B), the trunk is horizontal and the load is 400 mm from the spine; and in (C), the trunk is inclined at 45°, but the elbows are flexed so that the load is 600 mm from the spine. The calculations are based on anthropometric data from Pheasant (1986). The results are for a "standard person": 1675 mm tall, with a body weight of 70 kg and average bodily proportions. Note that the spinal loading which results from tilting the trunk through 90° (without lifting a weight at all) is equivalent to lifting about 30 kg in the most advantageous position.

### Stoop versus Crouch Lifting

The widely taught principle of "*bending the knees rather than bending the back*" serves to distinguish the two variations of lifting action shown in Figure 15.6:

- the *stoop lift* (or back lift), which is widely considered to be a bad thing;
- the *crouch lift* (or leg lift), which is widely considered to be a good thing.

"Keeping a straight back" may be interpreted in two ways: as the avoidance of trunk inclination and the avoidance of spinal flexion. In practice they generally go together and both are undesirable. (Although a perfectly vertical back may also be undesirable if it leads to an unstable starting position.) In a full stoop lift, the trunk may be inclined by as much as 90° and the lumbar spine may be flexed to its limit. In a correctly executed crouch lift, the inclination of the trunk is limited to about

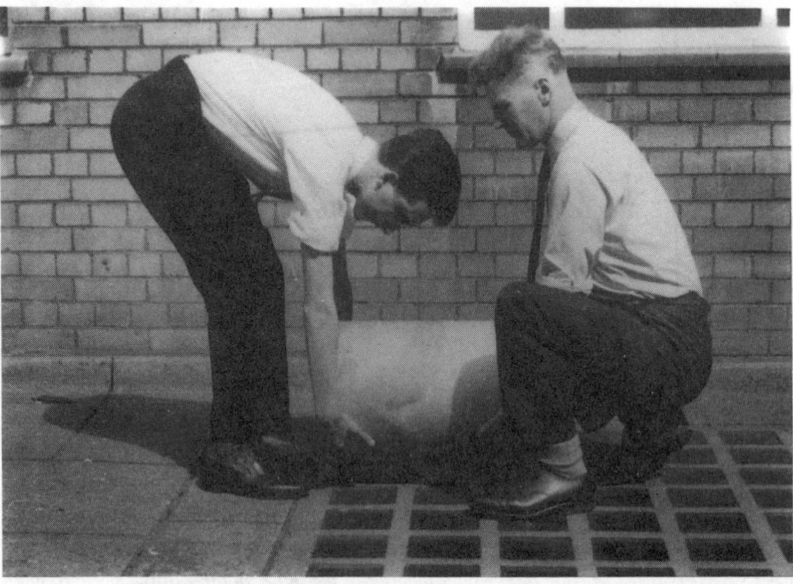

Figure 15.16 *Techniques for handling long loads. From Pheasant and Stubbs (1991); reproduced by kind permission of the National Back Pain Association*

30° and the lumbar spine remains closer to its mid-range of motion.

There are several good reasons why a flexed spine may be more susceptible to damage in a lifting action. The intervertebral disc is deformed, the posterior fibres of the annulus are under tension, and if these give way under pressure, the nucleus may be extruded. Cadaveric studies by Adams and Hutton (1982) confirm that in the normal spine, disc prolapse due to compressive loading is likely to occur only in positions of extreme flexion. Ligaments and other soft tissues may also be overstretched. Muscles in particular may be damaged if the lifting action includes an eccentric phase in which the spine is forced into further flexion (by the weight of the load) as it is attempting to extend. Grieve and Pheasant (1982) report studies which show that such eccentric phases do indeed occur in vigorous lifting actions (both crouch and stoop lifts).

The crouch lift is advantageous not only in that it will generally tend to keep the lumbar spine closer to the safer middle part of its range of motion, but also in that the centres of gravity of both the load and the upper parts of the body should (at least in principle) be closer to the fulcrum of the lumbosacral joint. Thus the strength of the lifting action may be as much as twice as great in a crouch lift position as compared with a stoop lift with similar foot placement (Damon *et al.*, 1971). The lower levels of spinal loading which we should expect in the crouch lift position have been confirmed in experimental studies using both direct measurement (Andersson *et al.*, 1976) and biomechanical analyses (Leskinen *et al.*, 1983). In the first of these studies, however, lifting techniques as such had less effect than load distance; and a subsequent study by Troup *et al.* (1983) failed to find any measurable difference between the techniques.

## Intra-abdominal Pressure (IAP)

Lifting actions (and other strenuous exertions) are associated with a transient increase of pressure within the abdominal cavity. This pressure may be measured using radio pills which the subject swallows (and which in due course may be retrieved and re-used). The pressure increase results from the combined actions of the diaphragm, the muscles of the anterior abdominal wall (particularly the transversus) and the pelvic floor. There may also be an increase in intrathoracic pressure—if the individual holds his breath against a closed glottis. The increase in intra-abdominal pressure (IAP), which seems to occur quite naturally as an intrinsic part of the lifting action, is probably a protective mechanism. The pressurized abdominal contents behave something like a fluid cushion, which helps to support the spine. The pressure (which may be regarded as acting upward on the diaphragm and downward on the pelvic floor) is thought to act as an additional extensor mechanism and it has been calculated that this mechanism may reduce the compressive loading on the spine by as much as 30% (Morris *et al.*, 1961; Davis and Troup, 1964), although this proportion has been challenged by McGill and Norman (1987a), who calculated that the compressive loading on the spine due to the activity of the abdominal muscles exceeded the reduction in loading resulting from the additional extensor mechanism. But the abdominal wall has certain weak points—particularly the inguinal canal in men. In susceptible individuals, the increases in IAP which occur during lifting may result in a progressive failure of the abdominal wall at these points, leading to hernia. (It is perhaps surprising that herniae due to lifting and handling have not been more extensively investigated epidemiologically.)

Early studies showed that the magnitude of the pressure increase was a function of both the weight of the load and its upward acceleration (Davis, 1959), and the angle of inclination of the trunk (Morris *et al.*, 1961). It was suggested therefore that intra-abdominal pressure could be used as an indirect index of spinal loading in the empirical evaluation of lifting and handling tasks.

Subsequent experimental studies by Andersson *et al.* (1977) have shown significant positive correlation between IAP and both the pressure within the nucleus of the disc (measured directly) and the tension in the erector spinae muscles (measured electromyographically). The correlation has been confirmed using biomechanical methods (Mairaux *et al.*, 1984), although there is some question as to the extent to which the relationship holds for flexed postures (Mairaux and Malchaire, 1988).

## Asymmetry and Twist

It is generally recognized that asymmetrical or twisted lifting actions are particularly hazardous. Back pain of sudden onset is often associated with twisting motions with or without lifting (see p. 70); and Kelsey *et al.* (1984c) have confirmed that lifting and twisting on the job is a strong risk factor for serious back problems (see below).

Asymmetry and twist are not the same thing. A lift which commences in an asymmetrical position—achieved by side bending, rotation or some combination of these—may or may not involve a subsequent twisting action. (It depends upon the subsequent trajectory of the load.)

The asymmetrical starting position will result in an increased loading on the spine (compared with a lift performed at an equal distance in front of the body), although it is difficult to calculate just how much greater, since the muscle actions involved are not well understood. Rotation of the lumbar spine can only be accomplished at the expense of an extensive deformation of the disc (because of the orientation of the facet joints in this region). Since the fibres of the annulus run around the disc in a spiral pattern, we should expect them to be at risk, especially if the rotation is combined with flexion.

Lifting strength falls off more or less equally rapidly with distance in front of the footbase and to the sides (Figure 15.7); and recent more detailed experiments, in which strength has been measured in different positions around the body at equal distances from the centre of the footbase, indicate that strength falls off relatively slowly with the angular displacement of the line of action of the lift from the straight-ahead position (David Sanchez, personal communication). This being the case, it may be that the risk associated with an asymmetrical lift which does not involve a turning motion is not very much greater than the risk associated with a

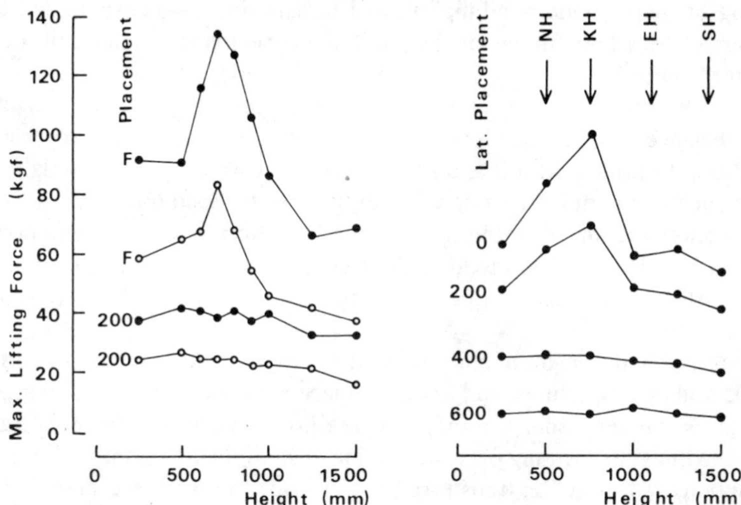


Figure 15.7 *Strength of a lifting action as a function of height above the ground and foot placement.* Left: *freestyle placement (F) and feet placed 200 mm behind the axis of the lift (● = men; ○ = women).* Right: *feet placed 20 mm behind the axis of the lift and various distances to the side. NH = knee height; KH = knuckle height; EH = elbow height; SH = shoulder height. From data kindly supplied by Ann-Marie Potts and Jane Dillon. Reproduced from Pheasant (1986)*

symmetrical lift at a similar distance from the centre of the footbase.

Lifting and turning actions, in which we pick up a load from one side of the body and swing it around to the other, are common in a wide range of everyday situations. The trunk turns about its own axis, with the load at a distance from the centre of rotation (approximately following an arc of a circle). In executing the turn, we characteristically "lead with the hips" and the shoulders (which are coupled to the inertia of the load) lag behind. By this means we maximize momentum transfer and increase the overall power output of the turn. We have at least two reasons to suppose that this might be hazardous—particularly if the lumbar spine commences in a position of flexion, as is often the case. The lumbar spine is subjected to a dynamic torsional loading—which may damage the fibres of the annulus (which are already under tension due to the starting position). And since the hips lead and the shoulders lag, there is an eccentric phase in the action of the muscles of the back and trunk.

In summary: asymmetry and twist are both likely to be associated with increased risk, but, of the two, it seems probable that the twisting action is the more hazardous.

## Use of Body Weight

Lifting trainers often talk about "using body weight", implying that by so doing, the effort required to move the load may be reduced. This concept is central to the teaching of the "kinetic handling" school of safe lifting—as expounded by the followers of McClurg Anderson (1951). The concept is surrounded by a certain amount of confusion.

When lifting outside the footbase, we must inevitably use body weight as a counterbalance—and correct foot placement is in many respects the key feature of good lifting technique. But it is difficult to see how we can use the weight of the body (which by definition acts downward) to move the load upward. Under some circumstances we may give the mass of the body an upward momentum, which may then be transferred to the load; but this is a somewhat different proposition, and the effect is only likely to be important for relatively light loads (Grieve and Pheasant, 1982).

The use of body weight to shift the load is a more realistic option in horizontal pushing and pulling actions; and again the placement of the feet, so as to maximize leverage, is the key issue. We may also use body weight in horizontal transfers across the body: by pivoting over the leading edge of the footbase, we can give the body momentum which is transferred to the load. But we noted above the risks that this may involve.

A curious misconception surrounds the rhythmic rocking action by which nurses are taught to lift patients from the sitting position. It is sometimes said that the patient acquires additional kinetic energy on each upswing of the rocking motion. In itself this is true; but the energy is lost again on each downswing. So this cannot be the reason why the rocking action makes it easier to lift the patient (as is certainly the case). Perhaps the rhythmic motion helps the nurse and the patient to co-ordinate their efforts more effectively.

## The Prevention of Lifting Injuries

In principle, we could think of three main strategies for dealing with the risks associated with lifting and handling tasks:

- (i) *selection*—attempting to identify those individuals who are particularly at risk and to eliminate them from the working population (or transfer them to other tasks);
- (ii) *training*—attempting to change people's behaviour by teaching them ways of lifting and handling loads which are considered to be safer;
- (iii) *work design (ergonomics)*—attempting to identify those aspects of the job which are particularly hazardous and to redesign them so as to make them safer.

All three approaches proceed from the assumption that lifting and handling injuries result from a mis-match between the demands of the task and the capacities of the person. Selection and training are aimed at solving the problem by fitting the person to the job (FPJ); work design is aimed at fitting the job to the person (FJP).

## Selection

A good selection criterion is one which will correctly identify those individuals who, by virtue of their personal characteristics, are at a high level of risk, without making recruitment needlessly difficult (and potentially discriminatory) by excluding job applicants who, in reality, could cope with the workload.

Can we achieve this in practice? With the exception of prior history and general unfitness, personal risk factors have relatively low predictive value for back pain. For other types of lifting injury, the predictive value of these factors would presumably be lower. It clearly makes sense to screen people for pre-existing back problems (and other conditions which would compromise their fitness for work) before allowing them to take up physically demanding jobs. But the available evidence suggests that the tangible benefits to be gained from medical screening are not as great as one might expect.

The most extensive study of these matters reported to date is that of Snook *et al.* (1978). The study involved a large sample of industrial concerns across the USA. Snook compared organizations which operated selection programmes based on (i) medical history alone; (ii) medical examinations (including history taking); (iii) medical examinations (including history taking) *plus* radiological examinations; (iv) none of these. There were no significant differences in back injury rate between the organizations in the four categories.

A certain amount of selection is commonly done on an informal basis—by allocating the heaviest jobs to the biggest and strongest-looking individuals. The possibility of formalizing this process has been investigated in two studies by Chaffin and co-workers (Chaffin and Park, 1973; Chaffin, 1974; Chaffin *et al.*, 1978). The manual jobs in a number of industrial plants were analysed. For each job, the weight of the load which was handled was compared with the average strength of the workers as measured in a comparable position. The ratio between the two was called the job strength rating (JSR). In both studies the subsequent incidence of back injuries increased with this ratio. In the first study there was a threefold increase in risk for jobs in which the weight of the load exceeded the average strength of the workers (JSR$>$1); but in the second study no such threshold effect was apparent (Figure 15.8). In the second study, however, the incidence of other musculoskeletal injuries and handling injuries both increased sharply in jobs in which the workers were frequently required to lift loads which approached or exceeded their maximum strength.

The analysis which was performed in these studies confounds differences between individuals with differences between jobs. Given the magnitudes of load which are commonly encountered in industry, and the average strengths we might expect to measure in different lifting positions, it would seem likely that those jobs in which the weight load exceeded the average strength of the workers would have been ones in which it was necessary to handle the load at a fair distance from the body. To clarify this position, we should need to know the injury rates for strong and weak individuals exposed to similar job demands.

Direct evidence in this respect is limited. Keyserling *et al.* (1980) found a lower

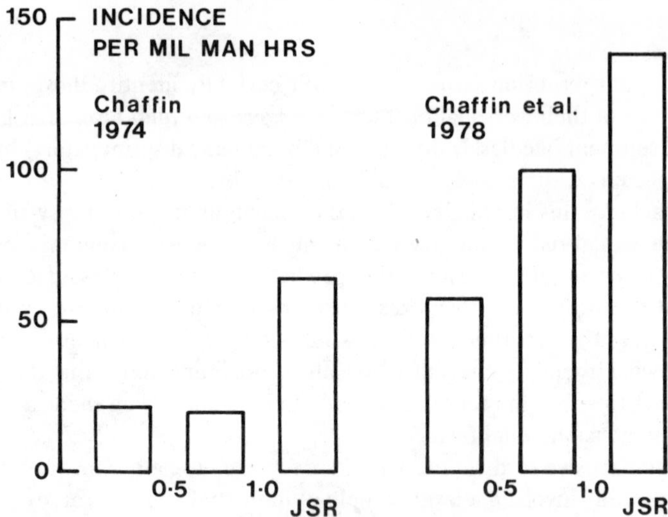


Figure 15.8 *Incidence of back injuries as a function of job strength rating (JSR) in two studies by Chaffin and co-workers (Chaffin 1974; Chaffin* et al., *1978), where JSR = load lifted on the job/average strength of workers*

injury rate in workers who had been selected on the basis of strength tests, as compared with those who had been subject only to medical screening. But the numbers involved were small and the statistical significance of the difference was only marginal.

## Sex Differences

There seems little reason to imagine that a woman engaged in a heavy manual occupation will be any more likely to injure herself than a man of the same size and strength. Sex differences in the shape of the pelvis are not great enough to make an appreciable amount of difference to the loading on the lumbar spine. But insomuch as women are on average a good deal weaker than men, we should expect a higher injury rate in women *than in men doing similarly heavy jobs*; and workstations which are suitable for male users may be less suitable for women.

There are a number of caveats to this general position. Uterine prolapse may be caused by lifting heavy weights. The risk is probably small, except where integrity of the pelvic floor is already seriously compromised (owing to an adverse obstetrical history). But because of this possible risk, some countries see fit to set the maximum weight of load which may be handled by a female worker at a level which is very much lower than that which is considered acceptable for men (ILO, 1990).

Heavy work is also thought to be undesirable during pregnancy. Goulet and Thériault (1987) concluded that heavy lifting and physical exertion carried an increased risk of spontaneous abortion during early pregnancy. In the later stages of pregnancy (from the 6th month onwards, or thereabouts) the prospective mother's exercise tolerance may begin to diminish rapidly and physical work becomes more arduous—although this is by no means universally the case. Slackening of the ligaments (in the sacro-iliac joint and elsewhere) may also make her back more vulnerable to injury, both during pregnancy and after her confinement. In some European countries, therefore, pregnant women may not be employed in heavy manual work. In Austria, for example, the maximum load for a pregnant woman is set at 10 kg for frequent lifting; and in Luxembourg it is set at 5 kg both during pregnancy and for 3 months after confinement (ILO, 1990).

## Training

Lifting training is based upon the implicit assumption that back injuries are characteristically caused by *faulty lifting technique* and that they may therefore be prevented by teaching people "*the correct way to lift*". In reality, there is relatively little direct evidence in support of either of these beliefs; and a certain amount of evidence to the contrary.

In principle, the content of a typical training course could be broken down into three components:

- knowledge;
- procedures;
- skill.

In practice, the distinctions between these components are not always clear; but they are worth exploring because they have a number of important consequences.

Most training programmes include a certain amount of background *knowledge*: concerning basic ergonomic principles, the anatomy and physiology of the spine, the "causes" of back pain, health and safety law, and so on. This background is important in that it helps put the practical side of the training programme (*procedures and skills*) into context.

Good lifting *technique* is partly a matter of following the correct *procedures* (given the nature of the load, etc.) and partly a matter of developing the necessary *skills*. An individual may follow the correct procedure but be unskilled in its execution (and therefore at risk); conversely, he may become skilled in the execution of potentially hazardous procedures (and perhaps even avoid injury thereby). If people are familiar with the correct procedures, but choose not to adopt them, they may sometimes be persuaded to mend their ways by *safety propaganda* (pep-talks, posters, and the like). But the acquisition of the skills which are required to execute a smoothly co-ordinated lifting action is a different matter altogether. It is more like learning a good tennis serve or a good golf swing. Skills of this kind can sometimes be taught as if they were procedures. This is the "doing it by numbers" approach. But the levels of skill which can be reached in this way are limited. The distinction between *safety training* and *safety propaganda* is worth emphasizing. Training is concerned with the learning of new knowledge, procedures and skills;

propaganda is concerned with persuading people to follow the correct procedures and to make the best use of the knowledge and skills they already have.

Can training be shown to result in a measurable reduction in the number of lifting injuries which occur, either in industry or elsewhere? As Brown (1972) points out, the basic principle of teaching people to lift weights by "bending their knees rather than by bending their backs" has been advocated since the 1930s. During this time there has been no overall reduction in the number of injuries attributable to lifting and handling.

The definitive study of the effectiveness of training, as a means of preventing industrial back injuries, is again that of Snook *et al.* (1978), which was based on a large sample of industrial concerns in the USA. Using a statistical approach of considerable ingenuity, Snook compared organizations which provided training programmes in safe lifting with those which did not. (No attempt was made to evaluate the content or quality of these programmes.) The analysis took into account the sizes of the organizations and the percentages of employees who were engaged in heavy lifting and handling tasks. On this basis it was possible to predict how many out of a sample of reported back injuries could be expected to occur in organizations that did and did not provide training—and thus to test the hypothesis that the injuries were distributed by chance alone (i.e. that training had no effect). *There was no evidence that those organizations that provided training programmes had fewer back injuries than those that did not.* In fact, the organizations who provided training programmes had marginally more back injuries than would be expected by chance ($0.05 > P > 0.01$).

As part of an extensive epidemiological investigation, Damkot *et al.* (1984) found that 70% of men who did not suffer from back pain had received instruction on safe lifting, compared with 83% of men with moderate back pain and 93% of men with severe back pain. A significantly higher proportion of the men with severe back pain reported that they habitually lifted with bent knees rather than with a bent back. We should expect that those men whose jobs placed them at risk would have been more likely to have been given lifting training. Perhaps they did indeed adopt the correct technique; but it did not prevent them from developing back problems. Alternatively, it might be that people who have developed severe back problems are more likely to adopt the correct technique.

Positive evidence for the effectiveness of training is limited—and what there is, is mainly anecdotal or based on small-scale, poorly documented studies. A study by Blow is widely quoted (e.g. by Davies, 1976; Troup and Edwards, 1985; etc.). This showed a 44% reduction in the man-hours lost owing to lifting and handling accidents in the year following the introduction of a training programme at an electrical goods factory in North London. Davies (1976) also cites unpublished studies in which a 30% reduction in injury rate was reported after the introduction of a training programme in a cigarette factory; and a 50% reduction was reported after a campaign of instruction in the safe handling of pit props. But of these three studies, Davies (1976) states that "none of the authors would claim that these figures are scientifically validated".

People in general are willing to tell the world about their successes; and less willing to report their failures. Many lifting trainers claim to have success stories to tell—which they believe to be evidence for the effectiveness of their methods—but few of these have reached the published literature.

## Back Pain in Nurses

Nurses are a high-risk group for low back pain. Table 15.1 is based upon the large-scale epidemiological survey of nurses reported by Stubbs *et al.* (1983a). The prevalence figures from this study are compared with data for physiotherapists (a broadly similar occupational group) and a control sample matched for age and sex, reported by Scholey and Hair (1989). The sickness absence figures are compared with DHSS data for the same year (quoted by Waddell, 1987) weighted to give a sex distribution similar to that of the nursing profession. The annual prevalence of back pain is a few per cent higher in the nurses; but there is a very much greater difference in the contribution which back pain makes to sickness absence. This is more striking in light of the low overall rates of sickness absence in the nurses. It seems unlikely that nurses are that much fitter than the rest of us; their low overall rates of sickness absence are probably due more to the high levels of stoicism which characterize the nursing profession. Note also that there was no significant difference between the prevalences of back pain in physiotherapists and controls—despite the fact that 91% of the physiotherapists in the sample had been directly involved in back care programmes for other people.

*Table 15.1 Back Pain in Nurses, Physiotherapists and the General Population*

| Back pain | Nurses | Physiotherapists | Controls |
|---|---|---|---|
| Point prevalence (%) | 17 | 14 | 15 |
| Annual prevalence (%) | 43 | 38 | 35 |
| **Sickness absence** | **Nurses** | | **General population** |
| Days/1000 persons/year | | | |
| all causes | 10,984 | | 18,131 |
| back pain | 1,774 | | 1,377 |
| Back pain | | | |
| % all causes | 16% | | 8% |

Data from Stubbs *et al.* (1983a); Scholey and Hair (1989); Waddell (1987).

It is widely believed that the back problems of nurses are attributable to the physically demanding nature of their work—and in particular to lifting patients. The data of Figure 15.9 (quoted from Stubbs *et al.*, 1980) support this belief—at least insomuch as the onset of symptoms is indicative of causation (see p. 71).

Lifting and handling sick people presents a special range of problems (not least of which is the weight and instability of the load), and special lifting and handling techniques are therefore required. The importance of training nurses in these techniques is often emphasized (e.g. RCN, 1979; Lloyd *et al.*, 1987). Does this

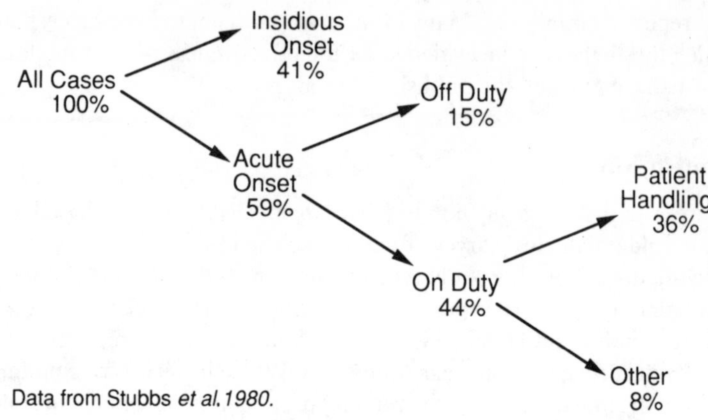


Figure 15.9 *Patterns of onset of back pain in nurses*

training lead to a reduction in risk? Stubbs *et al.* (1983b) investigated the prevalence of back pain in a sample of nurses as a function of the amount of lifting training they had received—both in the classroom and on the wards. The results are shown in Figure 15.10. No overall trend is discernible.

Figure 15.11 shows the quarterly back injury statistics over an 8 year period, for nursing staff in the Canterbury and Thanet area of South East England. (These unpublished data were kindly supplied by Janet Daws.) It is widely recognized that the quality of this training, given in this area, is very high. The injury rate fell dramatically following the commencement, in the summer of 1980, of an intensive programme, aimed at increasing awareness of the risks involved in lifting and

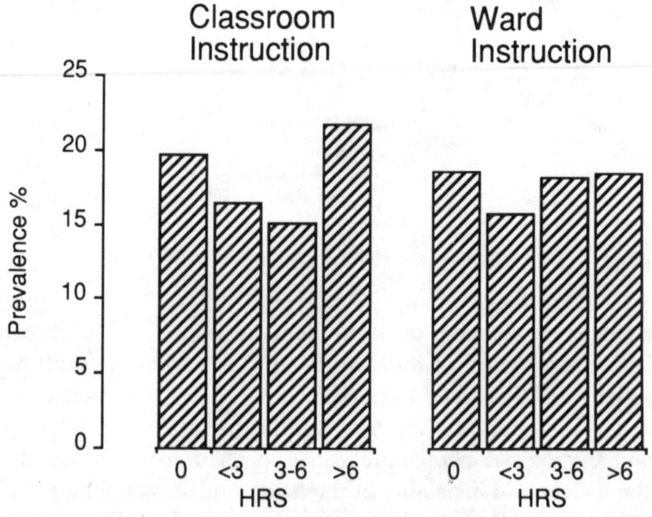


Figure 15.10 *Prevalence of back pain in nurses as a function of time spent in lifting training: in the classroom (*left*) and on the wards (*right*). Data from Stubbs* et al. *(1983b)*

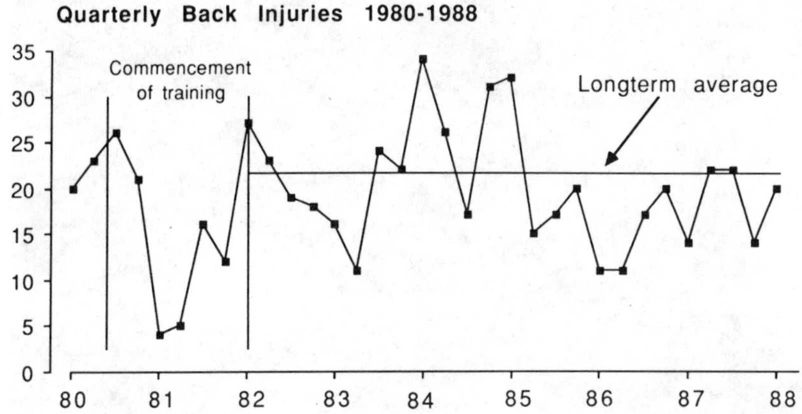

Figure 15.11 *Quarterly back injuries in Canterbury and Thanet area. Note short-term improvement after commencement of training, and subsequent return to original level*

handling. This improvement lasted for about 18 months. Since that time the quarterly figures have oscillated about an average level which is about the same as the average for the period immediately preceding the commencement of the programme. No overall long-term trend is discernible.

Stubbs *et al.* (1983b) used intra-abdominal pressure (IAP) as an indirect measure of the loading on the spine in an investigation of various methods for lifting and handling hospital patients, including the "shoulder (or Australian) lift" and the "orthodox lift", shown in Figures 15.12 and 15.13, respectively. The Australian lift has two important features: the patient's weight is taken on the nurse's left shoulder, thus bringing the line of thrust as close as possible to the spine; and the nurse presses down on the bed with her right hand (not visible in the picture), which reduces the loading on the spine. Stubbs and his co-workers found that the Australian lift results in significantly lower IAPs and may therefore be legitimately regarded as the safer procedure. In a second experiment, two nurses were given intensive training in the Australian and orthodox lifts and in certain other manoeuvres for transferring patients from beds to chairs and so on. Overall, no consistent trends were observable in the IAPs recorded before and after the intensive training period. This implies that, although some procedures for lifting patients are measurably better than others, there may be relatively little to be gained from attempting to enhance the individual's skill in the execution of these procedures by means of intensive training. This finding seems surprising. Even very experienced lifters (such as those responsible for training other people) may show considerable differences in technique when executing the same manoeuvre. For example, the angle of the trunk, when performing the Australian lift, may range from as little as 15° to almost 90° (i.e. horizontal). It is difficult to imagine that differences of this magnitude would have no measurable effect on spinal loading.

In principle, training is an appropriate strategy for dealing with those risks which arise specifically from faulty lifting technique rather than from other features of

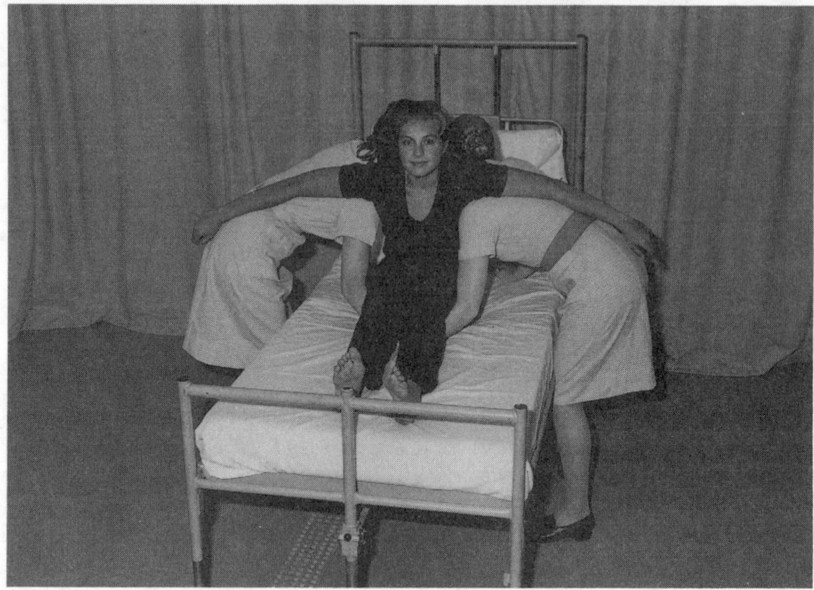

Figure 15.12 *The Australian lift. From Stubbs* et al. *(1983b), reproduced by kind permission of David Stubbs and the publishers of* Ergonomics, *Taylor and Francis Limited*

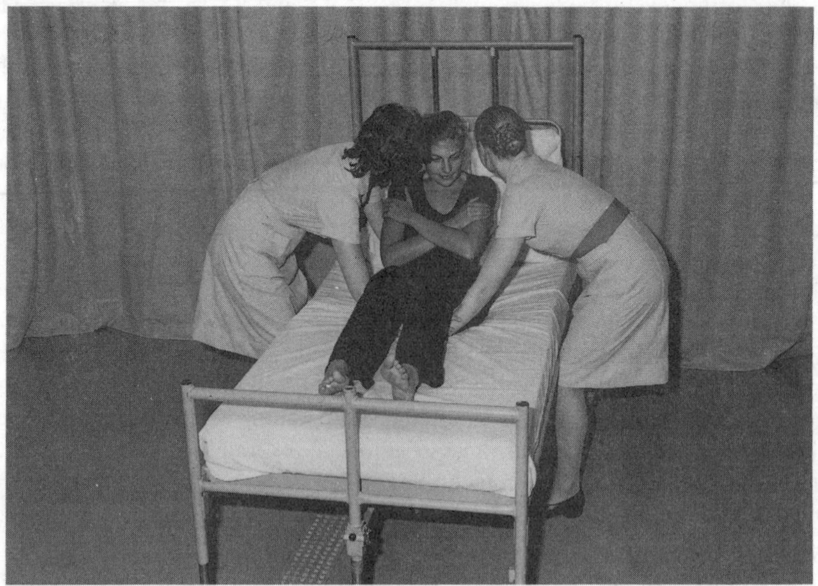

Figure 15.13 *The orthodox lift—now considered to be an unsafe working practice. From Stubbs* et al. *(1983b), reproduced by kind permission of David Stubbs and the publishers of* Ergonomics, *Taylor and Francis Limited*

the overall working situation which are outside the individual worker's control. The case of nursing is an informative one. In industry as a whole, it is unusual to find people handling loads in excess of 50 kg; but around 95% of adult human beings weigh more than this, and although nurses should always lift patients in pairs (or larger teams where required), the constraints of the working environment are such that the nurse will often be forced to support relatively heavy loads in extremely disadvantageous positions. The standard hospital bed is often 910 mm in width and it does not seem likely that this figure could reasonably be reduced. So, in practice, it is unlikely that the nurse attempting to lift a bedridden patient will be in a position much better than the one shown as Position C in Figure 15.5. Suppose she takes one-quarter of the body weight of a 90 kg patient. This would result in a spinal compression of the order of 550 kgf, which is well into the range which many people consider to be hazardous (see below). The postures which a nurse will be forced to adopt to womanhandle a patient into or out of a bath are even worse—probably not much better than Position B in Figure 15.5. The human body is a bulky and unstable load at best; and at worst the disorientated patient may writhe or thrash around as the nurse handles him. Other back pain risk factors are also present: studies have shown, for example, that, on average, a nurse may spend 22% of her working period in a stooped position; and in the hours between 9 and 11 (when she is making beds, etc.) she may be stooped 30% of the time (Baty and Stubbs, 1987). Nursing is also psychologically stressful.

Many people (both within the nursing profession and elsewhere) take the view that nurses have back problems because they are undertrained. The reality is that they are physically overloaded by their working activities. *In situations of this kind training alone is necessary but not sufficient.* To make further progress we need to identify the features of the working system which are responsible for the physical overload.

## The Limitations of Training

The available evidence suggests that successful programmes of lifting training are the exception rather than the rule. We should ask ourselves quite carefully why this is the case. A number of possibilities are set out, in the form of a checklist, in Table 15.2.

Consider the classic "good lifting action" shown in Figure 15.14. They key to the correct execution of this action lies in the person's foot placement in the starting position. The leading foot should be level with the front edge of the load, pointing in the direction in which the load will move. The other foot is placed behind the load. In principle, therefore, the upward thrust of the lifting action remains within the footbase. The lifter bends his knees and hips, and keeps his back in extension, with the trunk inclined no more than necessary (perhaps about 30° in practice). His arms are straight and he grips the load firmly by its corners, keeping it close to his body throughout the lifting action. As he begins to straighten up, he leans forward a little over the load, thus giving the upward thrust of the lifting action a forward component and keeping the line of action within the

*Table 15.2 Ten Questions Concerning Lifting Training Programmes*

| | |
|---|---|
| • Q1 | Are the techniques which are taught biomechanically correct? |
| • Q2 | Do the trainees (a) learn and remember the procedures they have been taught; (b) acquire the necessary skills? |
| • Q3 | Are the techniques which have been learned adequately transferred to the working situation? Do they become part of the trainee's instinctive repertoire of movement behaviour? |
| • Q4 | Does the training programme cover a wide enough range of practically relevant lifting and handling tasks? |
| • Q5 | Are there hidden disadvantages to the "correct" lifting technique which prevent it from being applied in the workplace? |
| • Q6 | Do the practical constraints of the working situation or the nature of the load *prevent* the student from using the "correct" lifting technique? |
| • Q7 | Is the weight of the load beyond reasonable limits? Is the load inherently unsafe for some other reason (bulky, unstable, poor grip, etc.)? |
| • Q8 | Is the overall physical workload beyond reasonable limits? |
| • Q9 | Is lifting and handling the principal cause of back pain in the working population concerned? |
| • Q10 | Is the social ethos of the workplace conducive to the maintenance of safe working practices? |

footbase. (Perhaps this is what people mean by "using body weight".) The lift is executed in a smoothly co-ordinated motion without jerking.

At least, that is how it is supposed to happen in theory. In practice, things may be very different. The well-established biomechanical advantages of the crouch lift technique are limited to loads which are compact enough to fit between the knees. If the load is wider than about 300–400 mm, it will be necessary to lift it *either* in front of the knees or *else* beside the knees. Both are undesirable: the former because the load must be handled at a considerable distance from the trunk; the latter because of the asymmetry involved. Park and Chaffin (1974) have shown that the loading on the spine in a crouch lift, in which the load passes in front of the knees, may actually be greater than in a stoop lift.

Furthermore, there are a very large number of practical circumstances in which it is quite impossible to adopt the starting position and foot placement of the classic "good lift", and in which the distance between the load and the body will inevitably be large, however the individual seeks to perform the task. Situations in which it is necessary to reach over obstacles or into containers are important cases in point (see below).

*In reality, therefore, the classic "good lift" is a feasible option only with compact loads in unobstructed spaces.*

In some cases it may be possible to devise special procedures for handling bulky or awkward loads, which incorporate some (at least) of the underlying principles of the classic good lift. Some examples are shown in Figures 15.15 and 15.16. If the content of the training programme does not accurately reflect the tasks which the trainees will accurately perform on the shop floor, then its impact will be negligible. But if people are forced to adopt unsafe lifting techniques, by

Figure 15.14 *The classic "good lift". Reproduced by kind permission of the National Back Pain Association*

circumstances which are outside their control, no amount of training is going to solve the problem.

Experience suggests that, left to their own devices, people rarely if ever lift loads by bending their knees—even when circumstances would permit them to do so.

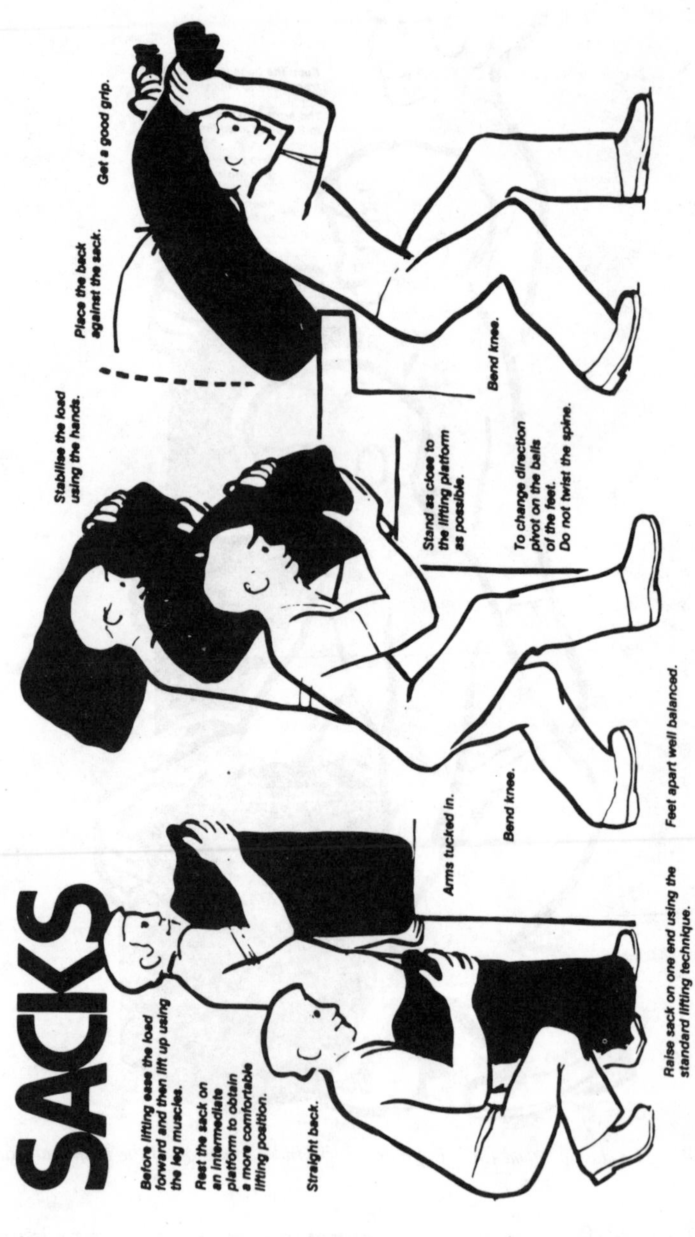


Figure 15.15 *Techniques for handling sacks. From Pheasant and Stubbs (1991); reproduced by kind permission of the National Back Pain Association*

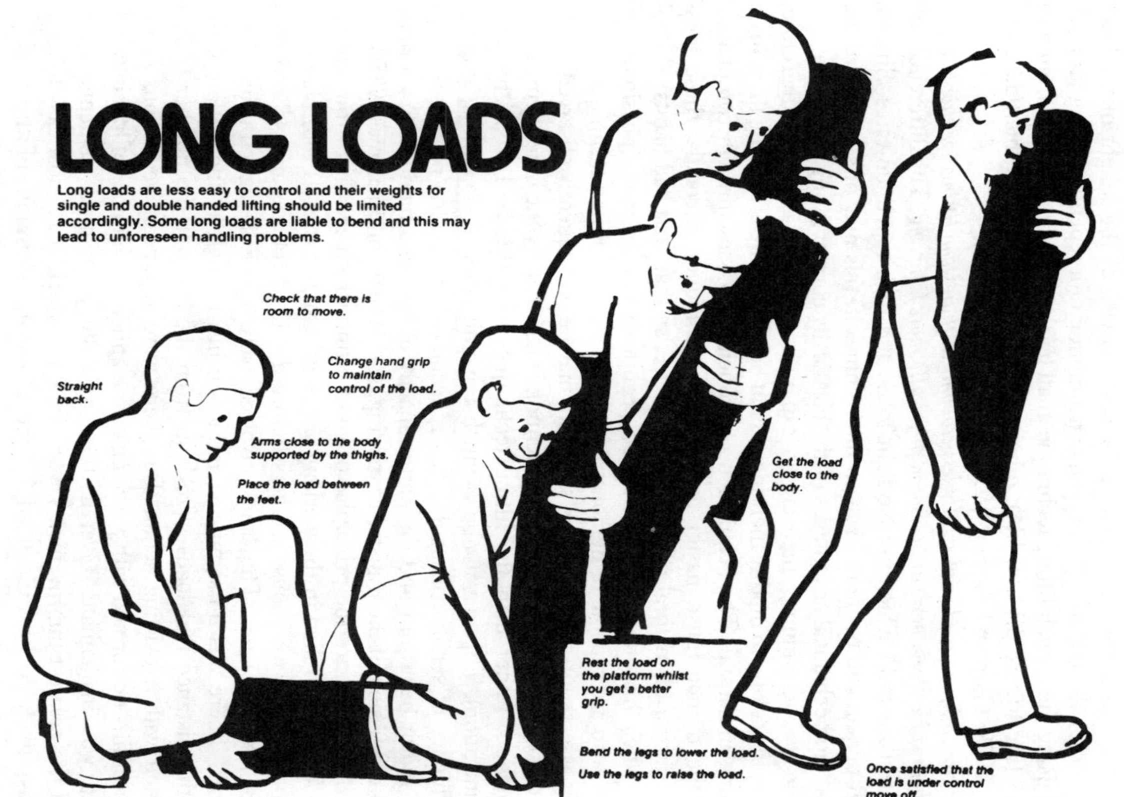


Figure 15.16 *Techniques for handling long loads. From Pheasant and Stubbs (1991); reproduced by kind permission of the National Back Pain Association*

The classic good lift must, therefore, have some hidden disadvantages. To execute it correctly, you require a fair degree of strength in the knee extensor muscles. This will be a problem for some people; and it has been suggested that too extreme a degree of knee flexion in the starting position may put the ligaments of the knee at risk. Crouch lifting is also more costly in terms of energy expenditure—since the muscles have to move the weight of the body as well as the weight of the load. Brown (1972) showed that the oxygen consumption of a repetitive lifting task was greater using the crouch lifting technique at all levels of loading ranging from 5 kg to 40 kg (see Figure 6.11, p. 134); other authors have confirmed this (Garg and Herrin, 1979; Garg and Saxena, 1979; but see also p. 311).

*Since crouch lifting is physiologically less economical than stoop lifting, we should expect it to be both subjectively more arduous and physically more fatiguing.* The difference will be greatest for tasks which involve light loads and high repetition rates. Since a fatigued person will be more likely to injure himself, this may offset some of the manifest biomechanical advantages of the classically correct lifting technique.

People seek to minimize their energy expenditure. There is a sense, therefore, in which, in seeking to modify their movement behaviour, we are asking them to do something unnatural. The acquisition of proficiency passes through two phases: firstly, the trainee must master the relevant techniques; secondly, these must become part of his natural movement repertoire, so that "old bad habits" do not intrude when his attention is otherwise occupied. For example, physiotherapy students who are capable of demonstrating proficiency in lifting technique, when formally assessed in the classroom situation, may be subsequently observed acting incorrectly on the wards, when their attention is taken up by the other demands of the situation, rather than the lifting technique itself (Fiona Turner, personal communication). This may not be so much a manifestation of sloppiness, or of the bad influence of senior colleagues, as a reflection of the neural processes which control motor behaviour and the finite capacity of human attention. (For a more general discussion of the transfer of training, see Annett and Sparrow, 1985.)

To what extent are the principles of good lifting which are taught on training courses actually applied in the working situation? Chaffin *et al.* (1986) filmed the lifting postures of warehouse workers before a 4 hour training course and then again 5–7 weeks later. The loads involved were relatively light by industrial standards. There was no significant reduction in the angle of inclination of the trunk, the distance that loads were held from the body or the number of lifts performed with a twisting action. But there were significant reductions in the number of lifts performed with a jerking action and in the number of lifts in which the grip was inadequate. Overall, therefore, the procedural elements which showed the greatest improvements were the ones which were most easy to learn and least likely to be influenced by the physical constraints of the working situation.

Does an improvement in lifting technique lead to a reduction in risk? Troup and Rauhala (1987) compared nurses who had been trained in the conventional way with a group who had attended a specially devised training course which emphasized a problem-solving approach based on ergonomic and biomechanical principles. The latter group were able to demonstrate significantly better patient handling technique in an examination situation. In a subsequent follow-up study, Videman *et al.* (1989) found no significant difference between the two groups, either in the frequency of "back injuries" (i.e. episodes having sudden onset in the workplace) or in the incidence of back pain in general. They then compared nurses who had been rated highly for lifting technique in the examination with nurses who had done badly. The nurses with good technique had a very much lower incidence of "back injuries" ($P<0.001$); but for other sorts of back pain, there was a slight trend in the opposite direction, although it did not achieve statistical significance. If we lump the two categories of back pain together (which the authors did not do), then overall there is no statistically significant difference between the two groups of nurses.

This extremely interesting finding may be interpreted in a number of ways. One possibility is that the acquisition of good lifting technique affects the context in which the onset of back pain occurs, rather than preventing back problems as such. Thus a nurse whose back is in a vulnerable state (by reason, for example, of postural stress or of her overall physical workload) may be able to avoid triggering an attack of back pain by lifting a patient awkwardly, if she has good lifting skills. But this may not prevent the symptoms of her underlying condition from becoming manifest in another context.

The ergonomist would argue that a training programme can only be effective if it is based upon the critical evaluation of existing working practices and it is part of a more extensive package of ergonomic improvements aimed at providing a safer system of work. Seen in these terms, training should be a participative process, whereby the trainee learns to plan his own working activities according to ergonomic principles. Technique as such is only part of the picture. The trainee must also learn to recognize his physical limitations and size up the load before attempting to move it (so as to get help where necessary). The importance of good housekeeping and an unobstructed working area cannot be overemphasized. In essence, the training programme becomes an exercise in self-help ergonomics. For a further discussion of the ergonomic approach to lifting training, see Pheasant and Stubbs (1991).

## Work Design

Lifting and handling tasks are characteristically unplanned, undesigned activities; they are often improvised solutions to avoidable problems. In many cases, the need to handle the load at all is a direct consequence of inefficient working practices or the inappropriate layout of the working area. Wherever possible, activities involving the manual handling of heavy loads should be designed out of the working system. Are there bottlenecks in the flow of work which increase the amount of handling which is required? Are storage facilities adequate? Can materials be delivered to the point at which they will be used? Are there suitable alternatives to carrying the load by hand? Is it possible to use cranes, hoists, jacks, conveyors, trucks, etc.? Can the load be palletized? Would it be easier to handle the load using a hook or a sling, or on a stretcher?

## Workstation Design

We noted earlier that certain types of lifting action carry a particularly high level of risk. These include:

- lifting at a distance;
- lifting actions involving asymmetry and twist.

These high-risk actions commonly result from the layout and physical constraints of the workspace, and in particular from:

- the need to reach over obstacles and/or into containers;
- confined working areas with limited headroom and/or horizontal clearance.

*The elimination of such problems is a high priority in the ergonomics of workstation design.*

Common industrial lifting tasks involved in forward or sideways reaching actions include palletization and depalletization operations (particularly when objects must be placed on a far corner of the pallet); the storage and retrieval of items from shelving; and the loading and unloading of trolleys and crates (Figures 15.17, 15.18). It will generally be better if crates have sides which can be opened. Examples from outside the occupational context include the lifting of loads from the boots and back seats of cars (Figure 15.19). Consider the loading on the spine

Figure 15.17 *Lifting at a distance—palletization task. From an original in the author's collection*

Figure 15.18 *Lifting at a distance—mobile storage bin for laundry bags. From an original in the author's collection*

Figure 15.19 *Lifting at a distance—the car boot. From an original in the author's collection*

which results from lifting a two- or three-year-old child from the back seat of a two-door car. Mechanics and maintenance men often have to work in confined spaces which constrain their postures—as do engineers who work in trenches, and people who load and unload vans. Some vehicle assembly tasks also fall into this category.

Asymmetry and twist often result from the need to swing the load across the body. This commonly occurs in loading and unloading tasks—for example, to and from conveyor belts and the backs of trucks. The action of turning with a load, so as to transfer it *from* a conveyor *into* a container or onto a pallet, is encountered particularly frequently. Trainers may attempt to deal with these problems by teaching people step turns—a bit like dancing a waltz with the load. Compliance with these procedures will be low because the dynamic lift and twist is very much more economical in terms of energy expenditure (see above).

## Vertical Height Range

When foot placement is optimal, the strength of the lifting action will be greatest at about half-way between knee height and hip height—that is, at about the height of the knuckles of the hand in the standing position. On average this is about 750 (±100) mm above the ground. The upper limbs are more or less straight, and the hips and knees are slightly flexed. The strong extensor muscles of the lower limb have the best possible mechanical advantage for generating the upward thrust of the lifting action.

The strength of the lifting action falls off steeply both above and below the optimum height (Figure 15.7). Below the optimum, increased flexion of the lower limb is required; above the optimum, increased flexion of the upper limbs is required. In both cases mechanical advantage is lost. When foot placement is disadvantageous, the peak is less apparent.

Lifts commencing at below knee level or thereabouts (about 550 ± 100 mm) may be regarded as undesirable, since they will require either a degree of lower limb flexion which greatly reduces the strength of the lifting action or a stooping action with a flexed spine.

A lift commencing at floor level (or above) can be continued without too much difficulty to a level somewhere near elbow height (about 1100 ± 150 mm). Beyond this point the lifter will *either* have to change grip (typically from one in which the palms are beneath the load *or* continue the lift by abducting the arms at the shoulder and leaning backward (hyperextending the spine). Both manoeuvres are potentially hazardous. A lift which *commences* above elbow height may be extended upward to shoulder height or more (about 1400 ± 150 mm) without too much difficulty, since no changes of grip are involved; but the loads which can be handled safely beyond this point are very much less, because the strength of the lifting action begins to decline very rapidly.

The easiest way of carrying a load like a crate or box is holding it by the front corners, with the arms straight, at hip height (about 900 ± 150 mm), so that it does not interfere with the movements of the lower limb. All things being equal, it makes sense for loads to be picked up from the level at which they will be carried.

These considerations define the vertical height ranges shown in Figure 15.20. The figures given for the various landmarks are based on anthropometric data from Pheasant (1986) rounded up to convenient whole numbers. The preferred height range for lifting and handling tasks extends from knuckle height to elbow height or thereabouts. The loads which people can reasonably be expected to lift will be greatest in this height range. The zones on either side of the preferred height range (from knee height to knuckle height and from elbow height to shoulder height) are acceptable for most purposes, provided that the weight is within reasonable limits for the person concerned. Lifting from below knee height or lifting to above shoulder height should be avoided where possible, and the weight of load we could reasonably regard as acceptable in these zones would be relatively low. Each height range may be divided up as shown into horizontal sections. Lifting actions performed in the zones marked "very poor" should be

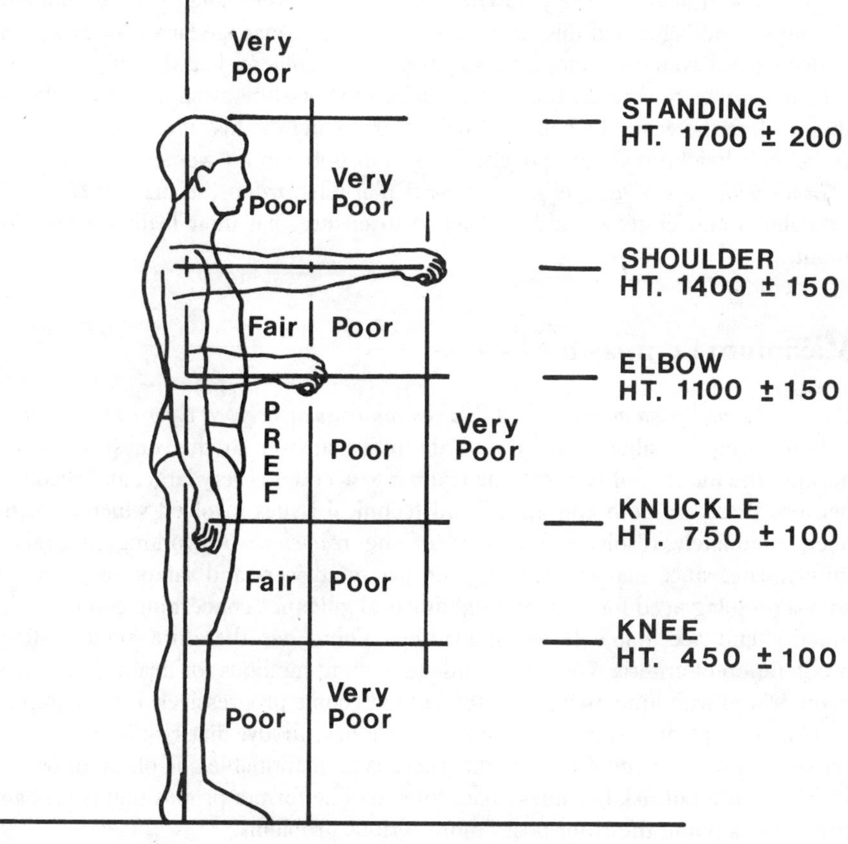


Figure 15.20 *Height ranges for lifting actions—after Pheasant and Stubbs (1991)*

avoided altogether. The weight which could be considered acceptable in these zones would be very low indeed.

### Load Characteristics

*Bulky loads* are more hazardous than compact loads (of the same weight)—not only because of the greater load moment (see above), but also because they will be more likely to put the person off-balance and they will require greater degrees of postural compensation.

*Wide loads* are more fatiguing to carry than narrow loads, because of the greater loading on the shoulder muscles; and loads which are more than about 300–400 mm in width are difficult to pick up from floor level (see above). A load which is more than about 600 mm in height will block forward vision, when carried at hip height by a short person; and if downward vision is required (to negotiate obstacles), the acceptable height will be less.

For general ease of handling, therefore (by the smaller person), the dimensions of a load such as a box or crate should not exceed about 350 mm wide by 350 mm deep (from front to back) by 450 mm high. (A few very small people may have difficulty handling a load this size; larger people may manage larger loads without difficulty; and even very compact loads may cause problems in awkward positions.)

*Loads carried by the sides* (i.e. like a suitcase) should be as slim as possible and should clear the ground when the carrier's arm hangs by his side. This effectively limits their height to about 650 mm for men or 600 mm for women.

*Loads which are difficult to grip* are particularly hazardous, as are *unstable loads*. Containers and crates should be packed to ensure that their loads do not shift about.

## Maximum Permissible Loads

*What is the maximum weight of load that you can reasonably expect a person to lift safely?*

It is rarely possible to provide a definitive answer to this question: firstly, because the number of factors to be taken into account is very large; and secondly, because there are both conceptual and technical issues involved which have not been adequately resolved. In terms of the *realpolitik* of working life this is unfortunate, since management, labour, safety advisers and safety inspectors all have a pressing need for clear and unequivocal guidance concerning good working practice and acceptable levels of loading. There has therefore been a steady proliferation of criteria, guidelines and assessment methods for dealing with these matters; and with time, these have tended to become progressively more complex.

The concept of a *maximum safe weight* implies, firstly, that risk increases as a function of weight; and secondly, that there is an identifiable cut-off point beyond which the level of risk becomes unacceptable. The former proposition is probably true up to a point; the latter poses more serious problems.

Suppose we understood the issues involved sufficiently well to plot a graph showing the risk of injury as a function of the heaviness of the workload. (The latter would presumably be defined by some composite measure incorporating the weight and nature of the load, the circumstances in which it is handled, and so on.) Two possible forms which this graph might take are shown in Figure 15.21. The complexities of the underlying mechanisms of damage (and the number of factors involved) suggest that it is extremely unlikely that the function which relates risk and workload will have any obvious points of discontinuity or threshold effects at which we could say "thus far and no further" (as shown in the left-hand curve). And even if there was such a threshold *for the individual*, we should expect such thresholds to be normally distributed in a working population (or approximately so). Thus the right-hand curve in which risk steadily increases with a workload seems the more probable of the two. (Although the function might well be expected to level off at very high workloads where everyone was at risk.) So there will inevitably be an extensive "grey area"—between loading levels which a reasonable person would consider "hazardous" and levels which a reasonable person would consider "safe". In other words, there is an extensive zone *of steadily increasing risk*—the upper and lower cut-off points of which are essentially arbitrary.

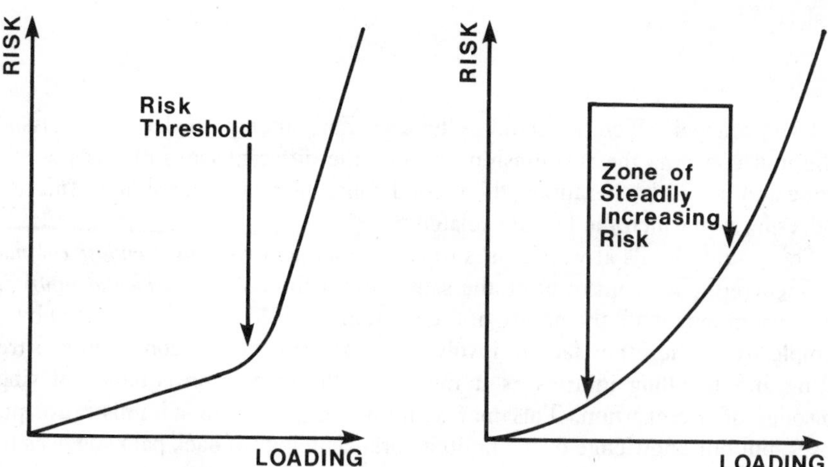


Figure 15.21 *Hypothetical risk curves:* left, *ideal;* right, *probable*

A few epidemiological studies have been reported which permit the plotting of *real risk* as a direct function of some measure of workload. Those of Chaffin and co-workers (Chaffin and Park, 1973; Chaffin *et al.*, 1978), shown in Figure 15.8, are a good example. Note that one of these showed a threshold effect, whereas the other showed a risk function which was more or less linear. Figure 15.22 shows another example, based on the data of Kelsey *et al.* (1984c). In this case the measure of workload is the number of lifts (of 25 lb or more) per day. For lifts which do not involve a twisting action there is a threshold effect, but even occasional twisting lifts carry high risks.

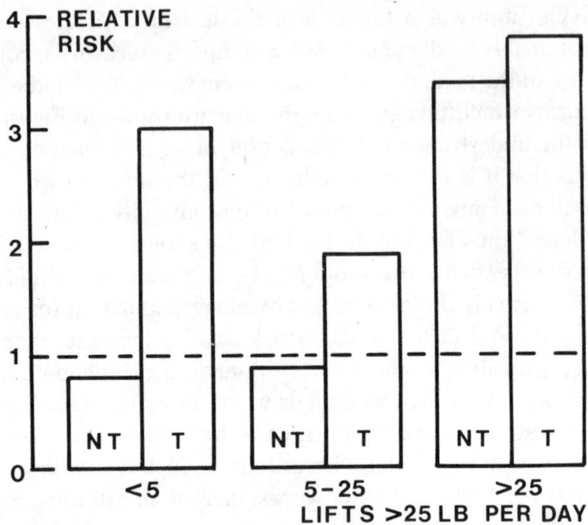


Figure 15.22 *Relative risk of low back pain, diagnosed as due to prolapsed intervertebral disc, as a function of lifts per day: T, with a twisting action; NT, without a twisting action. Data from Kelsey* et al. *(1984c)*

Two principal difficulties stand in the way of any attempt to clarify these issues. The first concerns the relationship between the different types of risk posed by lifting and handling operations; the second concerns human variability. And there are respects in which the two are related.

The loading levels at which risks of *overexertion* and risks of *cumulative damage* are incurred may or may not be the same; and if the risks of *accidental injury* are load-dependent at all, the nature of the relationship is likely to be obscured by the complexity of the other factors involved. At present, many people tend to treat lifting and handling injuries as if they were the direct consequence of single episodes of overexertion. This may simplify the matter for administrative purposes, but our knowledge of the multifactorial aetiology of back pain suggests that in many cases it may be far from the truth.

Many real working tasks are defined by the need to handle a certain total quantity of material: for example, building a wall out of concrete blocks or delivering a load of coal. Is it better to handle this load in large or small units? By decreasing the weight of each individual building block, or each sack of coal, we would presumably reduce the risk of overexertion. But would we get a pro rata decrease in the risk of cumulative damage? The decrease in unit weight is offset by an increase in the number or work cycles. In each of these cycles it will be necessary to lift both the weight of the load and some fraction of the weight of the body. So although the peak levels of loading to which the spine is exposed will decrease, the total quantity of work performed by the back muscles will increase.

At present we know relatively little about the levels of risk associated with high

*Table 15.3 The Vulnerable Back: Some Possible Causes*

**Subclinical back problems, resulting from:**
previous injuries from which recovery has been incomplete
cumulative overuse
"normal" degenerative processes

**General unfitness**

**Fatigue:** local and/or general
acute and/or chronic

**Virus infections**

peak loads and low repetition rates, compared with low peak loads and high repetition rates.

Consider now the problem of human variability. In Chapter 3 of this book we found evidence for the proposition that in any working population there are a certain number of individuals whose backs are in "a vulnerable state"—and who may injure themselves at levels of mechanical loading which are well within the limits of tolerance for other people. The vulnerable members of the population may constitute a distinct sub-group; or they may just be the tail of a continuous distribution of "at riskness". Some possible reasons for vulnerability are summarized in Table 15.3. Some of these conditions will tend to become more common with age (Figure 15.23), although they are by no means unknown in young people (and their presence need not necessarily indicate vulnerability in all cases).

Let us suppose that the limits of physical working capacity are normally distributed in the population. Suppose we follow the principle of the limiting user—and set our maximum acceptable workload at a level determined by the

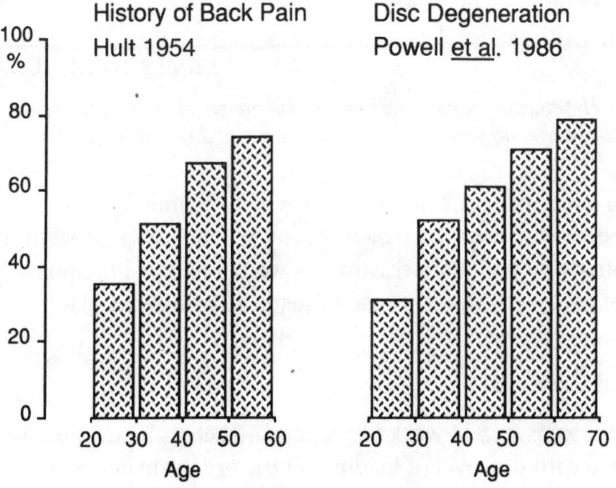


Figure 15.23 *The effects of age on:* left, *cumulative prevalence of back pain in male workers;* right, *prevalence of signs of disc degeneration in asymptomatic women*

capacity of an individual somewhere close to the bottom end of this distribution. This will minimize the number of people who are *overloaded* and therefore at risk; but it also means that the productive working capacities of the great majority of individuals will not be fully utilized. We could call this latter group "underloaded". The number of individuals falling into each category may be plotted as a function of the intensity of the workload, as shown in Figure 15.24. (The shape of the two curves in this figure follows from the nature of the normal distribution.) In principle we could perform some kind of cost–benefit analysis (as applied either to the working organization or to society as a whole) which took into account both overload and underload. The costs associated with overload arise only when an injury occurs; but the costs associated with failing to utilize the productive capacities of the workforce are incurred on a continuous basis. Thus the point at which this (hypothetical) cost–benefit function was optimized would presumably be one which placed a considerable percentage of the workforce at risk (perhaps even the majority).

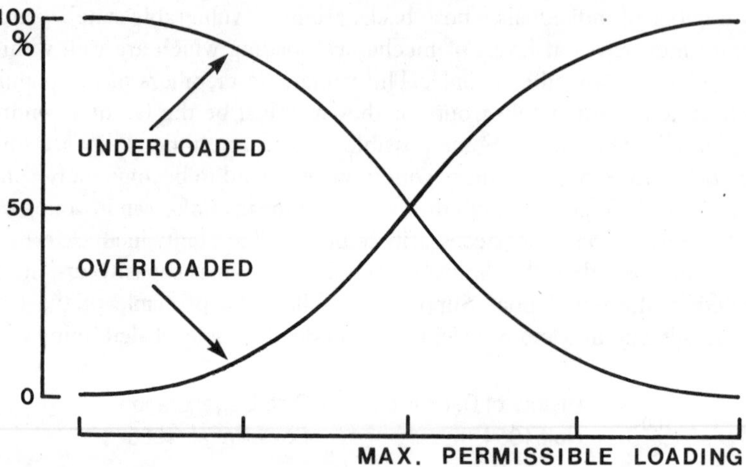


Figure 15.24 *Hypothetical curves showing the percentages of people who would be overloaded or underloaded if the maximum permissible loading was set at different levels*

We have a potential for conflict here which cannot be resolved on scientific grounds alone. Any consensus which can be reached in practice must necessarily reflect economic and political pressures as well. There is little point in proposing a limit which, if accepted, would cause industry to grind to a halt.

## Physiological Criteria

The traditional measures of work physiology—such as heart rate, energy expenditure, etc.—deal with the level of loading on the organism as a whole, averaged over a period of several minutes at least. Criteria of this kind are not applicable to lifting tasks of short duration. Furthermore, a task could be within acceptable limits as thus defined, but still involve localized overloading of specific muscle groups or soft tissue structures.

It is widely accepted that average energy expenditure during work should not exceed one-third maximum aerobic power—which for an average adult man is about 5 kcal/min (see p. 3). But acceptable limits, as thus defined, may be task-dependent. The maximum oxygen uptake during repetitive stoop lifting (i.e. using back muscles) may be as much as 15% less than it is during bicycling or running (Petrofsky and Lind, 1978a), whereas for repetitive crouch lifting (i.e. using leg muscles) there is no difference (Williams *et al.*, 1980). We should expect the level of energy expenditure associated with the onset of physiological fatigue to be low in tasks that have a low maximum oxygen uptake. On the basis of their physiological measurements, Petrofsky and Lind (1978b) proposed that the limiting workload for repetitive lifting tasks (over a 1–4 hour period) should not exceed 25% of the individual's "true" aerobic power (as determined during cycling, etc.); and Legg and Myles (1981) confirmed that the subjectively acceptable workload, for lifting and handling tasks over an 8 hour working day, was equivalent to 21% of aerobic power.

## Biomechanical Models

One approach to the problem of setting load limits where the traditional criteria of work physiology are inappropriate is based on the use of computerized biomechanical analyses which predict the mechanical loading on the spine (or, more specifically, the compressive loading on the L5/S1 disc). This approach is particularly influential at the present time (especially in the USA). The calculations which are required are relatively straightforward for static symmetrical two-handed exertions. They become more complex for asymmetric static exertions and for dynamic lifting actions (see Chaffin and Andersson, 1984, or Tracy, 1990, for an introduction to the technicalities of the subject).

As the various technical and computational problems which beset these models are solved, we may expect a steady increase in their accuracy and versatility as predictive tools. But even if the accuracy of the predictions were perfect (and we have by no means reached that point as yet), a knowledge of the loading on the spine in any given lifting task would not of itself tell us whether the task was safe. We should also need to know the levels of spinal loading at which injury is likely to occur.

Some indication concerning the latter may be derived from experiments in which cadaveric spines are tested to destruction. The data on the left-hand side of Figure 15.25 are based on experiments of this kind from various sources cited in NIOSH (1981). Set alongside these data are the epidemiological findings of Chaffin and Park (1973), in which the incidence of back injuries in industrial workers has been plotted as a function of the compressive loading on the spine in the jobs concerned (as predicted by the biomechanical model). A fair degree of agreement is apparent.

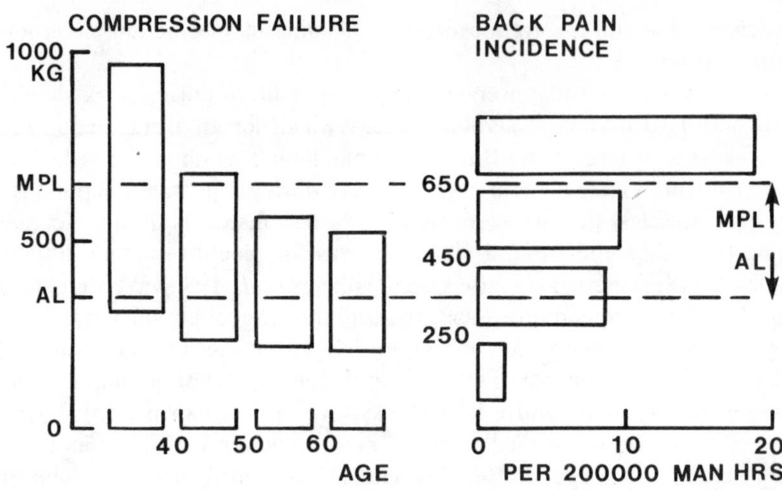


Figure 15.25 *Risk of injury as a function of spinal loading.* Left*: ultimate compression strengths of cadaveric specimens from subjects of different ages.* Right*: incidence of back injuries in the study by Chaffin and Park (1973)*

On the basis of this apparent agreement, it is widely accepted that:

(i) a compressive loading of more than 650 kgf should be regarded as hazardous for most people;
(ii) a compressive loading of less than 350 kgf should be regarded as safe for most people. (NIOSH, 1981)

Life is not this simple, however. Disc injuries are rare in these experiments. The characteristic mode of failure of the cadaveric spine, when it is subject to compressive loading in a testing laboratory, is a collapse of the end-plate of the vertebral body. In the clinical context, this type of injury is virtually unknown. Vertebral crush fractures are only likely to occur in young people as a result of sudden impact loading—such as a fall from a height or using an ejector seat. Similar fractures are common enough in elderly women as a result of osteoporosis—when they may often be caused by lifting. But in both cases the fracture involves the central part of the vertebral body rather than the end-plate. Microfractures of the end-plate are probably common enough at all ages—but it is not known whether they have any clinical significance.

More recent cadaveric experiments, using a different testing technique, have found lumbar spine compressive strengths ranging from 700 kgf to 1300 kgf in young male subjects, with an average value of about 1,000 kgf (Hutton and Adams, 1982). So the data upon which the NIOSH criteria for compressive loading were based may well *underestimate* the true strength of the spine by as much as 50%. Furthermore, the biomechanical model used by Chaffin and Park (1973) may well *overestimate* the compressive loading on the spine in the working situation by a similar amount—since it assumes that the back muscles act at an effective leverage of 50 mm, whereas more recent anatomical studies indicate that the true value would be around 75 mm (McGill and Norman, 1987b).

If these two sets of subsequent findings are correct, then most of the people injuring their backs in the epidemiological study upon which the criterion levels for compressive loading were based must have been doing so at levels of loading which were well within the compressive strengths of their lumbar spines. Given that most acute back injuries occurring during the course of lifting and handling tasks are probably muscular or soft tissue tears, this conclusion is not particularly surprising. Note also that the epidemiological study concerned was based on only 25 cases.

## The Psychophysical Approach

The psychophysical approach to defining good practice in the design of manual handling tasks is based on user trials in which samples of subjects (preferably industrial workers) estimate the levels of workload that they would find subjectively acceptable: that is, that they could sustain on a daily basis without an undue sense of strain. Snook (1978) has reported an extensive series of such experiments.

The principal advantage of the psychophysical approach lies in its directness and simplicity, and its potential for circumventing many of the limitations of the more analytical techniques of physiology and biomechanics; its principal weakness is the possibility that people might know what is good for them. Snook *et al.* (1978) investigated the loads which were handled, and the numbers of back injuries reported, in a range of industrial jobs. About one-quarter of the jobs which were analysed involved handling tasks which would be acceptable to less than 75% of the workers (according to psychophysical criteria); but half of the back injuries occurred in these particular jobs. Thus the risk of back injury is three times greater in jobs which are psychophysically acceptable to less than 75% of the workers.

## Statutory Limits

Some countries impose a statutory limit on the maximum weight to be lifted or handled at work; other countries do not.

At present, the UK imposes no such limit—except in the case of a certain few specific industries. These industry-specific limits are widely considered to be archaic—and it seems likely that they will be repealed in the near future. There is, however, a general requirement under the *Factories Act* (1961) that "a person shall not be employed to lift, carry or move any load so heavy as to be likely to cause injury to him". An HSE code of practice dealing with lifting and handling at work is in preparation at the time of writing (September, 1990). An early draft of this code of practice included specifications concerning maximum acceptable weights under various circumstances (HSC, 1982), but the more recently published draft does not (HSC, 1988).

The International Labour Office has recently published a survey of maximum permissible weights in various countries of the world (ILO, 1990). These figures

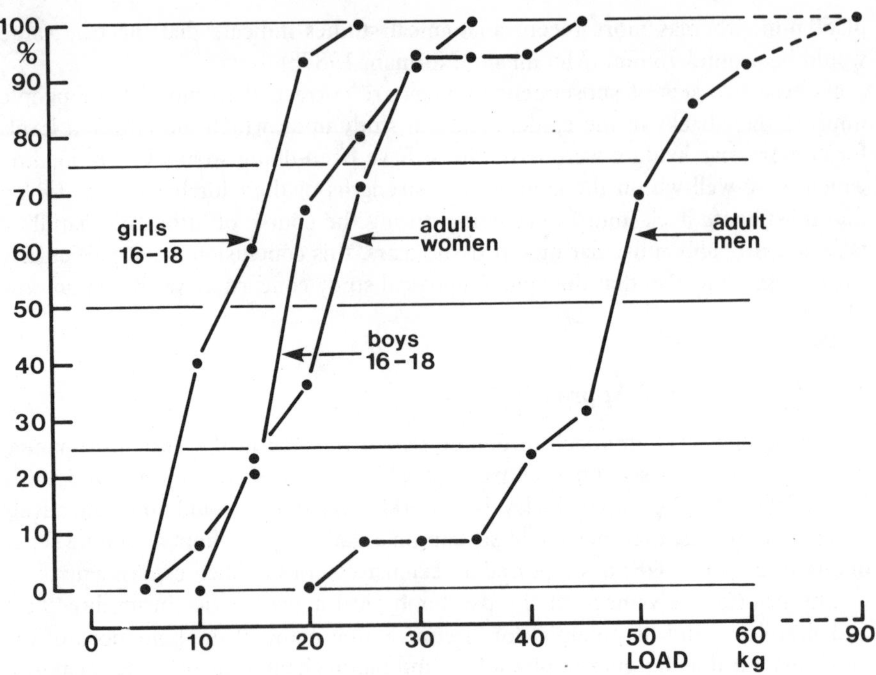


Figure 15.26 *Statutory limits on the weight to be handled by one worker in different countries of the world. Cumulative distributions based on information given in ILO (1990). See text for explanation*

are summarized in Figure 15.26. The data are plotted in cumulative form—in other words, the graphs show the percentage of countries where a load exceeding the weight concerned would be considered unacceptable. (The percentages only apply to countries having such regulations.) The median weight limit for adult men was 50 kg, with an inter-quartile range (IQR) of 45–55 kg (i.e. the middle 50% of values fell within this range). For adult women the median value was 25 kg (IQR = 20–25); for boys aged 16–18 the median value was 20 kg (IQR = 20–23); and for girls aged 16–18 the median value was 15 kg (IQR = 10–15).

Lifting and handling tasks are also the subject of a European Community directive adopted on 23 May 1990. This does not specify maximum permissible weights—but it contains a number of more general ergonomic provisions concerning the characteristics of the load, the working environment, the undesirability of lifting at a distance or with a turning action, and so on.

## ILO Recommendations

Table 15.4 shows the maximum weights for occasional lifting by men and women of different ages, as recommended in a guidance note published by the International Labour Office (ILO, 1962). The figures represent the consensus view of a working party of medical experts. For frequent lifting, it was recommended that

*Table 15.4 ILO Recommended Maximum Weights for Occasional Lifting*

| Age | Maximum weight (kg) | |
|---|---|---|
| | Men | Women |
| 14–16 | 15 | 10 |
| 16–18 | 19 | 12 |
| 18–20 | 23 | 14 |
| 20–35 | 25 | 15 |
| 35–50 | 21 | 13 |
| Over 50 | 16 | 10 |

the tabulated values should be reduced by 25%. These figures have been widely quoted (e.g. NIOSH, 1981; Chaffin and Andersson, 1984; etc.). A second publication by the International Labour Office quoted higher levels, however. Again it was based upon the deliberation of a panel of experts who recommended levels of 40–50 kg for men; 15–20 kg for women; 15–20 kg for boys aged 16–18 years; and 12–15 kg for girls aged 16–18 years (ILO, 1964).

The International Labour Office Maximum Weight Convention (ILO, 1967) restricts itself to general requirements only, stating that "no worker shall be required or permitted to engage in the manual transport of a load which, by reason of its weight, is likely to jeopardise his health or safety". But it adds the provisos that "the assignment of women and young workers to manual transport of loads other than light loads shall be limited" and that "where women and young workers are engaged in the manual transport of loads, the maximum weight of such loads shall be substantially less than that permitted for adult workers". According to ILO (1990), twenty-two member states had ratified this convention by January 1987.

## The NIOSH Guidelines

The guidelines published by the US *National Institute for Occupational Safety and Health* (NIOSH 1981) have been extremely influential—both in their country of origin and elsewhere. Features of these guidelines have been incorporated into the Code of Practice adopted by the Australian state of Victoria; and it is possible that some of these features may also be incorporated into the European (CEN) Standard which is in preparation at the time of writing.

The NIOSH guidelines are based on epidemiological, biomechanical, physiological and psychophysical considerations. NIOSH (1981) specifies two threshold limits: the action limit (AL) and the maximum permissible limit (MPL). The underlying criteria defining these limits are summarized in Table 15.5.

The version which is current at the time of writing (NIOSH, 1981) takes four features of a lifting task into account: the *horizontal location* of the load, the *vertical location* of the load at the starting point of the lift, the *vertical distance* which the load travels and the *frequency and duration* of the task.

If the four factors are known, for a given lifting task, then the action limit (AL)

*Table 15.5 NIOSH Criteria Defining the Action Limit (AL) and Maximum Permissible Limit (MPL)*

| | AL | MPL |
|---|---|---|
| *Epidemiological* | | |
| Increase in risk | Moderate | Significant |
| *Biomechanical* | | |
| Compressive loading (L5/S1 disc) | 350 kgf | 650 kgf |
| *Physiological* | | |
| Energy expenditure | 3.5 kcal/min | 5 kcal/min |
| *Psychophysical* | | |
| Acceptable to | 99% men<br>75% women | 25% men<br>1% women |

After NIOSH (1981).

and maximum permissible limit (MPL) for that task may be calculated by means of a mathematical procedure, the details of which need not concern us here. The load which is handled on the job is then compared with the limit values.

Lifting tasks may thus be placed into three categories:

(i) the "green zone"—if the load is within the action limit;
(ii) the "yellow zone"—if the load exceeds the action limit, but is within the maximum permissible limit;
(iii) the "red zone"—if the load exceeds the maximum permissible limit.

(The terms *green*, *yellow* and *red* do not actually appear in the text of the guidelines.)

Jobs in the green zone are considered to pose little or no risk of overexertion for the majority of people. Jobs in the red zone are considered to pose a high level of risk, and should therefore be ergonomically redesigned or eliminated altogether. Jobs in the yellow zone are considered to pose a moderate level of risk. For jobs in the yellow zone, an appropriate programme of selection or training is considered to be an acceptable alternative to changes in work design.

At the time of writing (November 1990) the NIOSH (1981) guidelines are being revised. It is not yet clear what changes will be made. It seems probable that the equation defining the limit values will be modified—possibly to include additional factors for *asymmetry* and for the *coupling* between the load and the hands. It is also conceivable that the upper limit (MPL) might be dropped.

For advantageous, but realistic, combinations of conditions, the NIOSH (1981) action limit for one-off lifting actions is rarely likely to go above 20 kg; and the maximum permissible limit is rarely likely to go above 60 kg. For disadvantageous (but also realistic) conditions (for example when lifting from the floor to waist height, at a distance from the body) the action limit can fall to 5 kg or less; and the maximum permissible limit can be 15 kg or less. The limit values decrease with repetition rate, reaching zero at between 12 and 18 lifts per minute depending upon the duration of the task and the height range.

The NIOSH guidelines are based upon an extensive body of research, but they have both theoretical and practical limitations. The underlying criteria on which they are based can be challenged in a number of respects. The validity of the biomechanical criteria is particularly open to question (see above); and there is some debate concerning the relationship between the physiological and psychophysical criteria, when it comes to tasks of long duration (Ciriello *et al.*, 1990).

In practical terms, the problems most commonly encountered in applying the NIOSH guidelines to real working situations are, firstly, that they only deal explicitly with lifting and lowering tasks, whereas real-world tasks often involve elements of holding, carrying, pushing, pulling, etc.; and secondly, that it may sometimes be difficult to assign realistic values to one or more factors in the equation. For example, if people are relatively free to take up different positions with respect to the load (as is often the case in practice), then it may be difficult to settle upon a single representative value for the horizontal position factor.

In practical terms also, the "yellow zone", between the action limit and the maximum permissible limit, is very extensive. Industrial jobs which exceed the maximum permissible limit are relatively rare—presumably because it would be difficult to persuade people to do such jobs. Where jobs do exceed the maximum permissible limit, they are often self-evidently hazardous on a common-sense basis. Conversely, jobs falling within the action limit tend to be self-evidently safe. So, in practice, most of the jobs which the ergonomist is called upon to analyse will tend to fall into the yellow zone. This will therefore tend to reinforce the status quo, in which training is seen as an acceptable alternative to more fundamental changes in the working system.

## The Robens Institute Guidelines

Peter Davis, David Stubbs and their co-workers at the *Robens Institute of Safety and Health* at the University of Surrey have devised a system of load limits based upon experimental and epidemiological studies in which intra-abdominal pressure (IAP) has been used as an index of spinal loading.

Studies of workers in the construction and telecommunications industries showed a high prevalence of back problems in jobs where the IAP commonly exceeds 100 mmHg (Stubbs, 1976, 1981; Davis and Sheppard, 1980; Nicholson *et al.*, 1981). An IAP of 90 mmHg was therefore adopted as the criterion level defining the safe limit for fit young men (Davis and Stubbs, 1977, 1978). Subsequent studies of injury rates in various groups of female workers suggested that for women the safe limit should be set at an IAP of 45 mmHg (David, 1987).

Experiments were conducted to determine safe limits for a wide variety of lifting, pushing and pulling actions according to the 90 mmHg criterion (Davis and Stubbs, 1977, 1978). The results were presented in the form of contour maps; an example is shown in Figure 15.27. The loads shown on the chart are considered to be within safe limits (according to the criterion) for 95% of men under the age of 50. The work of Davis and Stubbs (1977, 1978) forms the basis of a UK defence

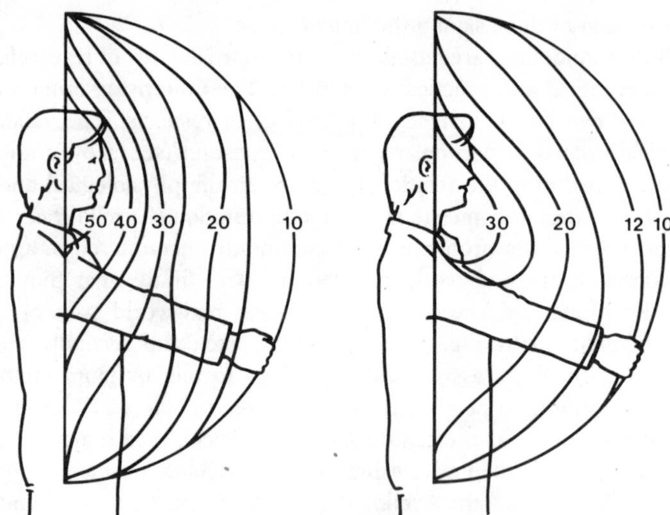


Figure 15.27 *Proposed safe limits (kg) for occasional lifting actions by fit young men, as given by Davis and Stubbs (1977, 1978):* left, *two-handed;* right, *one-handed. Reproduced from Pheasant (1986)*

standard (MoD, 1984). The correction factors given in Table 15.6 are quoted from this standard.

The principal limitation of these guidelines is that, in their original form, they only deal with lifts performed in front of the body, with an upright trunk. (It is assumed that a person will pick up a load from the floor by bending his knees, not his back.) The effects of asymmetry and trunk inclination have not been extensively documented.

Provisional estimates suggest that for some asymmetric lifts, the load limit could be reduced by about 40% (Ridd, 1985); and that the safe limit could be reduced by about 50% at a 45° trunk inclination (Nicholson, 1989). Further research concerning these matters is in progress at the time of writing (John Ridd, personal communication).

For a male worker under 50, performing a one-off lifting action under favourable (but realistic) conditions—compact load, close to body, optimum height range, etc.—the safe limit according to these guidelines is unlikely to exceed 30 kg.

*Table 15.6 Correction Factors for Load Limits*

*Reduce* the load limit given in the contour chart by the following amount:

| | Less than 1 lift per minute | More than 1 lift per minute |
|---|---|---|
| Men under 50 | – | 30% |
| Women under 50 | 40% | 58% |
| Men over 50 | 20% | 45% |
| Women over 50 | 52% | 66% |

But for a man over 50, lifting under unfavourable (but realistic conditions—from floor level, in an asymmetric position, at a distance from the body with an inclined trunk—the limit might be 5 kg (or less) for a single lifting action, and 2.5 kg (or less) for an action performed once per minute.

## "Safe Limits": A Personal Viewpoint

Given the extent of human variability, and the possible presence in the working population of a certain number of people who are especially vulnerable to injury, there is a very real sense in which we could say that there is no such thing as a "safe load", although with further research we could doubtless improve our understanding of the levels of loading which pose a "reasonably foreseeable risk".

The levels of loading defined by the NIOSH action limit and the Robens Institute guidelines are relatively low by industrial standards, and it is difficult to see how limits set at these levels could be rigorously enforced.

The "safe limits" predicted by these two sets of guidelines for lifts performed under disadvantageous (but realistic) circumstances are in some cases very low indeed. In many respects, therefore, it makes more sense to regard such a lifting action as inherently hazardous—whatever the load—and to take steps to eliminate it by redesigning the job according to ergonomic principles.

In other words, instead of asking "Is this load within safe limits?", we should really be asking "Is the design of this lifting and handling task ergonomically satisfactory?" To answer the latter question we need to consider the working system as a whole.

## Chapter Sixteen

# Clinical Ergonomics

Ask any assembled group of medical practitioners to pick one word to characterize the natural history of back pain and the chances are that the word they come up with will be "recurrent". A typical family practitioner could see around 20 patients suffering from back pain each week. Of these, 17 could be suffering from a condition which is work-related (see p. 69). But in how many of these cases will the patient's working activities ever be analysed in sufficient detail to identify those features of his working life which are placing him at risk? Instead he may well be told that his problems are due to the fact that he "has a long back" or that "back pain is the price that human beings have to pay for standing upright". (Any vet will tell you that quadrupeds have back problems too.)

This book has been mainly about primary prevention in the workplace—and it is in this area that the contribution of ergonomics has been most widely recognized. But ergonomics also has an important contribution to make in a much wider area of clinical practice. Every GP, every physiotherapist who deals with back and neck problems, every osteopath or chiropractor, sees patients who are suffering from work-related conditions. The epidemiological data indicate that they must be seeing such patients on a daily basis. But how often are the patient's symptoms treated in isolation from their aetiology?

Often enough, the patient's highest priority is to get back to work at the earliest opportunity. In essence he is asking his doctor (or physio, or osteopath) to fix up his symptoms—so that he can go back to doing those things which caused the problem in the first place, thus initiating the cycle of recurrence. In the worst cases, the patient's tragedy is that he can only be "cured" of his condition by depriving him of his livelihood.

The successful management of work-related musculoskeletal problems starts at the stage of the case history. The clinician must assess, as best he or she can, the relative importance of the various contributory factors in the aetiology of the patient's condition—so as to make a realistic appraisal of the extent to which these are under the patient's direct control—as against being imposed upon him by external circumstances. There is little point in telling someone to instruct someone to "sit up straight" at his desk, or to "bend the knees not the back" when lifting, if the physical constraints of the working situation are going to prevent this.

In the best of all possible worlds, the clinician would be in a position to visit the patient's workplace in order to see the problems for himself. This is rarely possible in practice—except in the context of the firm's occupational health department. But experience suggests that people are both willing and able to talk informatively about their work. Indeed, they often welcome the opportunity, and warm to the theme when they find someone who is prepared to take an interest in the subject. And although it is not quite as informative as the real thing, a great deal can be learned from asking the patient to demonstrate his working posture, mime the actions he performs, and so on. Recognizing the connections between these possible risk factors and the patient's clinical condition is as much as anything a matter of applied anatomy. A similar approach can be taken to the patient's non-occupational activities: housework, sports, hobbies, and so on.

There can be no single simple formula for gathering this information—since every case will obviously be different—but the checklist at the end of this section might be a useful starting point.

The clinician then needs to decide which of these risks can be eliminated by teaching the patient a different way of working or performing his other everyday activities; which require minor changes, such as the adjustment of chairs and workstations or the provision of a footrest; and which require more radical changes in the nature of the working system, which will necessarily involve a process of negotiation between the patient and his employer (perhaps via the intermediary of the patient's trade union).

Above all the patient has to go away with a better insight into the relationship between the symptoms which are troubling him and the activities which are causing them, and a positive attitude to finding ways and means for improving things.

## Checklist: Thirty Ergonomic Questions for the Patient's Case History

### A. General

1. Does the patient work in a fixed position or is he free to move around more or less at will?
2. Could the patient's working posture account for his symptoms?
3. Does the patient's working activity result in static muscle loading due to holding actions, sustained exertions, etc.?
4. Does the patient's job require him to exert himself to the limit of his physical capacities?
5. Does the patient's job require lengthy periods of intense concentration?
6. Does the patient find his work psychologically stressful or unrewarding?

### B. Standing

7. Does the patient stand continuously for more than half the working day?

   If he stands at all at work:

8. Does he stoop?
9. Is he required to lean over obstacles or into containers?

10. Is he required to make awkward reaching actions?
11. Does he work in an asymmetric or twisted position?
12. Can he shift his weight from foot to foot at will?
13. Would it help if he were provided with a stool?
14. Does he work with his shoulders elevated or his arms abducted to more than 60°?

## C. Sitting

15. Does the patient sit continuously for more than half the working day?

    If he sits at all at work:

16. Is the seat height correct
    (a) with respect to the working level?
    (b) with respect to the patient's leg length?
17. Does the patient know how to adjust his chair correctly?
18. If the patient is short, would he benefit from a foot rest?
19. If the patient is tall, would he benefit from a non-standard high desk?
20. Does the patient use a VDU? If so:
    (a) Is the screen located correctly with regard to the patient's angle of vision?
    (b) Is the reading distance satisfactory?
    (c) Are there glare problems?
21. Would the patient benefit from a reading stand?

## D. Repetitive Movements

22. Does the patient's job involve frequent or repetitive movements of the upper limb? If so, does he make more than 1,500 movements per hour or is the cycle time of the task less than 30 seconds?
23. Is he required to make repeated forceful gripping actions? Are these made with the wrist in a deviated position?
24. Does the pattern of movement involve extremes of
    (a) ulnar deviation?
    (b) radial deviation?
    (c) wrist flexion?
    (d) wrist extension?
    (e) pronation or supination?
    (f) upward reach?
    (g) backward reach?

## E. Lifting and Handling

25. Does the patient handle loads of more than 30 kg occasionally or 10 kg frequently?

26. Does he handle loads which are
    (a) bulky?
    (b) unstable?
    (c) difficult to grip?
27. Does he handle loads at a distance from the body?
28. Does he handle loads outside the preferred range of heights?
29. Does the task require him to lift with a twisting action?
30. Has he been trained in safe lifting?

## Back Schools

The potential value of an ergonomic approach to the management of musculoskeletal pain and dysfunction is illustrated by the success of "back schools". The underlying philosophy of the back school is that the patient needs to become an active participant in his or her own rehabilitation. The "curriculum" of the back school commonly includes basic instruction on the anatomy and physiology of the spine, the causes of back pain, posture, exercise and relaxation, as well as practical ergonomic advice aimed at reducing the loading on the spine both at work and in the performance of the patient's everyday activities (how to make the bed, how to clean the bath, etc.).

Controlled clinical trials have demonstrated the effectiveness of back schools in the management of both acute and chronic back problems. In a study of patients with acute back problems from the Volvo works at Goteborg, Berquist-Ullman and Larsson (1977) found that patients attending a back school were able to return to work earlier than those receiving conventional physiotherapy (and both groups fared better than a control group receiving a placebo treatment). Klaber-Moffet *et al.* (1986) compared patients suffering from chronic back pain who attended back schools with a control group who were treated with exercise only. Both treatments were effective in the short term—but for the patients in the exercise-only group the improvement only lasted a few weeks, whereas the back school patients continued to show long-term improvements with respect to both pain and functional disability.

The apparent success of back schools in the clinical management of back pain stands in contrast to the relatively poor track record of lifting training programmes. Why is this? In part it is probably because the back school programme addresses a much wider range of issues and risk factors. But it is probable also that the back school is more effective at initiating and maintaining behavioural change. The person who already has a back problem has a higher level of motivation and can see the relevance of what he is being taught—and he will get immediate feedback to reinforce the teaching as he finds ways of doing things which cause him less pain.

## Ageing and Disability

In the advanced industrialized countries, life expectancy is steadily increasing and the birth rate is lower than it was in the post-war period of the "baby boom". So both the absolute and relative numbers of elderly people are increasing. We live, it is often said, in an *ageing society*.

Figure 16.1 shows a recent set of demographic predictions for the UK population. In 1971 about one person in six was of retirement age (i.e. 65 in men, 60 in women); if present trends continue, it will be closer to one person in four by the year 2031. In 1971 the ratio between people of working age and people of retirement age stood at around 3.5 : 1; by the year 2031 it will be lower than 2.5 : 1 (*Social Trends*, **20**, 1990 edition, HMSO).

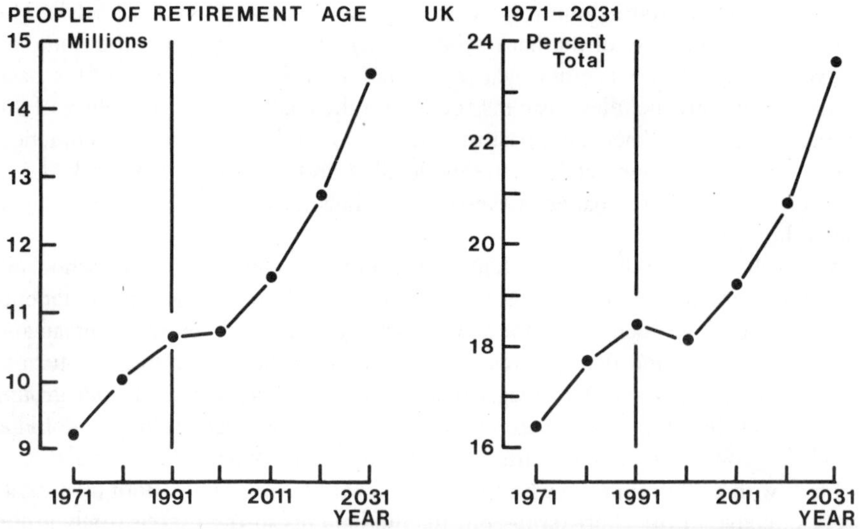


Figure 16.1 *The ageing population. People of retirement age in the UK:* left, *millions;* right, *as a percentage of the total population. Data from* Social Trends **20**, *1990 edition, HMSO*

The rate of increase is greatest in the older age groups—the number of people over 75 will increase dramatically by the end of the century. There are two ways of looking at this phenomenon: the optimistic and the pessimistic. The optimistic scenerio is that people will both live longer and stay fitter—as crippling diseases are defeated by the march of medical science. The pessimistic scenario is that medicine will merely succeed in keeping people alive in increasing states of decrepitude. In either case the demographic trend has important implications from the standpoint of ergonomics.

As we get older, our physical and mental capacities necessarily wane. The decline is partly due to the ageing process itself—and partly due to the effects of prior injury. (Statistically speaking, it is probably pointless to try to separate the two.)

Muscle strength, for example, declines with age (p. 43). Elderly people commonly show a pronounced wasting of the small muscles of the hand—perhaps because the T1 nerve root (which supplies these muscles) is compressed against the first rib, as the shoulders droop with age. The progressive weakness of grip which occurs, as a consequence of "normal ageing", may be exacerbated by the residual deformity and functional impairment which result from a chronic condition such as rheumatoid arthritis or an acute injury such as a Colles' fracture (a fracture of the lower end of the radius which is extremely common in post-menopausal women as a consequence of osteoporosis). Conversely, the ulnar-drift deformity of the fingers, which is usually regarded as a long-term consequence of rheumatoid arthritis, is actually quite common in elderly people with no history of the condition.

It is not uncommon, therefore, for an elderly person—perhaps an elderly woman who had a relatively weak grip in the first place—to find that she increasingly encounters problems in doing things like opening screwtop jars, turning doorknobs, turning keys in locks, and so on. Now it is easy enough to supply her with special devices—jar openers, key turners, etc.—which will help her to maintain her functional independence. You can replace her doorknobs with lever-type handles which are much easier to operate. But when you look at the problems in ergonomic terms, you may well discover that the products which are causing her problems are really quite badly designed from a functional standpoint. The caps of screwtop bottles and jars may be outside the optimal range of sizes, have smooth edges which make them difficult to grip, and so on. The door key is turned using a precision grip. For a person of "normal" strength this presents no real problem, but as strength declines, this inherently weak turning action may not be sufficient to operate the mechanism of the lock—particularly if the mechanism is in need of oiling or the door fits badly (as is often the case in the home of an elderly person). So we have a situation in which the product is by no means optimal in functional terms but it is just about good enough to be usable by a person of "normal abilities"—so there is relatively little commercial pressure for design improvements.

Consider the ordinary bread knife. You use it in a position of ulnar deviation. People with rheumatoid arthritis are sometimes given a special bread knife with its handle set at about 80° to the blade (like a carpenter's saw)—on the grounds that the ordinary bread knife tends to exacerbate their problems (increase ulnar drift, loading on the tendons, etc.). But, in reality, the redesigned bread knife is a much better tool for doing the job. The traditional design is only acceptable because most people have sufficient reserves of strength to overcome its deficiencies, and they don't cut bread for long enough for it to cause tenosynovitis.

The difficulties which elderly or infirm people experience in rising from easy chairs falls into exactly the same category (see p. 219), and there is no real evidence that people with back problems need special sorts of chairs—they just need well-designed chairs more acutely than the rest of us.

Kitchen equipment is another case in point. Kitchen worktops are currently manufactured to a standard height of 900 mm (BS 6222; ISO 3055)—and a variety of other products are manufactured to fit under the worktop or to be level with its surface. Anthropometric considerations and ergonomic studies of user preferences indicate that the standard worktop is too low for a substantial number of users (about 35% of women and 55% of men) and too high for a small number of women (about 4%). The kitchen sink is generally set into the worktop—so tasks performed in it will be at a working level which is a good deal too low for most users. Most people seem to tolerate these ergonomic deficiencies without any real difficulty, but for the tall back pain patient it can be a very serious problem—to the extent of making some kitchen tasks intolerably uncomfortable. In the early 1970s there was some interest in the possibility of providing kitchen fittings in a range of heights—but nothing has come of the idea, presumably because manufacturers think that it is "impractical" or that "there would not be enough demand to justify the expense". (For a further discussion see Pheasant, 1986.)

There are a number of reasons why it is useful to distinguish the concept of *disability* from the related concepts of *impairment* and *handicap*. According to current usage (as set out, for example, in Wood, 1975, and WHO, 1980), an *impairment* is a disturbance of normal structure or function (consequent upon some disease or injury). The impairment may lead to a *disability*—which is the inability to perform a task or tasks in the way that the majority of "normal" people are able to. And the disability may result in some kind of social or personal disadvantage for the individual concerned—which we would call a *handicap*. The whole process is referred to as one of *disablement*.

The concept of disability is, by its nature, both relative and task-specific (p. 120)—and any definition is necessarily arbitrary. Martin *et al.* (1988) conducted a survey aimed at establishing the numbers of people in the UK suffering from disabilities of various degrees of severity. The figures in Table 16.1 and Figure 16.2 are for the lowest criterion level (i.e. least severe). Overall some

*Table 16.1 Prevalence of Disabilities in UK Adults Living in British Households*

| Disability | Prevalence (%) |
|---|---|
| Locomotion | 9.3 |
| Hearing | 5.5 |
| Personal care | 5.0 |
| Dexterity | 3.7 |
| Seeing | 3.2 |
| Intellectual functioning | 2.8 |
| Behaviour | 2.7 |
| Reaching and stretching | 2.5 |
| Communication | 2.3 |
| Continence | 2.2 |
| Disfigurement | 0.9 |
| Eating, drinking, digesting | 0.5 |
| Consciousness | 0.4 |
| All Categories | 13.5 |

Data from Martin *et al.* (1988).

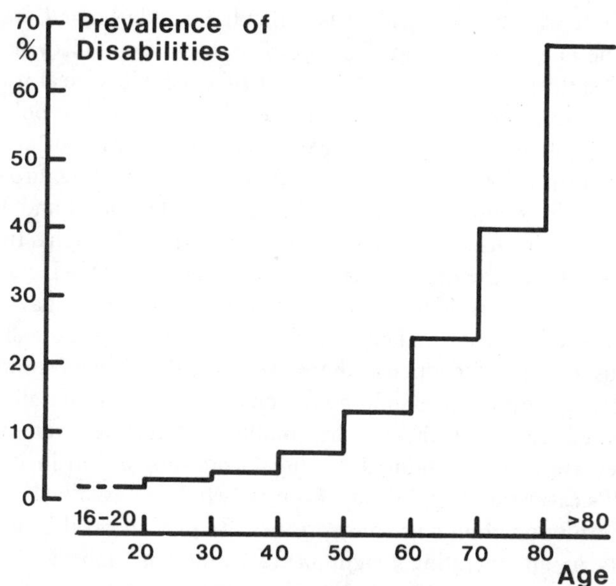


Figure 16.2 *Prevalence of disabilities in different age groups. Data from Martin* et al. *(1988)*

13.5% of British adults living in private households (i.e. not institutions) suffer from one or more measurable disabilities. The figure stands at 2% in the late teens. It climbs slowly at first and more rapidly after the age of 50 to a figure of 67% in the over-80 age group.

The extent to which these disabilities result in social or personal handicap is dependent in no small measure on ergonomic issues. The case of the wheelchair user is particularly informative—although you could tell a similar story for other sorts of disability. As Goldsmith (1976) observed, the wheelchair user is handicapped by three major sources of difficulty.

First, he has to cope with whatever problems put him in the wheelchair in the first place; secondly, his eye level and upward reach are on average about 400 mm less than that of other people; and thirdly, he trundles himself around in a cumbersome and relatively unmanoeuvrable machine which will not pass through gaps of less than about 800 mm in width (or 750 mm at a push) and cannot negotiate changes of level easily. (Indeed the effort of propelling the chair not uncommonly results in overuse disorders of the upper limbs or shoulders.)

As well as the obvious barriers to access—such as steps and kerbs—the wheelchair user may find himself defeated by all sorts of architectural trivia. For example, the critical feature, which determines whether he can use a lavatory, may be whether the door opens inwards or outwards. If it opens inwards, he may be able to get into the lavatory but not close the door behind him. But doors which open outwards may still cause him difficulties if they are set too close to a side wall—a clearance of 500 mm (or 300 mm at a push) is required to allow him room to manoeuvre as he opens the door.

With the best will in the world, it is difficult for able-bodied architects and designers to be empathetic about this sort of problem (see fallacy no. 5—p. 15). That is why we need guidelines and codes of practice which deal with access to buildings for disabled people (such as BS 5619, BS 5810, and so on).

A standing adult (in the normal range of stature and without any significant locomotor impairment) can reach and operate a wall-mounted control—such as a light switch or a lift control—at any height between floor level and 1800 mm or more. The wheelchair user can reach controls in a range of heights from 300 mm to 1400 mm—but only the middle part of this range (about 750–1250 mm) can be operated easily. So in positioning a lift control, for example, we have to trade off the convenience of the able-bodied majority against that of the small minority of wheelchair users. As it happens, there is no real conflict—controls in the 1000–1200 mm height range should be reasonably convenient for all concerned.

Wall-mounted electrical sockets are another interesting example. In older buildings they are usually mounted on the skirting board—an inch or so above floor level. (It's easiest to do it this way if you are wiring an existing building for the first time.) In modern buildings they are usually just above the skirting board, which results in an operating height of 150–200 mm. This is too low for a wheelchair user to reach (and exert the necessary force) without falling out of his chair. BS 5619 specifies a minimum height of 300 mm, which is just about satisfactory, but 500 mm would certainly be better and the best height would probably be somewhere around 800–900 mm. A socket at this height would have tangible advantages for ambulant users: it would reduce the amount of stooping required (a marked advantage for people with mobility difficulties or back problems) and it is out of reach for inquisitive small children. (You do, however, require a restraining clip to stop the plug falling on the floor.) So why do sockets remain close to floor level? It is partly a matter of the fact that it gives a neater appearance—but mainly probably that "we've always done it that way".

There is, however, at least one respect in which the requirements of the wheelchair user and the standing person are directly incompatible: working surfaces. There is no single worktop height which will be satisfactory for both standing and sitting people: the standing person requires a height of 900 mm plus; the wheelchair user requires a height of around 750 mm. This is a real problem in designing a kitchen which will be used both by a wheelchair-bound person and an ambulant spouse or helper—particularly if the spouse is elderly and perhaps suffers from locomotor difficulties which render him less able to cope with unsatisfactory working conditions.

# Epilogue: Macroergonomics

The problems we have dealt with in this book—such as back pain at work, RSI and the catastrophic failure of complex systems—are a microcosm of a broader set of issues, concerned with the humanization of technology. These broader issues are the subject area of *macroergonomics*.

We live in a consumer society—a self-reinforcing system of production and consumption. This has some important consequences for the design of our artefacts. Why should the standard three-pin plug have ceased its process of evolution at the point at which it is only just usable? Why are easy chairs so badly designed? Why do computer systems remain user-unfriendly? Why are people baffled by their video recorders and programmable domestic appliances? Why should you want pocket-sized computers to store information you would write in your diary? Why should anyone want a talking vacuum cleaner? There is a constant pressure leading to the design of innovative products (i.e. ones which create new markets) but not, it seems, for more *usable* products.

Some people think that man-made catastrophes are becoming more common. Statistically this would be difficult to prove either way—because of the small cluster effect. But, if it is true, it tells us something about the way that our demand for goods and services is outstripping the infrastructure which allows us to produce them safely. Air traffic control over South East England is a catastrophe waiting to happen. New technologies create new hazards. The computerization of industrial process control is a case in point. Software bugs may remain undetected for years, until an unforeseen combination of circumstances reveals them. A number of people have already been killed by industrial robots; and the production of genetically engineered micro-organisms has the potential for causing a catastrophic contamination of the ecosystem which would make Chernobyl look like a picnic.

We are fiercely competitive in our consumption of goods and services; and our sense of self-worth as individuals is tied up in our use of status symbols. This lies at the root of our high stress levels. Some people say that our lives have lost "the spiritual dimension". As a rationalist and a humanist, I take this to mean that we are in danger of losing touch with the part of our human nature which deals with things like compassion, the intuitive grasp of truth and beauty and the sense of unity with a higher purpose. Our cognitive tunnel vision leads us to neglect these things. We become like Eliot's "hollow men".

With technological progress, industrial production becomes progressively less dependent upon human effort. This has had far-reaching social consequences in the past and will continue to do so in the future. There is no historical inevitability. Our production technologies and our social institutions act as a coupled dynamic system. This system could become catastrophically unstable, but it is more likely that its equilibrium point will steadily drift in the same overall direction it has followed in the past. The industrial proletariat of classical Marxist theory has to all intents and purposes ceased to exist; the same long-term socio-technical trends which created our present consumer society contain the seeds of its destruction.

In general we may suppose that any job which *can* be computerized *will* be computerized. Progress in this direction has been slower than we expected. The paperless office remains in the future; and most of the repetitive work on industrial production lines is still done by human beings rather than robots. Early success stories in the design of expert systems—computer programs for playing chess, medical diagnosis, and so on—have not been followed up in other areas. The judgemental processes of the cognitive tasks, which were successfully computerized early in the game, lent themselves to computer simulation in ways that others have not. A few years ago, many people believed that artificial intelligence was just around the corner. Circumstances have forced us to modify that belief—the technical problems involved turned out to be greater than we had imagined. This must be one of the few examples in history of technological forecasting having turned out to be optimistic rather than conservative. But these technical problems will not hold us up forever.

As the production of goods and services for the consumer society requires progressively less human intervention, the economic link between production and consumption will weaken. As it stands, we reward people for productive effort by giving them the wherewithal for consumption. But when the need for work disappears, how shall we allocate the trappings of material success? What shall we use for status symbols? We shall increasingly be forced to ask ourselves what we actually need, rather than making do with those things which circumstances permit us to have. Unless, of course, we have poisoned the biosphere in the process or the disparity between the rich and poor countries of the world brings the whole edifice of our way of life toppling down about our ears. Our technology shapes our culture and our culture shapes us. What sorts of people do we wish to become? If we cannot solve these problems, we shall doubtless destroy ourselves.

> "And it ought to be remembered that there is nothing more difficult to take in hand, more perilous to conduct, or more uncertain in its success, than to take the lead in the introduction of a new order of things. Because the innovator has for enemies all those who have done well under the old conditions, and lukewarm defenders in those who may do well under the new."
>
> Niccolo Machiavelli, *The Prince* (1532)

# References


Aanonsen, A. (1964). *Shiftwork and Health.* Universitets Forlaget, Oslo

Aarås, A. (1987). Postural load and the development of musculo-skeletal illness. *Scandinavian Journal of Rehabilitation Medicine*, Supplement No. 18

Abbot, B. C., Bigland, B. and Ritchie, J. M. (1952) The physiological cost of negative work. *Journal of Physiology*, **117**, 380–390

Adams, J. G. U. (1985). *Risk and Freedom* (London: Transport and Publishing Projects)

Adams, J. G. U. (1988). Risk homeostasis and the purpose of safety regulation. *Ergonomics*, **31**, 407–428

Adams, M. A. and Hutton, W. C. (1982). Prolapsed intervertebral disc: a hyperflexion injury. *Spine*, **7**, 184–191

Adlauer, D. (1988). Shiftwork. In *Encyclopedia of Occupational Health and Safety* (Geneva: International Labour Office)

Åkerblom, B. (1948). *Standing and Sitting Posture* (Stockholm: AB Nordiska Bokhandeln)

Akerstedt, T. (1985). Adjustment of physiological circadian rhythms and the sleep–wake cycle to shiftwork. Ch. 15 in Folkard and Monk (1985b), pp. 185–198

Akerstedt, T. (1988). Sleepiness as a consequence of shiftwork. *Sleep*, **11**, 17–34

Akerstedt, T. and Torsvall, L. (1978). Experimental changes in shift schedules—their effect on well-being. *Ergonomics*, **21**, 849–856

Alderson, M. (1983). *Introduction to Epidemiology* (London: Macmillan)

Alexander, C. J. (1972). Chair-sitting and varicose veins. *The Lancet*, 15 April 1972, 822–824

Alkov, R. A. and Borowsky, M. S. (1980). A questionnaire study of psychological background factors in U.S. Navy aircraft accidents. *Aviation, Space and Environmental Medicine*, **51**, 860–863

Andersen, K. L., Masironi, R., Rutenfranz, J. and Seliger, V. (1978). *Habitual Physical Activity and Health.* World Health Organization, Regional Office for Europe, Copenhagen

Andersson, G. B. J., Murphy, R. W., Örtengren, R. and Nachemson, A. L. (1979). The influence of backrest inclination and lumbar support on lumbar lordosis. *Spine*, 4, 52–58

Andersson, G. B. J. and Örtengren, R. (1974). Lumbar disc pressure and myoelectric back muscle activity during sitting. II. Studies on an office chair. *Scandinavian Journal of Rehabilitation Medicine*, 3, 115–121

Andersson, G. B. J., Örtengren, R. and Nachemson, A. (1976). Quantitative studies of back loads in lifting. *Spine*, **1**, 178–185

Andersson, G. B. J., Örtengren, R. and Nachemson, A. (1977). Intradiscal pressure, intra-abdominal pressure and myoelectric back muscle activity related to posture and loading. *Orthopaedic Clinics of North America*, **8**, 85–96

Andersson, G. B. J., Örtengren, R., Nachemson, A. and Elfstrom, G. (1974). Lumbar disc pressure and myoelectric back, muscle activity during sitting. I. Studies on an experimental chair. *Scandinavian Journal of Rehabilitation Madicine*, **3**, 104–114

Andersson, G. B. J. and Pope, M. H. (1984). Occupational low back pain—the patients. Ch. 7 in Pope *et al.* (1984), pp. 137–156

Angle, J. and Wissman, D. A. (1980). The epidemiology of myopia. *American Journal of Epidemiology*, **111**, 220–228

Annett, J. and Sparrow, J. (1985). Transfer of training: a review of research and practical implications. *Programmed Learning and Educational Technology*, **22**, 116–124

Armstrong, T. J. and Chaffin, D. B. (1979). Carpal tunnel syndrome and selected personal attributes. *Journal of Occupational Medicine*, **21**, 481–484

Armstrong, T. J., Foulke, J. A., Bradley, J. S. and Goldstein, S. A. (1982). Investigation of cumulative trauma disorders in a poultry processing plant. *American Industrial Hygiene Association Journal*, **43**, 103–116

Armstrong, T. J., Radwin, R. G., Hansen, D. J. and Kennedy, K. W. (1986). Repetitive trauma disorders: job evaluation and design. *Human Factors*, **28**, 325–336

Asmussen, E. (1981). Similarities and dissimilarities between static and dynamic exercise. *Circulation Research*, **48**, Suppl. 1, 3–10

Asmussen, E. and Heebøll-Nielsen, K. (1962). Isometric muscle strength in relation to age in men and women. *Ergonomics*, **5**, 167–176

Asmussen, E. and Klausen, K. (1962). Form and function of the erect human spine. *Clinical Orthopaedics*, **25**, 55–63

Asmussen, E. and Mazin, B. (1978a). Recuperation after muscular fatigue by "diverting activities". *European Journal of Applied Physiology*, **38**, 1–7

Asmussen, E. and Mazin, B. (1978b). A central nervous component in local muscular fatigue. *European Journal of Applied Physiology*, **38**, 9–15

Asmussen, E. and Sørensen, N. (1971). The "wind up" movement in athletics. *Travail Humain*, **34**, 147–155

Åstrand, I. (1960). Aerobic work capacity in men and women with special reference to age. *Acta Physiologica Scandinavica*, **49**, (Suppl. 169)

Åstrand, I. (1967). Degree of strain during building work as related to individual aerobic capacity. *Ergonomics*, **10**, 293–303

Åstrand, I., Guharay, A. and Wahren, J. (1968). Circulatory responses to arm exercise with different arm positions. *Journal of Applied Physiology*, **25**, 528–532

Åstrand, P.-O. and Rodahl, K. (1986). *Textbook of Work Physiology—Physiological Basis of Exercise*, 3rd edition (New York: McGraw-Hill)

Atherton, J., Clarke, A. K. and Harrison, R. A. (1982). *Office Seating for the Arthritic and Low Back Patients*. Royal National Hospital for Rheumatic Diseases, Bath

Auliciems, A. (1989). Thermal comfort. Ch. 1 in *Building Design and Human Performance*, ed. N. C. Ruck (New York: Van Nostrand Reinhold)

Bailey, R. (1983). *Human Error in Computer Systems* (Englewood Cliffs, N. J.: Prentice-Hall)

Ballantine, M. (1983). Well how do children learn population stereotypes? In *Proceedings of the Ergonomics Society's Conference 1983*, ed. K. Coombes (London: Taylor and Francis), p. 1

Barber, P. (1987). *Applied Cognitive Psychology. An Information Processing Framework* (London: Methuen)

Bardana, E. J., Montanaro, A. and O'Hollaren, M. T. (1988). Building-related illness—A review of the available scientific data. *Clinical Reviews in Allergy*, **6**, 61–89

Barnard, P. J., Hammond, N. V., Morton, J., Long, J. B. and Clark, I. A. (1981). Consistency and compatibility in human–computer dialogue *International Journal of Man–Machine Studies*, **15**, 87–134

Barnes, R. M. (1963). *Motion and Time Study* (New York: Wiley)

Basmajian, J. V. and De Luca (1985). *Muscles Alive—Their Functions Revealed by Electromyography*, 5th edition (Baltimore: Williams and Wilkins)

Battié, M. C., Bigos, S. J., Fisher, L. D., Hansson, T., Nachemson, A. L., Spengler, D. M., Wortley, M. D. and Zeh, J. (1989). A prospective study of cardiovascular risk factors and fitness in industrial back pain complaints. *Spine*, **14**, 141–147

Baty, D. and Stubbs, D. A. (1987). Postural stress in geriatric nursing. *International Journal of Nursing Studies*, **24**, 339–344

Bauer, D. and Cavonius, C. R. (1980). Improving the legibility of visual display units through contrast reversal. In Grandjean and Vigliani (1980), pp. 137–142

Baxter, C. E. (1987). *Low Back Pain and Time of Day: a Study of the Effects on Psychophysical Performance*. PhD Thesis, University of Liverpool

Bear, J. C. and Richler, A. (1982). Re: Environmental influences on ocular refraction. *American Journal of Epidemiology*, **115**, 138–139

Bedale, E. M. (1924). Comparison of the energy expenditure of a woman carrying loads in eight different positions. In *The Effects of Posture and Rest in Muscular Work*, Report No. 29, Industrial Fatigue Research Board (London: HMSO)

Bendix, T. and Biering-Sørensen, F. (1983). Posture of the trunk when sitting on forward inclining seats. *Scandinavian Journal of Rehabilitation Medicine*, **15**, 197–203

Bendix, A., Jensen, C. V. and Bendix, T. (1988). Posture, acceptability and energy expenditure on a tiltable and knee-support chair. *Clinical Biomechanics*, **3**, 66–73

Bennet, R. M. (Ed.). (1986). The fibrositis/fibromyalgia syndrome. Current issues and perspectives. *American Journal of Medicine*, **81**, Suppl. 3A, 1–105

Bennet, R. M. (1987). Fibromyalgia. *Journal of the American Medical Association*, **257**, 2802–2803

Bennett, C. (1977). *Spaces for People—Human Factors in Design* (Englewood Cliffs, N.J.: Prentice-Hall)

Benson, H. (1975). *The Relaxation Response* (New York: William Morrow)

Bergh, H., Thorstensson, A., Sjödin, B., Hulten, B., Piehl, K. and Karlsson, J. (1978). Maximal oxygen uptake and muscle fibre types in trained and untrained humans. *Medicine and Science in Sports*, **10**, 151–154

Bergquist-Ullman, M. and Larsson, U. (1977). Acute low back pain in industry. *Acta Orthopaedica Scandinavica*, Suppl. 170

Bergqvist, V. O. (1984). Video display terminals and health: a technical and medical appraisal of the state of the art. *Scandinavian Journal of Work Environment and Health*, **10**, Suppl. 2, 1–87

von Bertalanffy, L. (1968). *General System Theory—Foundations, Development, Applications* (London: Penguin, 1973)

Bhatia, N. and Murrell, K. F. H. (1969). An industrial experiment in organized rest pauses. *Human Factors*, **11**, 167–172

Biering-Sørensen, F. (1982). Low back trouble in a population of 30-, 40-, 50- and 60-year-old men and women. *Danish Medical Bulletin*, **29**, 289–298

Biering-Sørensen, F. (1983a). A prospective study of low back pain in a general population. 1. Occurrence, recurrence and aetiology. *Scandinavian Journal of Rehabilitation Medicine*, **15**, 71–79

Biering-Sørensen, F. (1983b). A prospective study of low back pain in a general population. 2. Location, character, aggravating and relieving factors. *Scandinavian Journal of Rehabilitation Medicine*, **15**, 81–88

Biering-Sørensen, F. (1983c). Physical measurements as risk indicators for low back trouble over a one-year period. *Spine*, **9**, 106–119

Bigland-Ritchie, B. and Woods, J. J. (1976). Integrated electromyogram and oxygen uptake during positive and negative work. *Journal of Physiology*, **260**, 267–277

Bills, A. G. (1931). Blocking: a new principle of mental fatigue. *American Journal of Psychology*, **43**, 230–239

Birkbeck, M. and Beer, T. (1975). Occupation in relation to the carpal tunnel syndrome. *Rheumatology and Rehabilitation*, **14**, 218–221

Bjerner, B., Holm, A. and Swensson, A. (1955). Diurnal variation in mental performance. *British Journal of Industrial Medicine*, **12**, 103–110

Björksen, M. and Jonsson, B. (1977). Endurance limit of force in long term intermittent static contractions. *Scandinavian Journal of Work Environment and Health*, **3**, 23–27

Blake, M. J. F. (1971). Temperament and time of day. In *Biological Rhythms and Human Performance*, ed. W. P. Colquhoun (London: Academic Press), pp. 109–148

Bleed, A. S., Bleed, P., Cochran, D. J. and Riley, M. W. (1982). A performance comparison of Japanese and American handsaws. *Proceedings of the Human Factors Society*, **26**, 403–407

Bogduk, N. (1986). The anatomy of pathophysiology of whiplash. *Clinical Biochemechanics*, **1**, 92–101

Bohle, P. and Tilley, A. J. (1989). The impact of night work on psychological well-being. *Ergonomics*, **32**, 1089–1099

Böje, O. (1944). Energy production, pulmonary ventilation and length of step in well-trained runners working on a treadmill. *Acta Physiologica Scandinavica*, **7**, 362–371

Boyce, P. R. (1981). *Human Factors in Lighting* (Barking, Essex: Applied Science Publishers)

Branton, P. (1988). In praise of ergonomics—a personal perspective. *International Reviews of Ergonomics*, **1**, 1–20

Branton, P. and Grayson, G. (1967). An evaluation of train seats by observation of sitting behaviour. *Ergonomics*, **10**, 35–51

Bremner, J. M., Lawrence, J. S. and Miall, W. E. (1968). Degenerative joint disease in a Jamaican rural population. *Archives of Rheumatic Diseases*, **27**, 326–332

Broadbent, D. E. (1971). *Decision and Stress* (London: Academic Press)

Brooke, J. D., Toogood, S., Green, L. F. and Bagley, R. (1973a). Dietary pattern of carbohydrate provision and accident incidence in foundrymen. *Proceedings of the Nutrition Society*, **32**, 45A

Brooke, J. D., Toogood, S., Green, L. F. and Bagley, R. (1973b). Factory accidents and carbohydrate supplements. *Proceedings of the Nutrition Society*, **32**, 94A

Brown, I. D. (1982). Exposure and experience are a confounded nuisance in research on driver behaviour. *Accident Analysis and Prevention*, **14**, 345–352

Brown, J. R. (1972). *Manual Lifting and Related Fields—An Annotated Bibliography*. Labour Safety Council of Ontario

Brown, S. (1983). *Everyday Objects—Ergonomics and Evaluation* (Milton Keynes: The Open University Press)

Brunner, D., Manelis, G., Moldan, M. and Levin, S. (1974). Physical activity at work and the incidence of myocardial infarction, angina pectoris and death due to ischaemic heart disease: An epidemiological study in Israeli collective settlements (kibbutzim). *Journal of Chronic Diseases*, **12**, 217–233

Brunswic, M. (1981). *How Seat Design and Task Affect the Posture of the Spine in Unsupported Sitting*. MSc Dissertation, Ergonomics Unit, UCL, 26 Bedford Way, London, WC1

BS 5619 (1978). *Code of Practice for Design of Housing for Convenience of Disabled People*

BS 5810 (1979). *Code of Practice for Access for the Disabled to Buildings*

BS 5940 (1980). *Office Furniture. Part 1 (1980): Design and Dimensions of Office Workstations, Desks, Tables and Chairs*

BS 6222: Part 1 (1982). *Domestic Kitchen Equipment Specification for Co-ordinating Dimensions*

BS 6841 (1987). *British Standard Guide to Measurement and Evaluation of Human Exposure to Whole-body Mechanical Vibration and Repeated Shock*

BS 6842 (1987). *British Standard Guide to Measurement and Evaluation of Human Exposure to Vibration Transmitted to the Hand*

BS 8206 (1985). *Lighting for Buildings. Part 1 (1985): Code of Practice for Artificial Lighting*

Buckle, P. W. (1983). *A Multi-disciplinary Investigation of Factors Associated with Low-back Pain*. PhD Thesis, Cranfield Institute of Technology

Buckle, P. W., Kember, P. A., Wood, A. D. and Wood, S. N. (1980). Factors influencing occupational back pain in Bedfordshire. *Spine*, **5**, 254–258

Buckle, P. W. and Stubbs, D. A. (1990). Epidemiological aspects of musculoskeletal disorders of the shoulder and upper limbs. In *Contemporary Ergonomics*, 1990, ed. E. J. Lovesey (London: Taylor and Francis), pp. 75–78

Buckle, P. W., Stubbs, D. A. and Baty, D. (1986). Musculo-skeletal disorders (and discomfort) and associated work factors. In Corlett *et al.* (1986)

Buell, T. and Breslow, J. (1960). Mortality from coronary heart disease in California in men who work long hours. *Journal of Chronic Diseases*, **2**, 615–623

Burge, S., Hedge, A., Wilson, S., Bass, J. H. and Robertson, A. (1987). Sick building syndrome: a study of 4373 office workers. *Annals of Occupational Hygiene*, **31**, 493–504

Burton, A. K., Tillotson, K. M. and Troup, J. D. G. (1989a). Variation in lumbar saggital mobility with low-back trouble. *Spine*, **14**, 584–590

Burton, A. K., Tillotson, K. M. and Troup, J. D. G. (1989b). Prediction of low-back trouble frequency in a working population. *Spine*, **14**, 939–946

Cady, L. D., Bischoff, D. P., O'Connell, E. R., Thomas, P. C. and Allan, J. H. (1979). Strength and fitness and subsequent back injuries of firefighters. *Journal of Occupational Medicine*, **21**, 269–272

Caillet, R. (1968). *Low Back Pain Syndrome* (Philadelphia: F. A. Davis)

Cakir, A., Hart, J. J. and Stewart, T. F. M. (1980). *Visual Display Terminals* (Chichester: Wiley)

Caldwell, L. S. (1963). Relative muscle loading and endurance. *Journal of Engineering Psychology*, **2**, 155–161

Cannon, L. J., Bernacki, E. J. and Walter, S. D. (1981). Personal and occupational factors associated with carpal tunnel syndrome. *Journal of Occupational Medicine*, **23**, 255–258

Cannon, W. B. (1929). *Bodily Changes in Pain, Hunger, Fear and Rage* (Boston: Branford)

Carpentier, J. and Cazamian, P. (1977). *Night Work—Its Effects on the Health and Welfare of the Worker*. (Geneva: International Labour Office)

Castle, P. F. C. (1956). Accidents, absence and withdrawal from the work situation. *Human Relations*, **9**, 223–233

Cavagna, G. A., Dusman, E. and Margaria, R. (1968). Positive work done by a previously stretched muscle. *Journal of Applied Physiology*, **24**, 21–32

Cavanagh, P. R. and Kram, R. (1985). The efficiency of human movement—a statement of the problem. *Medicine and Science in Sports*, **17**, 304–308

Chaffin, D. B. (1974). Human strength capability and low back pain. *Journal of Occupational Medicine*, **16**, 248–254

Chaffin, D. B. and Andersson, G. B. J. (1984). *Occupational Biomechanics* (New York: Wiley)

Chaffin, D. B., Gallay, L. S., Woolley, C. B. and Kuciemba, S. R. (1986). An evaluation of the effect of a training program on worker lifting postures. *International Journal of Industrial Ergonomics*, **1**, 127–136

Chaffin, D. B., Herrin, G. D. and Keyserling, W. M. (1978). Preemployment strength testing—an updated position. *Journal of Occupational Medicine*, **20**, 403–408

Chaffin, D. B. and Park, K. S. (1973). A longitudinal study of low back pain as associated with occupational weight lifting factors. *American Industrial Hygiene Journal*, **34**, 513–525

Chapanis, A. (1988). 'Words, words, words' revisited. *International Reviews of Ergonomics*, **2**, 1–30

Chirico, A. M. and Stunkard, A. J. (1960). Physical activity and human obesity. *New England Journal of Medicine*, **263**, 935–941

CIE (1975). *Guide on Interior Lighting*. CIE Publication 29, Commission International de l'Éclairage

Ciriello, V. M., Snook, S. H., Blick, A. C. and Wilkinson, P. L. (1990). The effects of task duration on psychophysically-determined maximum acceptable weights and forces. *Ergonomics*, **33**, 187–200

Cleland, L. G. (1987). 'RSI': a model of social iatrogenesis. *Medical Journal of Australia*, **147**, 236–239

Cobb, C. R., de Vries, H. A., Urban, R. T., Leukens, C. A. and Bagg, R. J. (1975). Electrical activity in muscle pain. *American Journal of Physical Medicine*, **54**, 80–87

Coermann, R. R. (1962). The mechanical impedance of the human body in sitting and standing position at low frequencies. *Human Factors*, **4**, 227–253

Coermann, R. R. (1970). Mechanical vibrations. In *Ergonomics and Physical Environmental Factors*. Occupational Safety and Health Series No. 21 (Geneva: International Labour Office), pp. 17–41

Cohen, L. G. and Hallett, M. (1988). Hand cramps: clinical features and electromyographic patterns in a focal dystonia. *Neurology*, **38**, 1005–1012

Colquhoun, W. P. (1971). Circadian variation in mental efficiency. In *Biological Rhythms and Human Performance*, ed. W. P. Colquhoun (London: Academic Press), pp. 39–107

Colquhoun, W. P., Blake, M. J. J. and Edwards, R. S. (1986). Experimental studies of shift work I: A comparison of "rotating" and "stabilised" 4-hour shift systems. *Ergonomics*, **11**, 527–546

Conrad, R. and Hull, A. J. (1968). The preferred layout for numeral data entry keysets. *Ergonomics*, **11**, 165–174

Cooper, C. L., Cooper, R. D. and Eaker, L. H. (1988). *Living with Stress* (London: Penguin)

Corlett, E. N. (1978). The human body at work. *Management Sciences*, **22**, 20–53

Corlett, E. N. (1988). Cost-benefit analysis of ergonomic and work design changes. *International Reviews of Ergonomics*, **2**, 85–104

Corlett, E. N. and Bishop, R. P. (1978). The ergonomics of spot welders. *Applied Ergonomics*, **9**, 23–37

Corlett, E. N., Wilson, J. R. and Marenica, I. (1986). *Ergonomics of Working Postures* (London: Taylor and Francis)

Corrigan, D. L., Foley, V. and Widule, C. J. (1981). Axe use efficiency—a work theory explanation of an historical trend. *Ergonomics*, **24**, 103–109

Cox, M., Shephard, R. J. and Corey, P. (1981). Influence of an employee fitness programme upon fitness, productivity and absenteeism. *Ergonomics*, **24**, 795–806

Cox, T. (1978). *Stress* (London: Macmillan)

Cox, T. (1985). The nature and management of stress. *Ergonomics*, **28**, 1155–1163

Cox, T. (1987). Stress, coping and problem solving. *Work and Stress*, **1**, 5–14

Craig, A. and Richardson, E. (1989) Effects of experimental and habitual lunch size on performance, arousal, hunger and mood. *International Archives of Occupational and Environmental Health*, **61**, 313–319

Crisp, A. H. and Moldofsky, H. (1965). A psychosomatic study of writers' cramp. *British Journal of Psychiatry*, **111**, 841–858

Cruickshank, J. M., Gorlin, R. and Jennett, B. (1988). Air travel and thrombotic episodes: the economy class syndrome. *The Lancet*, 27 August, 497–498

Dainoff, M. J. and Dainoff, M. H. (1986). *People and Productivity—A Manager's Guide to Ergonomics in the Modern Office* (Toronto: Holt, Reinhardt and Winston)

Dalton, K. (1960). Menstruation and accidents. *British Medical Journal*, **2**, 1426–1427

Damkot, D. K., Pope, M. H., Lord, J. and Frymeyor, J. W. (1984). The relationship between work history, work environment and low back pain in men. *Spine*, **9**, 395–399

Damon, A. D. and Stoudt, H. W. (1963). The functional anthropometry of old men. *Human Factors*, **5**, 485–491

Damon, A., Stoudt, H. W. and McFarland, R. A. (1971). *The Human Body in Equipment Design* (Cambridge, Mass.: Harvard University Press)

David, G. C. (1987). Intra-abdominal measurements and load capacities for females. *Ergonomics*, **28**, 345–358

Davies, B. T. (1976). Training in manual handling and lifting. In *NIOSH Report on Safety In Manual Materials Handling, International Symposium* (National Institute of Occupational Safety and Health, USA)

Davis, P. R. (1959). The causation of herniae by weight lifting. *Lancet*, **2**, 155–157

Davis, P. R. and Sheppard, N. J. (1980). Pattern of accident distribution in the telecommunications industry. *British Journal of Industrial Medicine*, **37**, 175–179

Davis, P. R. and Stubbs, D. A. (1977, 1978). Safe levels of manual forces for young males. *Applied Ergonomics*, **8**, 141–150, 219–228, **9**, 33–37

Davis, P. R. and Troup, J. D. G. (1964). Pressures in the trunk cavities when lifting, pushing and pulling. *Ergonomics*, **7**, 465–474

Davis, P. R. and Troup, J. D. G. (1965). Effects on the trunk of handling heavy loads in different postures. In *Proceedings of the 2nd Conference of the International Ergonomics Association, Dortmund* (London: Taylor and Francis)

de Matteo, B. (1985). *Terminal Shock: The Health Hazards of Visual Display Terminals* (NC Press)

Dempsey, C. A. (1963). The design of body restraint systems. Ch. 10 in *Human Factors in Technology*, ed. E. Bennet, J. Degan and J. Spiegal (New York: McGraw-Hill), pp. 170–189

Dennet, X. and Fry, H. J. H. (1988). Overuse syndrome: a muscle biopsy study. *Lancet*, **i**, 905–908

de Vries, H. A. (1961). Prevention of muscular distress after exercise. *Research Quarterly AAHPER*, **32**, 177–185

Dillon, J. (1980). The role of ergonomics in the development of performance tests for furniture. *Applied Ergonomics*, **12**, 169–175

Dixon, N. F. (1976). *On the Psychology of Military Incompetence* (London: Jonathan Cape)

DoT (1987). *mv Herald of Free Enterprise.* Report of Court No. 8074, Department of Transport (London: HMSO)

Drillis, R. W. (1963). Folk norms and biomechanics. *Human Factors*, **5**, 427–441

Drinkall, J. N., Porter, R. W., Hibbert, C. S. and Evans, C. (1984). The value of ultrasonic measurement of the spinal canal in general practice. *British Medical Journal*, **288**, 121–122

Drury, C. G. and Francher, M. (1985). Evaluation of a forward-sloping chair. *Applied Ergonomics*, **16**, 41–47

Ducharme, R. E. (1975). Problem tools for women. *Industrial Engineering*, **7**, 46–50

Ducharme, R. E. (1977). Women workers rate male tools inadequate. *Human Factors Society Bulletin*, **20**, 1–2

Dupuis, H. and Zerlett, G. (1986). *The Effects of Whole-body Vibration* (Berlin: Springer-Verlag)

Durnin, J. V. G. A. and Passmore, R. (1967). *Energy, Work and Leisure* (London: Heinemann)

Dvorak, A., Merrick, N. F., Dealey, W. L. and Ford, G. C. (1936). *Typewriting Behaviour: Psychology Applied to Teaching and Learning Typewriting* (New York: American Book Co.)

Edholm, O. G. (1967). *The Biology of Work* (London: Weidenfeld and Nicholson)

Edwards, E. A. and Duntley, S. Q. (1939). The pigments and color of living human skin. *American Journal of Anatomy*, **65**, 1–33

Ehnström, G. (1981). Cited in Hagberg, M. (1981)

Eklund, J. A. E. (1988). Organization of assembly work: recent Swedish examples. In *Contemporary Ergonomics 1988*, ed. E. D. Megaw (London: Taylor and Francis), pp. 351–356

Engels, F. (1845). *The Condition of the Working Class in England* (Harmondsworth, Middlesex: Penguin, 1987)

English, C. J., Maclaren, W. M., Court-Brown, C., Hughes, S. P. F., Porter, R. W., Wallace, W. A., Graves, R. J., Pethick, A. J. and Soutar, C. A. (1989). *Clinical Epidemiological Study of Relations between Upper Limb Soft Tissue Disorders and Repetitive Movements at Work.* Report No. TM/88/19, Institute of Occupational Medicine, Edinburgh

Ericson, A. and Kallen, B. (1986a). An epidemiological study of work with video screens and pregnancy outcome. I. A register study. *American Journal of Industrial Medicine*, **9**, 447–457

Ericson, A. and Kallen, B. (1986b). An epidemiological study of work with video screens and pregnancy outcome. II. A case control study. *American Journal of Industrial Medicine*, **9**, 459–475

Fahrni, W. H. (1975). Conservative treatment of lumbar disc degeneration: our primary responsibility. *Orthopaedic Clinics of North America*, **6**, 93–103

Falk, B and Aarnio, P. (1983). Left-sided carpal tunnel syndrome in butchers. *Scandinavian Journal of Work Environment and Health*, **9**, 291–296

Fanger, P. O. (1973). *Thermal Comfort* (New York: McGraw-Hill)

Ferguson, D. A. (1987). 'RSI': putting the epidemic to rest. *Medical Journal of Australia*, **147**, 213–214

Finnegan, M. J., Pickering, C. A. and Burge, B. S. (1984). The sick building syndrome: prevalence studies. *British Medical Journal*, **289**, 1573–1575

Fisher, G. H., Brennan, M. G. and Binnall, D. H. J. (1978). After the children of

Benjamin: the increasing incidence of left handedness. Unpublished paper presented to the Ergonomics Society

Fitts, P. M. and Jones, R. E. (1947a). *Analysis of Factors Contributing to 460 "Pilot-Error" Experiences in Operating Aircraft Controls*. Report of Aeronautical Laboratory, Wright Patterson Air Force Base, Dayton, Ohio. Reprinted in *Selected Papers in the Design and Use of Control Systems*, ed. W. Sinako (New York: Dover)

Fitts, P. M. and Jones, R. E. (1947b). *Psychological Aspects of Instrument Displays*. Report of Aeronautical Laboratory, Wright Patterson Air Force Base, Dayton, Ohio. Reprinted in *Selected Papers in the Design and Use of Control Systems*, ed. W. Sinako (New York: Dover)

Fitzgerald, J. G. (1973). *The IAM Type Aircrew Lumbar Support: Fitting Manufacture and Use*. Aircrew Equipment Group Report No. 304; Institute of Aviation Medicine, Farnborough, Hampshire

Floyd, W. F. and Silver P. H. S. (1950). Electromyographic study of patterns of activity of the anterior abdominal wall muscles in man. *Journal of Anatomy*, **84**, 132–145

Floyd, W. F. and Silver, P. H. S. (1955). The function of the erectores spinae muscles in certain movements in postures in man. *Journal of Physiology*, **129**, 184–203

Folkard, S. and Monk, T. H. (1985a). Circadian performance rhythms. In Folkard and Monk (1985b), pp. 37–52

Folkard, S. and Monk, T. H. (Eds.) (1985b). *Hours of Work—Temporal Factors in Work-scheduling* (Chichester: Wiley)

Ford, H. (1922). *My Life and Work* (New York: Doubleday Page)

Fox, B. H. (1978). Premorbid psychological factors as related to cancer incidence. *Journal of Behavioural Medicine*, **1**, 1–12

Fox, E. L. and Matthews, D. K. (1981). *The Physiological Basis of Physical Education and Athletics*, 3rd edition (Philadelphia: Holt-Saunders)

Frankenhauser, M., Nordheden, B., Myrsten, A. L. and Post, B. (1971). Psychophysiological reactions to understimulation and overstimulation. *Acta Psychologica*, **35**, 298–308

Fraser, T. M. (1980). *Ergonomic Principles in the Design of Hand Tools*. Occupational Safety and Health Series, No. 44. (Geneva: International Labour Office)

Fraser, T. M. (1989). *The Worker at Work* (London: Taylor and Francis)

Freivalds, A. (1986a). The ergonomics of shovelling and shovel design—a review of the literature. *Ergonomics*, **29**, 3–18

Freivalds, A. (1986b). The ergonomics of shovelling and shovel design—an experimental study. *Ergonomics*, **29**, 19–30

Freivalds, A. (1987). The ergonomics of tools. *International Reviews of Ergonomics*, **1**, 43–76

Frese, M. and Semmer, N. (1986). Shiftwork, stress and psychosomatic complaints: a comparison between workers in different shifwork schedules, non-shiftworkers and former shiftworkers. *Ergonomics*, **29**, 99–114

Freud, S. (1901). *The Psychopathology of Everyday Life* (Harmondsworth, Middlesex: Penguin, 1975)

Friedman, M. and Rosenman, R. H. (1974). *Type A Behaviour and Your Heart* (New York: Knopf)

Fry, H. J. H. (1986). Overuse syndrome in musicians: prevention and management. *Lancet*, 27 September, 728–731

Frymoyer, J. W. (1988). Back pain and sciatica. *New England Journal of Medicine*, **318**, 219–300

Frymoyer, J. W., Pope, M. H., Clements, J. H., Wilder, D. G., MacPherson, B. and Ashikaga, T. (1983). Risk factors in low-back pain. *Journal of Bone and Joint Surgery*, **65A**, 213–218

Frymoyer, J. W., Pope, M. H., Constanza, M. C., Rosen, J. C. Coggin, J. E. and Wilder, D. G. (1980). Epidemiologic studies of low-back pain. *Spine*, **5**, 419–423

Garabant, D. H., Peters, J. M., Mack, T. M. and Bernstein, L. (1984). Job activity and colon cancer risk. *American Journal of Epidemiology*, **119**, 1005–1014

Garg, A. and Herrin, G. D. (1979). Stoop or squat: a biomechanical and metabolic evaluation. *AIIE Transactions*, **11**, 293–302

Garg, A. and Saxena, U. (1979). Effects of lifting frequency and techniques on physical fatigue with special reference to psychophysical methodology and metabolic rate. *American Industrial Hygiene Association Journal*, **40**, 894–903

Gerhardsson, M., Norell, S., Kiviranta, H., Pedersen, N. L. and Ahlbom, A. (1986). Sedentary jobs and colon cancer. *American Journal of Epidemiology*, **123**, 775–780

Glover, J. R. (1960). Back pain and hyperaesthesia. *Lancet*, 28 May, 1165–1169

Goldenberg, D. L. (1986). Fibromyalgia syndrome. An emerging but controversial condition. *Journal of the American Medical Association*, **257**, 2782–2787

Goldhaber, M. K., Polen, M. R. and Hiatt, R. A. (1988). The risk of miscarriage and birth defects among women who use visual display terminals during pregnancy. *American Journal of Industrial Medicine*, **13**, 695–706

Goldsmith, R. and Hale, T. (1971). Relationship between habitual physical activity and physical fitness. *American Journal of Clinical Nutrition*, **24**, 1489–1493

Goldsmith, S. (1976). *Designing for the Disabled*, 3rd edition (London: RIBA)

Goldthwaite, J. E. (1934). *Body Mechanics—In Health and Disease* (Lippincott)

Gollnick, P., Armstrong, R., Sembrowich, W., Shepherd, R. and Saltin, B. (1973). Glycogen depletion pattern in human skeletal muscle fiber after heavy exercise. *Journal of Applied Physiology*, **34**, 615–618

Goodenough, D. R. (1976). A review of individual differences in field dependence as a factor in auto safety. *Human Factors*, **18**, 53–62

Goodman, W. L. (1964). *The History of Woodworking Tools* (London: Bell and Hyman)

Gordon, A. M., Huxley, A. F. and Julian, F. (1966). The variation in isometric tension with sarcomere length in vertebrate muscle fibres. *Journal of Physiology*, **184**, 170–192

Gore, A. and Tasker, D. (1986). *Pause Gymnastics. Improving Comfort and Health at Work* (Sydney: CCH Australia Ltd)

Goulet, L. and Thériault, G. (1987). Association between spontaneous abortion and ergonomic factors. A literature review of the epidemiological evidence. *Scandinavian Journal of Work Environment and Health*, **13**, 399–403

Gowers, W. R. (1888). *A Manual of Diseases of the Nervous System* (London: Churchill)

Grandjean, E. (1973). *Ergonomics of the Home* (London: Taylor and Francis)

Grandjean, E. (Ed.) (1984). *Ergonomics and Health in Modern Offices* (London: Taylor and Francis)

Grandjean, E. (1987). *Ergonomics in Computerized Offices* (London: Taylor and Francis)

Grandjean, E. (1988). *Fitting the Task to the Man—a Textbook of Occupational Ergonomics* (London: Taylor and Francis)

Grandjean, E. and Burandt, H. U. (1962). Das sitzverhalten von Buroangestellten. *Industrielle Organisation*, **31**, 243–50. Cited in Grandjean (1987, 1988)

Grandjean, E., Hünting, W. and Piderman, M. (1983). VDT workstation design: preferred settings and their effects. *Human Factors*, **25**, 161–175

Grandjean, E. and Vigliani F. (Eds.) (1980). *Ergonomic Aspects of Visual Display Terminals* (London: Taylor and Francis)

Green, R. (1989). Human error on the flight deck. Paper presented at symposium on *Human Factors in High Risk Systems*. Royal Society, London

Greenwood, M. and Woods, H. M. (1919). *The Incidence of Industrial Accidents upon Individuals, with Special Reference to Multiple Accidents*. Industrial Fatigue Research Board Report No. 4, London

Grieco, A. (1986). Sitting posture: an old problem and a new one. *Ergonomics*, **29**, 345–362

Grieve, D. W. and Pheasant, S. T. (1982). Biomechanics. Ch. 3 in *The Body at Work*, ed. W. T. Singleton (Cambridge: Cambridge University Press), pp. 71–200

Griffin, M. J. (1978). The evaluation of vehicle vibration and seats. *Applied Ergonomics*, **9**, 15–21

Gyntelberg, F. (1974). One year incidence of low back pain among male residents of Copenhagen aged 40–59. *Danish Medical Bulletin*, **21**, 30–36

Hagberg, M. (1981). Muscular endurance and surface electromyogram in isometric and dynamic exercise. *Journal of Applied Physiology: Respiratory, Environmental and Exercise Physiology*, **51**, 1–7

Hagberg, M. (1986). Optimizing occupational muscular stress of the neck and shoulder. In Corlett *et al.* (1986), pp. 109–114

Haggard, H. W. and Greenberg, L. A. (1935). *Diet and Physical Efficiency* (Newhaven, Conn.: Yale University Press)

Haider, M., Kundi, M. and Weissenböck, M. (1980). Worker strain related to VDUs with differently coloured characters. In Grandjean and Vigliani (1980), pp. 53–64

Hale, A. R. and Hale, M. (1972) *A Review of the Industrial Accident Research Literature* (London: HMSO)

Hansson, T., Bigos, S., Beecher, P. and Wortley, M. (1985). The lumbar lordosis in acute and chronic back pain. *Spine*, **10**, 154–155

Hart, J. and Dillon, J. (1987). *Forward Sloping Seats—A Comparative Study*. Unpublished Internal Report. Post Office Research, Swindon

Hawkins, L. H. (1981). The influence of air ions, temperature and humidity on subjective well-being and comfort. *Journal of Environmental Psychology*, **1**, 279–292

Hawkins, L. H. and Barker, T. (1978). Air ions and human performance. *Ergonomics*, **21**, 273–278

Helander, M. G. and Rupp, B. A. (1984). An overview of standards and guidelines for visual display terminals. *Applied Ergonomics*, **15**, 185–195

Heliövaara, M. (1987). Body height, obesity and risk of herniated lumbar intervertebral disc. *Spine*, **12**, 469–472

Hendriksson, K. G. (1988). Muscle pain in neuromuscular disorders and primary fibromyalgia. *European Journal of Applied Physiology and Occupational Physiology*, **57**, 348–352

Herberts, P. and Kadefors, R. A. (1976). A study of painful shoulder in welders. *Acta Orthopaedica Scandinavica*, **47**, 381–387

Hersey, R. B. (1936). Emotional factors in accidents. *Personnel Journal*, **15**, 59–65

Hertzberg, H. T. E. (1955). Some contributions of applied physical anthropology to human engineering. *Annals of the New York Academy of Sciences*, **63**, 616–629

Herzberg, F. (1968). *Work and the Nature of Man* (London: Staples Press)

Herzberg, F., Mausner, B. and Snyderman, B. (1959). *The Motivation to Work* (New York: Wiley)

Hettinger, T. (1961). *Physiology of Strength* (Springfield, Ill.: Charles C. Thomas)

Hietanen, E. (1984). Cardiovascular response to static exercise. *Scandinavian Journal of Work Environment and Health*, **10**, 397–402

Hill, J. M. M. and Trist, E. L. (1953). A consideration of industrial accidents as a means of withdrawal from the work situation. *Human Relations*, **6**, 357–380

Hira, D. S. (1980). An ergonomic appraisal of educational desks. *Ergonomics*, **23**, 213–221

HMCIF (1924–1974). *Annual Report of HM Chief Inspector of Factories* (London: HMSO)

Hobday, S. W. (1985). Keyboards designed to fit hands and reduce postural stress. In *Ergonomics International 85*, ed. I. D. Brown, R. Goldsmith, K. Coomes, and M. A. Sinclair (London: Taylor and Francis), pp. 457–459

Holdstock, D. J., Misiewicz, J. J., Smith, T. and Rowlands, E. N. (1970). Propulsion (mass movements) in the human colon and its relationship to meals and somatic activity. *Gut*, **11**, 91–99

Homans, J. (1954). Thrombosis of the deep leg veins due to prolonged sitting. *New England Journal of Medicine*, 28 January, 148–149

Hopkinson, R. G. and Collins, J. B. (1970). *The Ergonomics of Lighting* (London: Macdonald Technical and Scientific)

Hornibrook, F. A. (1934). *The Culture of the Abdomen* (New York: Doubleday)

HSC (1982). *Proposals for Health and Safety (Manual Handling of Loads) Regulations and Guidance*. Health and Safety Commission Consultative Document (London: HMSO)

HSC (1988). *Handling Loads at Work—Proposals for Regulations and Guidance*. Health and Safety Commission Consultative Document (London: HMSO)
HSE (1975–1977). *Health and Safety—Manufacturing and Service Industries*. Health and Safety Executive (London: HMSO)
HSE (1978–1982). *Health and Safety Statistics*. Health and Safety Executive (London: HMSO)
HSE (1983). *Visual Display Units*. Health and Safety Executive (London: HMSO)
HSE (1989). Health and Safety Statistics 1986–1987. *Employment Gazette*, Vol. 97, No. 2, Occasional Supplement No. 1 (London: HMSO)
HSE (1990). *Noise at Work. Guidance of Regulations* (London: HMSO)
Hult, L. (1954). Cervical, dorsal and lumbar spinal syndromes. *Acta Orthopaedica Scandinavica*, Suppl. 17.
Hunter, D. (1978). *The Diseases of Occupations* (London: Hodder and Stoughton)
Hünting, W. and Grandjean, E. (1976). Sitzverhalten und subjektives Wohlbefinden auf schwenkbaren und fixierten Formsitzen. *Zeitschrift der Arbeitswissenschaft*, **30**, 161–164. Cited in Grandjean (1987, 1988)
Hünting, W., Läublit, T. and Grandjean, E. (1981). Postural and visual loads at VDT workstations, Part 1. Constrained postures. *Ergonomics*, **24**, 917–931
Hutchinson, R. C. (1954). Effect of gastric contents on mental concentration and production rate. *Journal of Applied Physiology*, **7**, 143–147
Hutton, W. C. and Adams, M. A. (1982). Can the lumbar spine be crushed in heavy lifting? *Spine*, **7**, 586–590
Ikai, M. and Steinhaus, A. H. (1961). Some factors modifying the expression of human strength. *Journal of Applied Physiology*, **16**, 157–163
ILO (1962). *Manual Lifting and Carrying*. International Occupational Safety and Health Information Sheet No. 3 (Geneva: International Labour Office)
ILO (1964). *Maximum Permissible Weight to be Carried by One Worker*. Occupational Safety and Health Series, No. 5 (Geneva: International Labour Office)
ILO (1967). *Convention Concerning the Maximum Permissible Weight to be Carried by One Worker*. Convention 127 (Geneva: International Labour Office)
ILO (1978). *Management of Working Time in Industrialised Countries*. (Geneva: International Labour Office)
ILO (1990). *Maximum Weights in Load Lifting and Carrying*. Occupational Safety and Health Series No. 59 (Geneva: International Labour Office)
Inman, V. T., Ralston, H. J. and Todd, F. (1981). *Human Walking* (Baltimore: Williams and Wilkins)
ISO 2631 (1985). *Evaluation of Human Exposure to Whole-body Vibration*
ISO 3055 (1975). *Kitchen Equipment—Co-ordinating Sizes*
ISO 7243 (1982). *Hot Environments—Estimation of the Heat Stress on Working Man, Based on the WBGT-index (Wet Bulb Globe Temperature)*
ISO 7726 (1985). *Thermal Environments—Instruments and Methods for Measuring Physical Quantities*
ISO 7730 (1984). *Moderate Thermal Environments—Determination of the PMV and PPD Indices and Specification of the Conditions for Thermal Comfort*
ISO 7933 (1989). *Hot Environments—Analytical Determination and Interpretation of Thermal Stress Using Calculation of Required Sweat Rate*
ISO 8995 (1989). *Principles of Visual Ergonomics—The Lighting of Indoor Work Systems*
Jacobson, E. (1944). *Progressive Relaxation: A Physiological and Clinical Investigation of Muscular States and their Significance in Psychology and Medical Practice* (Chicago: Chicago University Press)
James, W. (1890). *The Principles of Psychology* (New York: Henry Holt)
Janis, I. L. (1972). *Victims of Groupthink* (Boston: Houghton Mifflin)
Jaschinski-Kruza, W. (1988). Visual strain during VDU work: the effect of viewing distance and the dark focus. *Ergonomics*, **31**, 1449–1466

Johnson, L. C., Tepas, D. I., Colquhoun, W. P. and Colligan, M. J. (1981). *Biological Rhythms, Sleep and Shift Work* (New York: Spectrum)

Johnson, M. A., Polgar, J., Weightman, D. and Appleton, D. (1973). Data on distribution of fibre types in thirty-six human muscles. An autopsy study. *Journal of Neurological Science*, **18**, 111–129

Jonsson, B. (1988). The static load component in muscle work. *European Journal of Applied Physiology and Occupational Physiology*, **57**, 305–310

Jørgensen, K., Falletin, N., Krogh-Lund, C. and Jensen, B. (1988). Electromyography and fatigue during prolonged low level contractions. *European Journal of Applied Physiology and Occupational Physiology*, **57**, 316–321

Joseph, J. (1960). *Man's Posture: Electromyographic Studies* (Springfield, Ill.: Charles C. Thomas)

Kamon, E. and Belding, H. S. (1971). The physiological cost of carrying loads in temperate and hot environments. *Human Factors*, **13**, 153–161

Kapanji, I. A. (1974). *The Physiology of the Joints* (Edinburgh: Churchill Livingstone)

Karasek, R. (1979). Job demands, job decision latitude and mental strain: implications for job redesign. *Administrative Science Quarterly*, **24**, 285–306

Karasek, R., Baker, D., Marxer, F., Ahlbom, A. and Theorell, T. (1981). Job decision latitude, job demands and cardiovascular disease: a prospective study of Swedish men. *American Journal of Public Health*, **71**, 694–705

Karvonen, M. J. (1984). Physical activity and cardiovascular mortality. *Scandinavian Journal of Work Environment and Health*, **10**, 389–395

Kasl, S. V., Evans, A. F. and Niederman, J. C. (1979). Psychosocial risk factors in the development of mononucleosis. *Psychosomatic Medicine*, **41**, 455–466

Kay-Shuttleworth, J. (1832). *The Moral and Physical Condition of the Working Class Employed in the Cotton Manufacture in Manchester*

Kelsey, J. L. (1975a). An epidemiological study of herniated lumbar intervertebral discs. *Rheumatology and Rehabilitation*, **14**, 144–159

Kelsey, J. L. (1975b). An epidemiological study of the relationship between occupations and acute herniated lumbar intervertebral disks. *International Journal of Epidemiology*, **4**, 197–205

Kelsey, J. L., Githens, P. B., O'Connor, T., Weil, V., Calogero, J. A., Holford, T. R., White, A. A., Walter, S. D., Ostfield, A. M. and Southwick, W. O. (1984a). Acute prolapsed lumbar intervertebral disc. An epidemiologic study with special reference to driving automobiles and cigarette smoking. *Spine*, **9**, 608–613

Kelsey, J. L., Githens, P. B., Walter, S. D., Southwick, W. O., Weil, V., Holford, T. R., Ostfield, A. M., Calogero, J. A. and O'Connor, T. (1984b). An epidemiological study of acute prolapsed cervical intervertebral disc. *Journal of Bone and Joint Surgery*, **66A**, 907–914

Kelsey, J. L., Githens, P. B., White, A. A., Holford, T. R., Walter, S. D., O'Connor, T., Ostfield, A. M., Weil, W., Southwick, W. O. and Calogero, J. A. (1984c). An epidemiological study of lifting and twisting on the job and risk for acute prolapsed lumbar intervertebral disc. *Journal of Orthopaedic Research*, **2**, 61–66

Kerkhoff, G. (1985). Individual differences in circadian rhythms. Ch. 3 in Folkard and Monk (1985b), pp. 29–36

Kersley, G. D. (1979). Back pain: its problems and treatments. *Current Medical Research and Opinion*, **6**, Suppl. 2, 27–32

Keyserling, W. M., Herring, C. D. and Chaffin, D. B. (1980). Isometric strength testing as a means of controlling medical incidents on strenuous jobs. *Journal of Occupational Medicine*, **22**, 332–336

Klaber-Moffet, J. A., Chase, S. M., Portek, I. and Ennis, J. R. (1986). A controlled prospective study to evaluate the effectiveness of a back school in the relief of chronic low back pain. *Spine*, **11**, 120–122

Kleberg, I. G. and Ridd, J. E. (1987). An evaluation of office seating. In *Contemporary Ergonomics 1987*, ed. E. D. Megaw (London: Taylor and Francis), pp. 203–208

Kleitmann, N. (1939, 1963). *Sleep and Wakefulness* (Chicago: University of Chicago Press)

Klockenberg, E. A. (1926). *Rationalisierung der Schreibundschine und ihrer Bedienung* (Berlin: Springer-Verlag)

Knave, B. G., Wibom, R. I., Bergqvist, V. O., Carlsson, L. L. W., Levin, M. I. B. and Nylen, P. R. (1985a). Work with video display terminals among office employees II. Physical exposure factors. *Scandinavian Journal of Work Environment and Health*, **11**, 467–474

Knave, B. G., Wibom, R. I., Voss, M., Hedström, L. D. and Bergqvist, V. O. (1985b). Work with video display terminals among office employees. I. Subjective symptoms and discomfort. *Scandinavian Journal of Work Environment and Health*, **11**, 457–466

Knutsson, A., Akerstedt, T., Jonsson, B. G. and Orth-Gomer, K. (1989). Increased risk of ischaemic heart disease in shift workers. *Lancet*, 12 July, 89–92

Koller, M., Kundi, M. and Cervinka, R. (1978). Field studies of shiftwork in an Austrian oil refinery. 1: Health and psychosocial wellbeing of workers who drop out of shiftwork. *Ergonomics*, **2**, 835–847

Komi, P. V. and Buskirk, F. R. (1972). Effect of eccentric and concentric muscle conditioning on tension and electrical activity of human muscle. *Ergonomics*, **15**, 417–434

Kroemer, K. H. E. (1971). Foot operation of controls. *Ergonomics*, **14**, 333–361

Kroemer, K. H. E. (1972). Human engineering: the keyboard. *Human Factors*, **14**, 51–63

Kroemer, K. H. E. and Hill, S. G. (1986). Preferred line of sight angle. *Ergonomics*, **29**, 1129–1134

Kucera, J. D. and Robins, T. G. (1989). Relationship of cumulative trauma disorders of the upper extremity to degree of hand preference. *Journal of Occupational Medicine*, **31**, 17–22

Kukkonen, R., Luopajarvi, T. and Riihimäki, V. (1983). Prevention of fatigue amongst data entry operators. In *Ergonomics of Workstation Design*, ed. T. O. Kvålseth (London: Butterworths), pp. 28–34

Kuopajarvi, T., Kuorinka, I., Virolainen, M. and Holmberg, M. (1979). Prevalence of tenosynovitis and other injuries of the upper extremities in repetitive work. *Scandinavian Journal of Work Environment and Health*, **5**, Suppl. 3, 48–55

Kurppa, K., Holmberg, P. C., Rantala, K., Nurmine, T. and Saxén, L. (1985). Birth defects and exposure to video display terminals during pregnancy. A Finnish case-referent study. *Scandinavian Journal of Work Environment and Health*, **11**, 353–356

Kurppa, K., Waris, P. and Rokkanen M. D. (1979a). Peritendinitis and tenosynovitis—a review. *Scandinavian Journal of Work Environment and Health*, Suppl. 3, 19–24

Kurppa, K., Waris, P. and Rokkanen, P. (1979b). Tennis elbow—lateral elbow pain syndrome. *Scandinavian Journal of Work Environment and Health*, Suppl. 3, 15–18

Lance, J. W. (1969). *The Mechanism and Management of Headache* (London: Butterworths)

Lander, C., Korbon, G. A., de Good de, and Rowlingson, J. C. (1987). The Balans chair and its semi-kneeling position: an ergonomic comparison with the conventional sitting position. *Spine*, **12**, 269–272

Laporte, W. (1966). The influence of a gymnastic pause upon recovery following post office work. *Ergonomics*, **9**, 501–506

Larsson, L., Grimby, G. and Karlsson, J. (1979). Muscle strength and speed of movement in relation to age and muscle morphology. *Journal of Applied Physiology*, **46**, 451–456

Läubli, T., and Grandjean, E. (1984). The magic of control groups in VDT field studies. In Grandjean (1984), pp. 105–112

Lawrence, J. S. (1961). Rheumatism in cotton operatives. *British Journal of Industrial Medicine*, **18**, 270–276

Lazarus, R. S. (1976). *Patterns of Adjustment* (New York: McGraw-Hill)

Ledgard, H., Singer, A. and Whiteside, J. (1981). *Directions in Human Factors for Interactive Systems* (New York: Springer-Verlag)

Legg, S. J. and Myles, W. S. (1981). Maximum acceptable repetitive lifting workloads for an 8 hour workday using psychophysical and subjective rating methods. *Ergonomics*, **24**, 907–916

Lehmann, G. (1958). Physiological measurements as a basis of work organization in industry. *Ergonomics*, **1**, 328–344

Lehmann, G. (1962). *Praktische Arbeitsphysiologie* (Stuttgart: Thieme Verlag)

Leithead, C. S. and Lind, A. R. (1964). *Heat Stress and Heat Disorders* (London: Cassell)

Leskinen, T. P. J., Stålhammar, H. R., Kuorinica, J. A. A. and Troup, J. D. G. (1983). A dynamic analysis of spinal compression with different lifting techniques. *Ergonomics*, **26**, 595–604

Levi, L. (1972). Stress and distress in response to psychosocial stimuli. *Acta Medica Scandinavica*, **191**, Suppl. 528

Lewicki, R., Tchorzewski, H., Denys, A., Kowalska, M. and Golinska, A. (1987). Effects of physical exercise on some parameters of immunity in conditioned sportsmen. *International Journal of Sports Medicine*, **8**, 309–314

Liebowitz, H. W. and Owens, D. A. (1975). Anomalous myopias and the intermediate dark focus of accommodation. *Science*, **189**, 646–648

Life, M. A. and Pheasant, S. T. (1984). An integrated approach to the study of posture in keyboard operation. *Applied Ergonomics*, **15**, 83–90

Lippold, O. C. J. (1973). *The Origin of the Alpha Rhythm* (Edinburgh: Churchill Livingstone)

Lloyd, D. C. E. F. and Troup, J. D. S. (1983). Recurrent back pain and its prediction. *Journal of the Society of Occupational Medicine*, **33**, 66–74

Lloyd, P., Tarling, C., Troup, J. D. G. and Wright, B. (1987). *The Handling of Patients: A Guide for Nurses*. The Back Pain Association in collaboration with the Royal College of Nursing, London

Loveless, N. E. (1962). Direction of motion stereotypes: a review. *Ergonomics*, **5**, 357–384

Lundervold, A. (1951). Electromyographic investigations of position and manner of working in typewriting. *Acta Physiologica Scandinavica*, **24**, Suppl. 84, 1–17

Lundervold, A. (1958). Electromyographic investigations during typewriting. *Ergonomics*, **1**, 226–233

McArdle, W. D., Katch, F. I. and Katch, V. L. (1986). *Exercise Physiology—Energy, Nutrition and Human Performance* (Philadelphia: Lea and Febiger)

McClurg Anderson, T. (1951). *Human Kinetics and Analysing Body Movements* (London: Heinemann)

McCormick, E. S. (1970). *Human Factors Engineering* (New York: McGraw-Hill)

McDonald, A., Cherry, N., Delorme, C. and McDonald, J. C. (1986). Visual display units and pregnancy: evidence from the Montreal survey. *Journal of Occupational Medicine*, **28**, 1126–1131

Macdonald, E. B., Porter, R., Hibbert, C. and Hart, J. (1984). The relation between spinal canal diameter and back pain in coal miners. *Journal of the Society of Occupational Medicine*, **26**, 23–28

McGill, C. M. (1968). Industrial back problems: a control program. *Journal of Occupational Medicine*, **10**, 174–178

McGill, S. M. and Norman, R. W. (1987a). Reassessment of the role of intra-abdominal pressure in spinal compression. *Ergonomics*, **30**, 1565–1588

McGill, S. M. and Norman, R. W. (1987b). Effects of an anatomically detailed erector spinae model on L4/L5 disc compression and shear. *Journal of Biomechanics*, **20**, 591–600

Mackay, C. (1987). The alleged reproductive hazards of VDUs. *Work and Stress*, **1**, 49–58

McKenna, F. P. (1982). The human factor in driving accidents: an overview of approaches and problems. *Ergonomics*, **25**, 867–877

McKenna, F. P. (1983). Accident proneness: a conceptual analysis. *Accident Analysis and Prevention*, **15**, 65–71

McKenna, F. P. (1985). Do safety measures really work? An examination of risk homeostasis theory. *Ergonomics*, **28**, 489–498

McKenna, F. P. (1987). Behavioural compensation and safety. *Journal of Occupational Accidents*, **9**, 107–121

Maeda, K. (1977). Occupational cervicobrachial disorder and its causative factors. *Journal of Human Ergology*, **6**, 193–202

Magora, A. (1970a). Investigations of the relation between low back pain and occupation I. *Industrial Medicine*, **39** (11), 31–37

Magora, A. (1970b). Investigations of the relation between low back pain and occupation II. *Industrial Medicine*, **39** (12), 28–34

Magora, A. (1972). Investigations of the relation between low back pain and occupation III. Physical requirements: sitting, standing and weight lifting. *Industrial Medicine*, **41** (12), 5–9

Magora, A. (1973a). Investigations of the relation between low back pain and occupation IV. Physical requirements: bending, rotation, reaching and sudden maximal effort. *Scandinavian Journal of Rehabilitation Medicine*, **5**, 186–190

Magora, A. (1973b). Investigations of the relation between low back pain and occupation V. Psychological aspects. *Scandinavian Journal of Rehabilitation Medicine*, **5**, 191–196

Mairaux, Ph., Davis, P. R., Stubbs, D. B. and Baty, D. (1984). Relation between intra-abdominal pressure and lumbar movements when lifting weights in the erect position. *Ergonomics*, **27**, 883–894

Mairaux, Ph. and Malchaire, J. (1988). Relation between intra-abdominal pressure and lumbar stress: effects of trunk posture. *Ergonomics*, **9**, 1331–1342

Mandal, A. C. (1976). Work chair with tilted seat. *Ergonomics*, **19**, 157–164

Mandal, A. C. (1981). The seated man (Homo sedens). The seated work position, theory and practice. *Applied Ergonomics*, **12**, 19–26

Manning, D. P., Mitchell, P. G. and Blanchfield, L. P. (1984). Body movements and events contributing to accidental and nonaccidental back injuries. *Spine*, **9**, 734–739

Marey, E. J. (1895). *Movement*, Trans. E. Pritchard (New York: Appleton, Century, Crofts)

Martin, J., Meltzer, H. and Eliot, D. (1988). *The Prevalence of Disability Among Adults* (London: HMSO)

Maslow, A. H. (1954). *Motivation and Personality* (New York: Harper and Row)

Matthews, J. and Knight, A. A. (1971). *Ergonomics in Agricultural Equipment Design*. National Institute of Agricultural Engineering, Silsoe, Bedfordshire, UK

Mayer, J., Roy, P. and Mitra, K. P. (1956). Relation between caloric intake, body weight and physical work in an industrial male population in West Bengal. *American Journal of Clinical Nutrition*, **4**, 169–175

Mekky, S., Schilling, R. S. T. and Walford, J. (1962). Varicose veins in women cotton workers. An epidemiological study in England and Egypt. *British Medical Journal*, **2**, 591–595

Melzack, R. and Wall, P. D. (1988). *The Challenge of Pain*, 2nd edition (London: Penguin)

Miller, A. (1988). Stress on the job. *Newsweek*, 25 April, 28–33

Minors, D. S. and Waterhouse, J. M. (1985). Introduction to circadian rhythms. Ch. 1 in Folkard and Monk (1985b), pp. 1–14

Mitler, M. M., Carskadon, M. A., Czeisler, C. A., Dement, W. C., Dinges, D. F. and Graeber, R. C. (1988). Catastrophes, sleep and public policy: consensus report. *Sleep*, **11**, 100–109

Miwa, T. (1963). Evaluation method for vibration effect. Part 3. Measurements of threshold and equal sensation contours on hand for vertical and horizontal sinusoidal vibrations. *Industrial Health*, **5**, 213–220

MoD (1984). *Defence Standard. Human Factors for Designers of Equipment Part 3: Body Strength and Stamina*. MoD 00-25, Pt 3—Ministry of Defence, London

Moldofsky, H. (1986). Sleep and musculoskeletal pain. *American Journal of Medicine*, **81**, Suppl. 3A, 85–89

Monk, T. H. and Folkard, S. (1983). Circadian rhythms and shiftwork. Ch. 4 in *Stress and Fatigue in Human Performance*, ed. R. Hockey, pp. 97–122

Monk, T. H. and Folkard, S. (1985). Individual differences in shiftwork adjustment. Ch. 18 in Folkard and Monk (1985b), (Chichester: Wiley), pp. 227–238

Monod, H. and Scherrer, J. (1967). The work capacity of a synergic muscle group. *Ergonomics*, **8**, 329–333

Montoye, H. J. and Lamphier, D. E. (1977). Grip and arm strengths in males and females. *Research Quarterly AAHPER*, **48**, 109–120

Morris, J. N., Chave, S. P. W., Adam, C., Sirey, C., Epstein, L. and Sheehan, D. (1973). Vigorous exercise in leisure time and the incidence of coronary heart disease. *Lancet*, **1**, 333–339

Morris, J. N., Heady, J., Raffle, P., Roberts, C. and Parks, J. (1953). Coronary heart diesease and physical activity at work. *Lancet*, **2**, 1053–1057

Morris, J. N., Lucas, D. B. and Bresler, B. (1961). Role of the trunk in stability of the spine. *Journal of Bone and Joint Surgery*, **43A**, 327–351

Murrell, H. (1976). *Men and Machines* (London: Methuen)

Murrell, K. F. H. (1969). *Ergonomics—Man and His Working Environment* (London: Chapman and Hall)

Nachemson, A. (1966). Electromyographic studies on the vertebral portion of the psoas muscle. *Acta Orthopaedica Scandinavica*, **37**, 177–190

Nachemson, A. and Elfström, G. (1979). Intravital dynamic pressure measurements in lumbar discs. *Scandinavian Journal of Rehabilitation Medicine*, Suppl. 1

Naitoh, P. (1981). Circadian cycles and restorative power of naps. In *Johnson et al.* (1981), pp. 553–580

Nakaseko, N., Grandjean, E., Hünting, N. and Gierer, R. (1985). Studies in ergonomically designed alphanumeric keyboards. *Human Factors*, **27**, 175–187

Napier, J. R. (1956). The prehensile movements of the human hand. *Journal of Bone and Joint Surgery*, **38B**, 902–913

NBOSH (1987). *Occupational Injuries in Sweden*. National Board of Occupational Safety and Health, Swedish Work Environment Fund, Sweden

Newbold, E. M. (1926). *A Contribution to the Study of the Human Factor in the Causation of Accidents*. Industrial Health Research Board, Report No. 34, London

Newham, D. J. (1988). The consequences of eccentric contractions and their relationship to delayed onset muscle pain. *European Journal of Applied Physiology and Occupational Physiology*, **57**, 353–359

Newsholme, E. and Leech, T. (1988). Fatigue stops play. *New Scientist*, 22 September, 39–43

Nicholson, A. S. (1989). A comparative study of methods for establishing load handling capabilities. *Ergonomics*, **32**, 1125–1144

Nicholson, A. S., Davis, P. R. and Sheppard, N. J. (1981). Magnitude and distribution of trunk stresses in telecommunications engineers. *British Journal of Industrial Medicine*, **38**, 364–371

Nicholson, A. S. and Ridd, J. E. (Eds.) (1988). *Health, Safety and Ergonomics* (London: Butterworths)

NIOSH (1981). *Work Practices Guide for Manual Lifting*. National Institute for Occupational Safety and Health, Cincinnati

Nixon, P. G. F. (1982). Stress and the cardiovascular system. *Practitioner*, **226**, 1589–1598

Norman, D. A. (1988). *The Psychology of Everyday Things* (New York: Basic Books)

Oborne, D. J. (1985a). *Computers at Work. A Behavioural Approach* (Chichester: Wiley)

Oborne, D. J. (1985b). Whole-body vibration and ISO 2631: A critique. *Human Factors*, **25**, 55–70

Oborne, D. J. (1987). *Ergonomics at Work* (Chichester: Wiley)

Ong, C. N. (1984). VDT workplace design and physical fatigue—a case study in Singapore. In Grandjean (1984), pp. 484–489

Orth-Gomer, K. (1983). Intervention on coronary risk factors by adapting a shiftwork schedule to biological rhythmicity. *Psychosomatic Medicine*, **45**, 407–415

Östberg, O. (1980). Accommodation and visual fatigue in display work. In Grandjean and Vigliani (1980), pp. 41–52

OTA (1985). *Automation of America's Offices*. Report OTA-CIT-287, US Congress, Office of Technology Assessment (Washington, D.C.: US Government Printing Office)

Paffenbarger, R. S., Gima, A. S., Laughlin, M. E. and Black, R. A. (1971). Characteristics of longshoremen related to fatal coronary heart disease and stroke. *American Journal of Public Health*, **61**, 1362–1370

Paffenbarger, R. S., Wing, A. L. and Hyde, R. T. (1978). Physical activity as an index of heart attack risk in college alumni. *American Journal of Epidemiology*, **108**, 161–175

Palmore, E. (1969). Physical, mental and social factors in predicting longevity. *Gerontologist*, **9**, 103–108

Park, K. S. and Chaffin, D. B. (1974). A biomechanical evaluation of two methods of manual load lifting. *AIIE Transactions*, **6**, 105–113

Perera, J. (1988). The hazards of heavy breathing. *New Scientist*, 3rd December, 46–48

Perrow, C. (1984). *Normal Accidents: Living with High Risk Systems* (New York: Basic Books)

Person, R. S. (1956). An electromyographic investigation on co-ordination of the activity of antagonist muscles in man during the development of a motor unit. *Zhurnal Vysshei Nervnoi Deyatel 'nosti imeni IP Pavlova*, **1**, 17–27

Petrofsky, J. S. (1982). *Isometric Exercise and its Clinical Implications* (Springfield, Ill.: Charles C. Thomas)

Petrofsky, J. S. and Lind, A. R. (1978a). Comparison of metabolic and ventilatory responses of men to various lifting tasks and bicycle ergometry. *Journal of Applied Physiology, Respiratory, Environmental and Exercise Physiology*, **45**, 60–63

Petrofsky, J. S. and Lind, A. R. (1978b). Metabolic, cardiovascular and respiratory factors in the development of fatigue in lifting tasks. *Journal of Applied Physiology: Respiratory, Environmental and Exercise Physiology*, **45**, 64–68

Pheasant, S. T. (1977). *A Biomechanical Analysis of Human Strength*. PhD Thesis, University of London

Pheasant, S. T. (1983). Sex differences in strength—some observations on their variability. *Applied Ergonomics*, **14**, 205–211

Pheasant, S. T. (1986). *Bodyspace—Anthropometry, Ergonomics and Design* (London: Taylor and Francis)

Pheasant, S. T. (1987). *Ergonomics: Standards and Guidelines for Designers*. PP 7317, British Standards Institution, London

Pheasant, S. T. (1988a). The Zeebrugge–Harrisburg syndrome. *New Scientist*, 21 January, 55–58

Pheasant, S. T. (1988b). User-centred design. Ch. 7 in Nicholson and Ridd (1988), pp. 73–96

Pheasant, S. T. (1991). *Anthropometrics—An Introduction*, 2nd edition (London: British Standards Institution)

Pheasant, S. T. and Harris, C. M. T. (1982). Human strength in the operation of tractor pedals. *Ergonomics*, **25**, 53–63

Pheasant, S. T. and O'Neill, D. (1975). Performance in gripping and turning—a study of hand/handle effectiveness. *Applied Ergonomics*, **6**, 205–208

Pheasant, S. T. and Scriven, J. G. (1983). Sex differences in strength—some implications for the design of hand tools. In *Proceedings of the Ergonomics Society's Conference, 1983*, ed. K. Coombes (London: Taylor and Francis), pp. 303–315

Pheasant, S. T. and Stubbs, D. A. (1991). *Lifting and Handling—An Ergonomic Approach*. National Back Pain Association, Teddington, Middlesex

Pope, M. H., Bevins, T., Wilder, D. G. and Frymoyer, J. W. (1985). The relationship between anthropometric postural, muscular and mobility characteristics of males ages 18–55. *Spine*, **10**, 644–648

Pope, M. H., Frymoyer, J. W. and Andersson, G. (Eds.) (1984). *Occupational Low Back Pain* (New York: Praeger)

Porter, C. S. and Corlett, E. N. (1989). Performance differences of individuals classified by questionnaire as accident prone or non-accident prone. *Ergonomics*, **32**, 317–334

Porter, S. (1988). Accident proneness: a review of the concept. *International Reviews of Ergonomics*, **2**, 177–206

Pottier, M., Dubreuil, A. and Monod, H. (1969). The effects of sitting posture on the volume of the foot. *Ergonomics*, **12**, 753–758

Poulsen, E. and Jørgensen, K. (1971). Back muscle strength, lifting and stooped working postures. *Applied Ergonomics*, **2**, 133–137

Poulton, E. C. (1970). *Environment and Human Efficiency* (Springfield, Ill.: Charles C. Thomas)

Poulton, E. C. (1971). Skilled performance and stress. Ch. 3 in *Psychology at Work* (1st edition), ed. P. B. Warr (Harmondsworth, Middlesex: Penguin), pp. 55–75

Poulton, E. C., Hunt, E. M., Carpenter, A. and Edwards, R. S. (1978). The performance of junior hospital doctors following reduced sleep and long hours of work. *Ergonomics*, **21**, 279–296

Powell, M. C., Wilson, M., Szypryt, P., Symonds, E. M. and Worthington, B. S. (1986). Prevalence of lumbar disc degeneration observed by magnetic resonance in symptomless women. *Lancet*, 13 December, 1366–1367

Powell, P. I., Hale, M., Martin, J. and Simon, M. (1971). *2000 Accidents* (London: National Institute of Industrial Psychology)

Putz-Anderson, V. (Ed.) (1988). *Cumulative Trauma Disorders—A Manual for Musculoskeletal Diseases of the Upper Limbs* (London: Taylor and Francis)

Pyykko, I. (1986). Clinical aspects of hand arm vibration syndrome—a review. *Scandinavian Journal of Work Environment and Health*, **12**, 4427–447

Pyykko, I., Farkkila, M., Tiovanen, J., Korhonen, O. and Hyvarinen, J. (1976). Transmission of vibration in the hand–arm system with special reference to changes in compression force and acceleration. *Scandinavian Journal of Work Environment and Health*, **2**, 87–95

Quick, J. C. and Quick, J. D. (1984). *Organizational Stress and Preventative Management* (New York: McGraw-Hill)

Radl, G. W. (1980). Experimental investigations for optimal presentation mode and colour of symbols on the CRT-screen. In Grandjean and Vigliani (1980), pp. 271–276

Ramazzini, B. (1713). *De Morbis Artificum*. Trans. W. C. Wright (1940), *Diseases of Workers* (Chicago: University of Chicago)

Randle, I. P. M. (1988). *The Development of a Model of the Human Responses to Load Carriage*. PhD Thesis, University of Surrey

RCN (1979). *Avoiding Low Back Injury Among Nurses*. The Royal College of Nursing, London

Reason, J. (1987). The Chernobyl errors. *Bulletin of the British Psychological Society*, **40**, 201–206

Reason, J. (1989). *Human Error* (Cambridge: Cambridge University Press)

Reason, J. T. and Brand, J. J. (1975). *Motion Sickness* (London: Academic Press)

Reason, J. T. and Mycielska, K. (1982). *Absent Minded? The Psychology of Mental Lapses and Everyday Errors* (London: Academic Press)

Reilly, T. and Rothwell, J. (1988). Adverse effects of overtraining in females. In *Contemporary Ergonomics 1988*, ed. E. D. Megaw (London: Taylor and Francis)

Reynolds, D. D. and Angevine, E. N. (1977). Hand–arm vibration, Part II: Vibration transmission characteristics of the hand and arm. *Journal of Sound and Vibration*, **51**, 255–265

Ridd, J. E. (1985). Spatial restraints and intra-abdominal pressure. *Ergonomics*, **28**, 149–166

Rodahl, K. (1989). *The Physiology of Work* (London: Taylor and Francis)

Roethlisberger, F. J. and Dickson, W. J. (1949). *Management and the Worker* (Cambridge, Mass.: Harvard University Press)

Rogers, S. H. (1987). Recovery time needs for repetitive work. *Seminars in Occupational Medicine*, **2**, 19–24

Rohmert, W. (1960). Ermittlung von Erhlungpausen für statische Arbeit des Menschen. *Internationale Zeitschrift für Angewandte Physiologie Einschliesslich Arbeitsphysiologie*, **18**, 123–164

Rohmert, W., Wangenheim, M., Manzer, J., Zipp. P. and Lesser, W. (1986). A study stressing the need for a static postural force model for work analysis. *Ergonomics*, **29**, 1235–1249

Roland, M. O. (1986). A critical review of the evidence for a pain–spasm–pain cycle in spinal disorders. *Clinical Biomechanics*, **1**, 102–109

Rosegger, R. and Rosegger, S. (1960). Health effects of tractor driving. *Journal of Agricultural Engineering Research*, **5**, 241–275

Roth, M. (1861). *The Prevention of Spinal Deformities, Especially of Lateral Curvatures, with Notes on the Causes, the Artificial Production and the Injurious Modes of Treatment of Many of These Complaints* (London: Groombridge). Also in *British Journal of Homoeopathy*, **LXXVI**, 217–293

Rowe, M. L. (1969). Low back pain in industry—a position paper. *Journal of Occupational Medicine*, **11**, 161–169

Rowe, M. L. (1971). Low back pain disability in industry: updated position. *Journal of Occupational Medicine*, **13**, 476–478

Rummer, K., Berggrund, V., Jernberg, P. and Ytterbom, U. (1976). Driver reaction to a technical safety measure—studded tyres. *Human Factors*, **18**, 443–454

Russek, H. I. and Zohman, B. L. (1958). Relative significance of heredity, diet and occupational stress in CHD of young adults. *American Journal of Medical Sciences*, **235**, 266–275

Rutenfranz, J., Haider, M. and Koler, M. (1985). Occupational health measures for nightworkers and shiftworkers. Ch. 16 in Folkard and Monk (1985b), pp. 119–210

Rutenfranz, J., Knauth, P. and Angersbach, D. (1981). Shift work research issues. In Johnson *et al.* (1981), pp. 165–196

Rutenfranz, J., Knauth, P. and Colquhoun, W. P. (1976). Hours of work and shiftwork. *Ergonomics*, **19**, 331–340

Ryan, G. A. and Bampton, M. (1988). Comparison of data process operators with and without symptoms. *Community Health Studies*, **12**, 63–68

Saarhi, J. T. and Lautela, J. (1979). Characteristics of jobs in high and low accident frequency companies in the light metal working industry. *Accident Analysis and Prevention*, **11**, 51–60

Salman, R. A. (1975). *Dictionary of Tools Used in the Woodworking and Allied Trades c. 1700–1970* (London: Allen and Unwin)

Sandover, J. (1983). Dynamic loading as a possible source of low back disorders. *Spine*, **6**, 652–657

Sauter, S. L., Chapman, L. J., Knutson, S. J. and Anderson, H. A. (1987). Case example of wrist traumas in keyboard use. *Applied Ergonomics*, **18**, 183–186

Schmitke, H. (1980). Ergonomic design principles of alphanumeric displays. In Grandjean and Vigliani (1980), pp. 267–270

Scholey, M. and Hair, M. (1989). Back pain in physiotherapists involved in back care education. *Ergonomics*, **32**, 179–190

Selye, H. (1956). *The Stress of Life* (New York: McGraw-Hill)

Selzer, M. L. and Vinokor, A. (1974). Life events, subjective stress and traffic accidents. *American Journal of Psychiatry*, **13**, 903–906

Shaffer, L. H. (1973). Latency mechanisms in transcription. In *Attention and Performance IV*, ed. S. Kornblum (New York: Academic Press)

Shaper, A. G. (1988). *Coronary Heart Disease Risks and Reasons* (London: Current Medical Literature)

Sheehy, M. P. and Marsden, C. D. (1982). Writer's cramp: a focal dystonia. *Brain*, **107**, 461–480

Shephard, R. J. (1968). Intensity, duration and frequency of exercise as determinants of the response to a training regime. *Internationale Zeitschrift für Angewandte Physiologie Einschliesslich Arbeitsphysiologie*, **26**, 272–278

Sheridan, T. (1980). Human error in nuclear power. *Technology Review*, **82**, 22–33

Shotton, M. A. (1986). Computer dependency—a survey. In *Contemporary Ergonomics 1986*, ed. D. J. Oborne (London: Taylor and Francis), pp. 300–304

Shute, S. J. and Starr, S. J. (1984). Effects of adjustable furniture on VDT users. *Human Factors*, **26**, 157–170

Silverstein, B. A., Fine, L. J. and Armstrong, T. J. (1986). Hand wrist cumulative trauma disorders in industry. *British Journal of Industrial Medicine*, **43**, 779–784

Silverstein, B. A., Fine, L. J. and Armstrong, T. J. (1987). Occupational factors in carpal tunnel syndrome. *American Journal of Industrial Medicine*, **11**, 343–358

Simons, D. G. (1976). Muscle pain syndromes—Part II. *American Journal of Physical Medicine*, **55**, 15–42

Simpson, G. C. (1988a). Is ergonomics cost-effective? Ch. 12 in Nicholson and Ridd (1988), pp. 154–170

Simpson, G. C. (1988b). Hazard awareness and risk perception. Ch. 5 in Nicholson and Ridd (1988), pp. 48–57

Simpson, K. (1940). Shelter deaths from pulmonary embolism. *Lancet*, 14 December, 744

Slovic, P., Fischoff, B. and Lichtenstein, S. (1981). Perceived risk: psychological factors and social implications. *Proceedings of the Royal Society of London A*, **376**, 17–34

Smith, A. (1776). *An Inquiry into the Nature and Causes of the Wealth of the Nations* (Harmondsworth, Middlesex: Penguin, 1970)

Smith, A. (1989). Respiratory virus infections and performance. Paper presented at symposium on *Human Factors in High Risk Systems*. Royal Society, London

Smith, J. W. (1953). The act of standing. *Acta Orthopaedica Scandinavica*, **23**, 153–168

Smith, M. J., Cohen, B. G. F., Stammerjohn, L. W. and Happ, A. (1981). An investigation of health complaints and job stress in video display operations. *Human Factors*, **23**, 387–399

Smith, M. J., Stammerjohn, L. W., Cohen, G. F. and Lalich, N. R. (1980). Job stress in video display operations. In Grandjean and Vigliani (1980), pp. 201–210

Snook, S. H. (1978). The design of manual handling tasks. *Ergonomics*, **21**, 963–986

Snook, S. H., Campanelli, R. A. and Hart, J. W. (1978). A study of three preventative approaches to low back pain. *Journal of Occupational Medicine*, **20**, 478–481

Solly, S. (1864). Clinical lectures on scrivener's palsy or the paralysis of writers. Cited in Sheehy and Marsden (1982), 709–711

Speroff, B. and Kerr, W. (1952). Steel mill 'hot strip' accidents and interpersonal desirability values. *Journal of Clinical Psychology*, **8**, 89–91

Spilling, S., Eiterheim, J. and Aarås, A. (1986). Cost-benefit analysis of work environment investment at STK's telephone plant at Kongsvinger. In *The Ergonomics of Working Postures*, ed. E. N. Corlett, J. Wilson, and I. Manenica (London: Taylor and Francis), pp. 380–397

Spitzer, H. (1951). Physiologische grundlagen fur den erhdungszchlag bei Schwerarbeit. REFA—Nachrichten, Heft, Darmstadt

Stammerjohn, L. W., Smith, M. J. and Cohen, B. G. F. (1981). Evaluation of workstation design factors in VDT operations. *Human Factors*, **23**, 401–412

Stranden, E., Aarås, A., Anderson, D. M., Myhre, H. O. and Martinsen, K. (1983). The effects of working posture on musculo-skeletal load and circulatory condition. In *Proceedings of the Ergonomics Society's Conference*, ed. K. Coombes (London: Taylor and Francis)

Stubbs, D. A. (1976). *Trunk Stresses in Construction Workers*. PhD Thesis, University of Surrey

Stubbs, D. A. (1981). Trunk stresses in construction and other industrial workers. *Spine*, **6**, 83–89

Stubbs, D. A., Buckle, P. W., Hudson, M. P., Rivers, P. M. and Batty, D. (1986). Backing out: nurse wastage associated with pack pain. *International Journal of Nursing Studies*, **23**, 325–336

Stubbs, D. A., Buckle, P., Hudson, M., Rivers, P. and Worringham, C. (1983a). Back pain in the nursing profession. Part I. Epidemiology and pilot methodology. *Ergonomics*, **26**, 755–765

Stubbs, D. A., Buckle, P., Hudson, M., Rivers, P. and Worringham, C. (1983b). Back pain in the nursing profession. Part II. The effectiveness of training. *Ergonomics*, **26**, 767–779

Stubbs, D. A., Hudson, M. P., Rivers, P. M. and Worringham, C. J. (1980). Patient handling and truncal stress in nursing. In *Prevention of Back Pain in Nursing*, Proceedings of the Conference organized by the Nursing Practice Research Unit, Northwick Park Hospital, Back Pain Association and DHSS, pp. 14–27

Symington, I. A. and Stack, B. H. R. (1977). Pulmonary thromboembolism after air travel. *British Journal of Diseases of the Chest*, **71**, 138–140

Tabary, J. C., Tabary, C., Tardieu, C., Tardieu, G. and Goldspink, E. (1972). Physiological and structural changes in the cat's soleus muscle due to immobilisation at different lengths of plaster casts. *Journal of Physiology*, **24**, 231–244

Taylor, D. J. and Pocock, S. J. (1972). Mortality of shift and day workers 1956–68. *British Journal of Industrial Medicine*, **29**, 201–207

Taylor, F. W. (1911). *The Principles of Scientific Management* (New York: Harper)

Taylor, H. R. (1981). Racial variations in vision. *American Journal of Epidemiology*, **113**, 62–80

Taylor, H. R. (1983). The Author replies. *American Journal of Epidemiology*, **115**, 139–142

Taylor, W. (1974). The vibration syndrome—an introduction. In *The Vibration Syndrome*, ed. W. Taylor (London: Academic Press), pp. 1–12

Taylor, W. (1985). Vibration white finger: a newly prescribed disease. *British Medical Journal*, **291**, 921–922

Taylor, W. (1988). Biological effects of the hand–arm vibration syndrome: historical perspective and current research. *Journal of the Acoustical Society of America*, **82**, 415–422

Taylor, W. (1989). Bad vibrations. *New Scientist*, 14 January, 41–47

Taylor, W., Pelmear, P. L. and Pearson, J. C. G. (1975). Vibration-induced white finger epidemiology. In *Vibration White Finger in Industry*, ed. W. Taylor and P. L. Pelmear (London: Academic Press), pp. 1–13

Teisinger, J. (1972). Vascular disease disorders resulting from vibrating tools. *Journal of Occupational Medicine*, **14**, 129–133

Thackrah, C. T. (1832). *The Effects of the Principal Arts, Trades and Professions and of Civic States and Habits of Living on Health and Longevity*

Thompson, A. R., Plewes, L. W. and Shaw, E. G. (1951). Peritendinitis crepitans and simple tenosynovitis: a clinical study of 544 cases in industry. *British Journal of Industrial Medicine*, **8**, 150–158

Tichauer, E. R. (1978). *The Biomechanical Basis of Ergonomics—Anatomy Applied to the Work Situation* (New York: Wiley)

Tiffin, J. and McCormick, E. J. (1970). *Industrial Psychology* (London: Allen and Unwin)

Townsend, P., Davidson, N. and Whitehead, M. (1988). *Inequalities in Health* (Harmondsworth, Middlesex: Penguin)

Tracy, M. F. (1990). Biomechanical methods in posture analysis. Ch. 23 in *Evaluation of Human Work*, ed. J. R. Wilson and E. N. Corlett (London: Taylor and Francis)

Travell, J. (1967). Mechanical Headache, *Headache*, **7**, 23–29

Travell, J. and Rinzler, S. H. (1952). The myofascial genesis of pain. *Postgraduate Medicine*, **11**, 425–434

Travell, J., Rinzler, S. and Herman, M. (1942). Pain and disability of the shoulder and arm. *Journal of the American Medical Association*, **120**, 417–422

Travell, J. E. and Simons, D. E. (1983). *Myofascial Pain and Dysfunction: The Trigger Point Manual* (Baltimore: Williams and Wilkins)

Troup, J. D. G. (1988). The perception of musculoskeletal pain and incapacity for work: prevention and early treatment. *Physiotherapy*, **74**, 435–439

Troup, J. D. G. and Edwards F. C. (1985). *Manual Handling and Lifting. An Information and Literature Review with Special Reference to the Back.* The Health and Safety Executive (London: HMSO)

Troup, J. D. G., Foreman, T. K., Baxter, C. E. and Brown, D. (1987). The perception of back pain and the role of psychophysical tests of lifting capacity. *Spine*, **12**, 645–657

Troup, J. D. G., Leskinen, T. P. J., Stålhammer, H. R. and Kuorinka, I. A. A. (1983). A comparison of intra-abdominal pressure increases, hip torque, and lumbar vertebral compression in different lifting techniques. *Human Factors*, **25**, 517–525

Troup, J. D. G., Martin, J. W. and Lloyd, D. C. E. F. (1981). Back pain in industry: a prospective survey. *Spine*, **6**, 61–69

Troup, J. D. G. and Rauhala, H. H. (1987). Ergonomics and training. *International Journal of Nursing Studies*, **24**, 325–330

Turner, J. P. and Buckle, P. W. (1987). Carpal tunnel syndrome and associated risk factors. In *Musculoskeletal Disorders at Work*, ed. P. W. Buckle (London: Taylor and Francis), pp. 124–132

Vällfors, B. (1985). Acute, subacute and chronic low back pain—clinical symptoms, absenteeism and working environment. *Scandinavian Journal of Rehabilitation Medicine*, Suppl. 11, 1–98

Vena, J. E., Graham, S., Zielezny, M., Swanson, M. K., Barnes, R. F. and Nolan, J. (1985). Lifetime occupational exercise and colon cancer. *American Journal of Epidemiology*, **122**, 357–365

Verhaegen, P., Strubbe, J., Vonk, R. and van den Abeele, J. (1985). Absenteeism, accidents and risk taking. *Journal of Occupational Accidents*, **7**, 177–186

Verhaegen, P., Vanhalst, B., Derijcke, H. and van Hoeke, M. (1979). The value of some psychological theories on industrial accidents. *Journal of Occupational Accidents*, **1**, 39–45

Vernon, H. M. (1921). *Industrial Fatigue and Efficiency* (London: Routledge)

Vernon, H. M. (1936). *Accidents and Their Prevention* (Cambridge: Cambridge University Press)

Videman, T., Rauhala, H., Asp, S., Lindström, K., Cedercreutz, G., Kämpi, M., Tola, S. and Troup, J. D. G. (1989). Patient handling skill, back injuries and back pain—an intervention study in nurses. *Spine*, **14**, 148–156

Viikari-Juntura, A. (1984). Tenosynovitis, peritendinitis, and the tennis elbow syndrome. *Scandinavian Journal of Work Environment and Health*, **10**, 443–449

Viitisalo, J. T., Era, P., Leskinen, A.-L. and Heikkinen, E. (1985). Muscular strength profiles and anthropometry in random samples of men aged 31–35, 51–55 and 71–75 years. *Ergonomics*, **28**, 1563–1574

Village, J., Morrison, J. B. and Leong, D. K. N. (1989). Whole-body vibration in load-haul-dump vehicles. *Ergonomics*, **32**, 1167–1183

Vøllestad, N. K. and Sejersted, O. M. (1988). Biochemical correlates of fatigue. *European Journal of Applied Physiology and Occupational Physiology*, **57**, 336–347

Waddell, G. (1987). A new clinical model for the treatment of low-back pain. *Spine*, **12**, 632–644

Wallace, M. and Buckle, P. (1987). Ergonomic aspects of upper limb disorders. *International Reviews of Ergonomics*, **1**, 173–200

Waris, P. (1980). Occupational cervicobrachial syndromes—A review. *Scandinavian Journal of Work Environment and Health*, **6**, Suppl. 3, 3–14

Warr, P. and Wall, T. (1973). *Work and Well-Being* (Harmondsworth, Middlesex: Penguin)

Welford, A. T. (1968). *Fundamentals of Skill* (London: Methuen)

Welford, A. T. (1973). Stress and performance. *Ergonomics*, **5**, 567–580

Westgaard, R. H. and Aarås, A. (1984). Postural muscle strain as a causal factor in the development of musculo-skeletal illness. *Applied Ergonomics*, **15**, 162–174

Westgaard, R. H. and Aarås, A. (1985). The effect of improved workplace design on the environment of work-related musculo-skeletal illness. *Applied Ergonomics*, **16**, 91–97

Westgaard, R. H. and Bjørklund, R. (1987). Generation of muscle tension additional to postural load. *Ergonomics*, **30**, 911–924

Weston, H. C. (1953). Visual fatigue with special reference to lighting. In *Symposium on Fatigue*, ed. W. F. Floyd and A. T. Welford (London: H. K. Lewis)

Wever, R. A. (1985). Man in temporal isolation: basic principles of the circadian system. Ch. 2 in Folkard and Monk (1985b), pp. 15–28

WHO (1980). *International Classification of Impairments, Disabilities and Handicaps* (Geneva: World Health Organization)

Wickström, G. (1978). Effect of work on degenerative back disease. *Scandinavian Journal of Work Environment and Health*, **4**, Suppl. 1, 1–12

Widule, C. J., Foley, V. and Demo, F. (1978). Dynamics of the axe swing. *Ergonomics*, **21**, 925–930

Wiener, N. (1954). *The Human Use of Human Beings* (Sphere edition, 1968)

Wiener, N. (1961). *Cybernetics: or Control and Communication in the Animal and the Machine*, 2nd edition (Cambridge, Mass.: MIT Press)

Wilde, G. J. S. (1982). The theory of risk homeostasis: implications for safety and health. *Risk Analysis*, **2**, 209–225

Wilde, G. J. S. (1989). Accident countermeasures and behavioural compensation: the position of risk homeostasis theory. *Journal of Occupational Accidents*, **10**, 267–292

Wilder, S. G., Woodworth, B. B., Frymoyer, J. W. and Pope, M. H. (1982). Vibration and the human spine. *Spine*, **7**, 243–254

Wilkins, A. (1986). Why are some things unpleasant to look at? In *Contemporary Ergonomics 1986*, ed. D. J. Oborne (London: Taylor and Francis), pp. 259–263

Wilkins, A. J. and Nimmo-Smith, I. (1984). On the reduction of eye-strain when reading. *Ophthalmic and Physiological Optics*, **4**, 53–59

Wilkins, A. J. and Nimmo-Smith, I. (1987). The clarity and comfort of printed text. *Ergonomics*, **30**, 1705–1720

Wilkins, A. J., Nimmo-Smith, I., Slater, A. I. and Bedocs, L. (1988). Fluorescent lighting, headaches and eye strain. Paper presented at *National Lighting Conference, 1988*

Wilkins, A., Nimmo-Smith, I., Tait, A., McManus, C., Della Salla, S., Tilley, A., Arnold, K., Barrie, M. and Scott, S. (1984). A neurological basis for visual discomfort. *Brain*, **107**, 989–1017

Williams, C. A., Petrofsky, J. S. and Lind, A. R. (1980). Physiological responses of women during lifting exercise. *European Journal of Applied Physiology*, **50**, 133–144

Williams, R. (1976). *Keywords—A Vocabulary of Culture and Society* (London: Fontana)

Willis, E. (1986). RSI as a social process. *Community Health Studies*, **10**, 210–219

Winkel, J. (1985). *On Foot Swelling During Sedentary Work and the Significance of Leg Activity*. National Board of Occupational Safety and Health, Stockholm, Sweden

Winsemius, W. (1965). Some ergonomic aspects of safety. *Ergonomics*, **8**, 151–162

Winter, D. A. (1979). *Biomechanics of Human Movement* (New York: Wiley)

Wood, C. (1986). Good behaviour for a heart attack. *New Scientist*, 13 March, 31–34

Wood, J. (1987). Lighting for control rooms. In *Contemporary Ergonomics 1987*, ed. E. D. Megaw (London: Taylor and Francis), pp. 309–314

Wood, P. H. N. (1975). *Classification of Impairments and Handicaps* (Geneva: World Health Organization)

Wurtman, R. J. and Wurtman, J. J. (1989). Carbohydrates and depression. *Scientific American*, **260**, 50–57

Yerkes, R. M. and Dodson, J. D. (1908). The relation of strength of stimulus to rapidity of habit formation. *Journal of Comparative Neurology and Psychology*, **18**, 459–482

Yousef, G. E., Bell, E. J., Mann, G. F., Murugesan, V., Smith, D. E., McCartney, R. A. and Mowbray, H. J. (1988). Chronic enterovirus infection in patients with postviral fatigue syndrome. *Lancet*, 23 January, 146–149

Yunnus, M. B., Kalyan-Raman, U. P., Kalyan-Raman, K. and Masi, A. T. (1986).

Pathologic changes in muscle in primary fibromyalgia syndrome. *American Journal of Medicine*, **81**, Suppl. 3A, 38–42

Zipp, P., Haider, E., Halpern, N. and Rohmert, W. (1983). Keyboard design through physiological strain measurements. *Applied Ergonomics*, **14**, 117–122

Zulley, J. and Campbell, S. S. (1985). Napping behaviour during "spontaneous internal desynchronization": sleep remains in synchrony with body temperature. *Human Neurobiology*, 4, 123–126

# Index